PRAISE FOR ROGER N. LANCASTER'S *Th*

"Sex, lies, and videotape. Culture wars, scienc[...] stupid. With interdisciplinary brilliance and biting wit, Lancaster makes sad sense of sociobiology's contemporary renaissance while dissecting its popular and scholarly practitioners. But Lancaster's decidedly queer perspective connects science to shifting sexuality, family, and economic inequality in the cultural stew of the present. E.M. Forster famously adjured us to 'only connect.' What does *Will and Grace* have to do with post-Fordist economies? Journalists' just-so stories about ducks and sex with the 9/11 terror? *The Trouble with Nature* connects us all, in surprisingly new ways." **MICAELA DI LEONARDO,** author of *Exotics at Home: Anthropologies, Others, American Modernity*

"What is a woman, man, the homosexual, or heterosexual? asks Roger Lancaster in this lively, engaging new book. And well may he ask as, once again, academic pop stars hawk their biodetermined creations. Eschewing simplistic caricatures, he offers vivid examples of the ambiguities, contradictions, and complexities that characterize real people, while showing how the biomyths serve to revivify constricting ideologies about sex and family. An original and fascinating book." **RUTH HUBBARD,** author of *Exploding the Gene Myth* and *The Politics of Women's Biology*

"*The Trouble with Nature* is a brilliant, wide-ranging, masterful critique of the cultural impact of sociobiology and evolutionary psychology on popular as well as scholarly understandings of gender, sexuality, and political economy. There are no cheap, trendy shots at science here, nor grandstand gestures to the prejudices of cultural relativists. Lancaster displays the skills of a science journalist while producing a major cultural studies opus." **JUDITH STACEY,** author of *In The Name of the Family: Rethinking Family Values in the Postmodern Age*

"Provocative, witty, illuminating, and politically pointed, *The Trouble with Nature* shows us how the flat-footed fixities of biological reductionism limit and constrain us and why we need an expansive progressive political imagination to free us." **LISA DUGGAN,** coauthor of *Sex Wars: Sexual Dissent and Political Culture*

"*The Trouble with Nature* will be a valuable addition to my library. It is a book I will want to share with colleagues and students. A pleasure to read, it is full of insights about the place of sexuality in popular consciousness. Lancaster has written a personal and a political study while avoiding many of the clichés too common in contemporary cultural criticism." **LAWRENCE GROSSBERG,** author of *Dancing in Spite of Myself: Essays on Popular Culture*

The Trouble with Nature

The Trouble with Nature

Sex in Science and Popular Culture

ROGER N. LANCASTER

University of California Press

BERKELEY   LOS ANGELES   LONDON

University of California Press
Berkeley and Los Angeles, California

University of California Press, Ltd.
London, England

© 2003 by the Regents of the University of California

Library of Congress Cataloging-in-Publication Data

Lancaster, Roger N.
    The trouble with nature : sex in science and popular culture /
Roger N. Lancaster.
        p.   cm.
    Includes bibliographical references and index.
    ISBN 0-520-20287-2 (alk. paper).—ISBN 0-520-23620-3
(pbk. : alk. paper)
    1. Sex in popular culture.   2. Science news.   3. Pseudoscience.
4. Sexual orientation—Physiological aspects.   I. Title.

HQ23 .L29   2003
306.7—dc21   2002002311

Manufactured in Canada
12   11   10   09   08   07   06   05   04   03
10 9 8 7 6 5 4 3 2 1

The paper used in this publication is both acid-free and totally chlorine-
free (TCF). It meets the minimum requirements of ANSI/NISO
Z39.48–1992 (R 1997) (Permanence of Paper).

My suspicion is that the universe is not only queerer than we suppose, but queerer than we *can* suppose.

J. B. S. HALDANE

Naturalism has served as deceptively in the modern world as supernaturalism ever did in the past.

KENNETH BURKE

We have to understand that with our desires, through our desires, go new forms of relationship, new forms of love, new forms of creation.

MICHEL FOUCAULT

Contents

Illustrations

Acknowledgments

Let me say right off that I feel like I've been wading through a toxic waste dump of ideas for the better part of several years, pressing little exhibits into my media scrapbook like Doris Lessing on a bad day. Actually, it didn't start off that way. A materialist with a bit of a vulgar streak, I began this project more than a little weary of the airy idealist conceits that have dominated recent discussions of sex in anthropology and cultural studies. Looking for approaches that might put some *body* into discussions of the body, I had hoped to discover sophisticated new biological perspectives on sex and sexuality. What I encountered instead was the same old reductivism warmed over: the belief that complex social identities are scripted in strands of DNA. The notion that desires are coded, microscopically, in the genes. Ideas about social good (and ill), staged as media-savvy evolutionary just-so stories.

Lots of scholars feign sophistication on these matters, but my research methods for this project were simple enough: I read newspapers and news weeklies to monitor what kinds of science studies made headlines, and to track how scientific claims about gender and sexuality entered into public discourse. For a time, I almost imagined subtitling this book *Reading the* New York Times *Science News* (with apologies to Catherine Lutz and Jane Collins), for it seemed that if a science story was deemed newsworthy in that influential paper, chances were good you'd also see it in the *Washington Post* or in *Newsweek* or in some other national news source. In addition to reading the papers, I watched network news and, *à contretemps*, TV entertainment programs, taking notes on treatments of gender and sexuality in those very different venues. (Okay, I admit it: I would probably have watched a certain amount of TV anyway—probably not ABC's *Home Improvement*, but definitely WB's *Buffy the Vampire Slayer*.) Over time, I've tried to tame a free-flowing series of impromptu notes into something more closely resembling book chapters. The ensuing text follows the flickering line between media studies and science studies, lesbigay and gender studies— as informed by my more conventional academic training and research in anthropology and specifically by my interest in cultural studies.

I cannot say whether this book is a work of the auto-ethnographic sort, or whether it belongs to the genre of cultural criticism. I hope it represents a bit of both. My role models are Roland Barthes, who appropriately camped up a reading of nature at the confluence of Marx, Lukács, and Saussure in *Mythologies;* Marshall Sahlins, who dispatched the first wave of sociological ideas with considerable panache in *The Use and Abuse of Biology* (a book whose basic arguments have not been superseded in twenty-five years); and the "Left biologists"—Anne Fausto-Sterling, Stephen Jay Gould, Ruth Hubbard, and Richard Lewontin. I've drawn special inspiration from gay-studies scholars like Lisa Duggan, Lauren Berlant, and Michael Warner, authors of exceptional talent who've written about mass culture and have sometimes attempted to mediate between the often recondite field of queer theory and the public sphere. (I cribbed this book's title from Warner's *The Trouble with Normal.*)

A number of readers have indulged me with helpful and selflessly critical readings of earlier drafts of various chapters: Andy Bickford, Susan Bordo, William Byne, Samuel Colón, Marcial Godoy-Anativia, Deborah Heath, David Kaufmann, Bill Leap, Micaela di Leonardo, Michael Higgins, Jonathan Marks, David McBride, Ann Palkovich, Richard Parker, Adolph Reed, Nancy Scheper-Hughes, Naomi Schneider, Paul Smith, Susan Sperling, Judith Stacey, Suzanna Walters, and Eric Zinner. Alan Dundes generously pointed me toward some profitable readings in folklore and mythology. Anne Fausto-Sterling was most helpful, sending me in draft form a portion of her new book, *Sexing the Body*, while Bill Byne, Jon Marks, and Suzanna Walters each supplied me with drafts of material then in press. Gina Maranto engaged me in a delightful and informative online conversation about the social construction of normality, and my research assistants, Lynne Constantine and Joanne Clarke Dillman, lent considerable help with portions of the research and editing. At the University of California Press, my editor, Naomi Schneider, shepherded this project through the usual hazards. Erika Büky capably oversaw production. Bud Bynack and Sue Carter did a great job with copyediting, while staffers Annie Decker and Sierra Filucci kept things moving along (and did more than their share of hand-holding with an anxious author). Staff seldom get the credit they deserve for the work they do. Zina Santos, the Cultural Studies Program's administrative assistant, ran the office with admirable efficiency, which freed up my time to write and do research. Zina also volunteered to help locate illustrations and request permissions. Special thanks, then, to Zina, for service beyond the call of duty, to my partners in crime, Micaela and Paul, for several close readings of the text, and to my partner in life, Samuel, for the same—and for putting up with all that goes into the production of a book. Obstinate and perverse creature that I am, I have not always followed my fellow-travelers' advice. Hence, the traditional self-curse: Any resulting omissions, mistakes, or misrepresentations are, of course, my own.

RNL

Introduction:
Culture Wars, Nature Wars

A Report from the Front

Nature is, first of all, a wish.
HARVIE FERGUSON, *The Science of Pleasure*

A series of TV greeting-card commercials broadcast in the mid-1990s followed the conversations of a young, attractive, and presumably married couple on a variety of special occasions: holidays, anniversaries, and birthdays. Over the course of these ads, the woman has let her husband in on one of her sex's many little secrets. Women always—and, if they are well practiced, *discreetly*—turn the card around to inspect its logo, to see if it's a Hallmark®. So went an entire genre of mass media representations at the time: coy expositions on the role of sexual difference in modern courtship, romance, and family life. Vignettes on the true nature of men, women, and desire.

In 1997, Hallmark aired a variant of this ad in which the young man presents his infant niece—his wife's sister's daughter—with her very first greeting card. TV viewers hear the baby's cooing and giggling. From a perspective at the side of the bassinet, we see the card rear up, then turn around. With the young man's shocked face on screen, the voice-over queries: *"Could it be this need to look for Hallmark on the back is . . . genetic?"* At the same time, molecular geneticist Dean Hamer—a National Institutes of Health scientist famous for genetic research on the causes of male homosexuality—was lending gravitas to scenarios not far removed from Madison Avenue whimsy.[1] At a Harvard conference on biology and sexuality, Hamer announced that his present research indicates the existence of two genes, each with two variants. One gene determines the frequency with which people like to have sex: once a week or more, versus less than once a week. The other gene, as Simon LeVay puts it, determines whether we prefer multiple partners, "unusual" positions, and sexual "experimentation" or whether we prefer monogamy and sexual repetition.[2] Alas, it becomes difficult to parody the claims of science on the "nature" of desire.

1

MEANWHILE, BACK AT THE SEXUAL REVOLUTION

Even though popular science of the past decade tends to deliver emphatic parables about a timeless human nature, empirical happenings on the social ground suggest little in the way of certainty about sexual identities or gender roles. Quite the contrary. At the unfolding of the new millennium, as in the waning years of the twentieth century, American society is in turmoil over sex, and presumably will remain so for a long time to come.

The cultural (as opposed to economic) issues that came to the fore in the 1960s continue to dominate politics in the United States, where elections more often than not turn on conflicts over abortion, homosexuality, marriage, family life, and relations between the sexes. The old feminist slogan "The personal is political" finds new life in a consequently altered social landscape: in the "tabloidization" of news, in sex scandals of the rich and famous, in John and Jane Q. Public "human interest" stories that purport to monitor the health and definition of modern families, and in general, in what Lauren Berlant has described as a "privatization" or "downsizing" of citizenship—the notion "that the core context of politics should be the sphere of private life."[3] This is to say nothing of recent technological advances in assisted reproduction and cloning, from which have ensued not only bio-ethical quandaries, but also existential dilemmas over the meanings of sex, conception, and personhood.

All said, in that long interregnum between the close of the Gulf War in 1991 and the terrorist attacks on New York and the Pentagon in 2001, the defining news topics of the decade amount to a now hidden, now open struggle over what it means to be a man or a woman. These struggles continue unabated today—not only in contentions over family law, welfare benefits, pornography, or the right to choose, but also in deliberations over the aims of foreign policy or the justifications for military intervention, and in controversies about funding for international institutions that deal with sexual health and reproduction. But after eighty years of universal suffrage, decades after the first salvos of the sexual revolution, and after thirty years of modern feminism and gay liberation, the sides have diversified, the issues have grown more complex, and just what it might mean to be male or female is anybody's guess. Basic meanings and practices are, as it were, up for grabs.

Is a real man tough or nurturing, self-reliant or relationship-oriented? Does he eat steak, or does he eat quiche? Should a woman, in Henry Higgins's quaint phrase (as in the recurring quandaries of feminism), be "more like a man"—or decidedly less so? Is the properly politicized feminist anti-

Figure 1. Abortion activists pro and con confront each other outside the U.S. Supreme Court Building in Washington, June 30, 1992. Photo by Marcy Nighswander. AP/World Wide Photos.

pornography, pro-sex, or (I think here of the so-called babe feminists) downright sleazy? And what is the point of ongoing struggles for sexual equality: to claim a legitimate place for everyone within expanded definitions of male and female, church and family, community and state? Or is it time for a queer coalition of fags, dykes, sissies, butches, misfits, and rebels to secede forever from True Man, Real Woman, and other fake universals?

Indeed, just what *is* a man or a woman? Contestation, confusion, and equivocation dramatically manifest themselves around these basic questions in public discourses today—and not only in American society.

When British Army officer Sergeant Major Joe Rushdon announced his plans to undergo male-to-female gender reassignment surgery, it was unclear whether he would be discharged from the service. Homosexuals were (at the time) still barred from military service in Britain. In response to Rushdon's announcement, the Defense Ministry hastily cobbled together a new policy allowing transsexuals to serve in the military—while continuing to bar homosexuals. "There is a clear difference between homosexuality and transsexuality," a Defense Ministry spokesperson told BBC News, hypothecating on the nature of desire and claiming clarity where others have seen murk. "Transsexuality is a gender issue, not one of sexuality," he continued. Invoking an analytical distinction between "gender" and "sexuality," the Defense Ministry cited one of the sturdier dualisms of feminism and gay liberation. But by asserting a wall of separation between the two, the improvised policy implied anything but liberatory effects. Instead, it suggested a nervous patrolling of unstable identities. Christine Burns, speaking for the transgender civil rights group Press for Change, underscored the weird ironies and implausible consequences of the government's position. Rushdon, she pointed out, "will physically be a woman but legally be a man. Given the Army's ban on homosexuality, it is not clear whether she will be allowed to have sexual relations with a man or a woman—if anyone."[4]

THE CHANGING CONTEXT

Such confusions about basic rules illuminate wider conflicts and uncertainties in the changing, ever so interconnected worlds of gender and sexuality. Nowadays, as in the past, men and women come of age in a society dominated by male prerogatives and heterosexual values. But unlike past generations, young people now come to adulthood in a society where those values are openly questioned from every possible angle and in every conceivable forum. From bedroom to boardroom, in classrooms and on the streets, the "culture wars" have become an inescapable backdrop of modern life: a many-sided struggle in which old allies no sooner achieve some small victory than they part company, an ongoing revolution in which every loss renews the fight on another front, an all-around crisis in which even individuals are torn between hostile camps.

In the aftermath of oppositional contest and institutional compromise, a certain social transformation has already occurred. Family, marriage, and community, by all accounts, are not what they used to be, which has been a cause for celebration on one side and lamentation on the other. Gay institutions, which three decades ago occupied a shadowy realm on the margins of public life, now anchor a substantial portion of cultural and economic life in every major American city. It is difficult to imagine the modern metropolis without its Village/Chelsea complex, its West Hollywood, its greater Castro, its South Beach, or its Dupont Circle area.[5] The cultural values expressed in mainstream mass entertainment, too, have changed. To the consternation of cultural conservatives, Disney cartoons strive to teach multicultural, inclusive, and tolerant values, while an up-to-date Popeye and a thoroughly modern Olive Oyl solve their problems together using brains, not brawn. Feminist issues (fat, body image, domestic violence, sexual freedom, and sexual coercion) are robustly debated for the masses on afternoon talk shows, while suburban parents agonize over the social effects of traditional, gendered toys—guns for boys and dolls for girls—and conspire to mute gender disparities among their sons and daughters.

In a time of contested, changing values, even moralistic finger wagging takes on an oddly nuanced tone. Opining in *Newsweek* on the Clinton sex scandals, senior editor Jonathan Alter tilts against permissiveness with the following precariously balanced sentence: "There's something to the idea that the welcome and long-overdue tolerance of blacks, women, Jews, foreigners and gays has left us indiscriminately tolerant."[6] Push such a rickety contention one way or the other and it might well say something intolerant. But because it is caught in the tow of its own historical moment, it merely expresses the author's uneasiness about his own moralizing.

In more ways than one, the Monica Lewinsky affair proved a sensitive barometer of ongoing changes in American sexual culture. The Starr Report made it clear that President Bill Clinton and White House intern Lewinsky enjoyed a wide gamut of nonreproductive sex acts (manual manipulation, digital penetration, fellatio, cunnilingus, anilingus, and penetration by phallic objects). These revelations triggered virtually no public talk about "unnatural sex"—despite the fact that a good number of the couple's steamy passions remain illegal under many states' sodomy laws.[7] In the end, Clinton survived impeachment for lying under oath about the affair and staged an obligatory, brief, and not very convincing performance of spiritual contrition. Voters punished pro-impeachment Republicans in the 1998 off-year elections, and public opinion proved so blasé about the entire affair that leaders of the religious Right conceded that "Moral Minority" might be a more appropriate term for Christian conservatism than "Moral Majority." Subsequently, Hillary Rodham Clinton's 2000 run for a United States Senate seat in New York provided the opportunity for yet another round of speculation about the first couple's private lives: Republican commentators attempted to revive the characterization of Hillary Clinton as "The Great Enabler," while tabloids gossiped about the first lady's flings with Lesbian "gal pals"... to no avail. HRC won handily.

Meanwhile, even that bulwark of social retrenchment, evangelical Protestantism, seems unable to resist the flow of certain social currents, having appropriated much of the style and some of the substance of post-sixties social movements: youth programs, hip music, quasi-feminist "consciousness-raising" discussions, a rhetorical emphasis on therapeutic introspection and self-actualization.[8] A quick perusal of the home schooling section of a Christian bookstore in my area turned up a surprisingly large representation of progressive ideas—notions not altogether consonant with the aim of educating children within the constraints of a narrow and sometimes bigoted tradition (e.g., tailoring instruction to the unique individual, making everyday happenings into learning experiences, encouraging creativity, empowering little girls to learn, etc.). In the 1970s and 1980s, Christian fundamentalists had built political capital by tapping widespread homophobia; by the late 1990s, it had become clear that sexual intolerance ran contrary to social tides and alienated even conservative voters. So in 1999, the Reverend Jerry Falwell, founder of the aforementioned Moral Majority, and supporters at Liberty University in Lynchburg, Virginia—ground zero in the culture wars—met with gay and lesbian Christians in an effort to "lower the rhetoric" and to encourage a change of tone in discussions of religion, homosexuality, and gay civil rights.[9]

In short, times have changed. They continue to change, and people with them. A great mass of people—gays, lesbians, bisexuals, and transgenders, surely, but also live-in couples, working mothers, single mothers, even apparently "conventional" married couples—no more recognize their lives in the old nostrums and received wisdom than working-class families of Marx's day could recognize themselves in bourgeois mores.[10]

THE ALL-AMERICAN QUEST FOR A SELF:
QUEER DILEMMAS

Recognition, precisely—re-cognition—is the heart of the matter: the wound that hurts, the vital link between personal turmoil and public troubles. For how do we catch a glimpse of ourselves in the midst of revolutionizing metamorphoses? How might we make sense of our changing practices, to ourselves and to others? And how could we ever hope to attain certainty about something so multifarious and evanescent as a self?

Such are the exhilarating and vertiginous questions usually associated with gay and lesbian "coming-out" experiences—a dramatic contest for existence that pits a heroic self against so many social obstacles, a search for authenticity that opens up fundamental (and ultimately unresolvable) questions about manhood, womanhood, selfhood. But today, it is not just gays who are party to such existential questions. Everyone is.

I still remember the jolt of surprise I felt reading a newspaper article about the thirtieth anniversary of the 1963 March on Washington (a feeling not so different from the psychic bump I experienced some time around the twenty-fifth anniversary of the Stonewall Riots when in a send-up of the investigative interview technique used by some TV talk shows to conceal a speaker's identity, Tony the Tiger began urging adult enthusiasts of Kellogg's Frosted Flakes to come out of the closet). A young woman attending the commemorative march described the original civil rights event as a "national coming-out day" for black people. Perhaps the speaker's analogy got the historical process backward (the gay/lesbian movement borrowed tactics from the civil rights movement, not vice versa—historically, black pride comes before gay pride), but her conflation underscores a simple and more immediate truth: coming out has become a general trope in American popular political culture. Anyone can emerge from denial, evasion, and secrecy to embrace a public identity. We all have to decide for ourselves what it means to be a man or a woman, to be black or white, to have this body or that. Everyone "comes out" today.

Perhaps this was always so, for much is old about our current dilemmas. When the new social movements politicized identity—when feminists first raised consciousness and gays first came out—they broke with institutional structures and ideological strictures. But they also acted as heirs of a recurring and peculiarly American injunction to find ourselves, to know ourselves, to be true to ourselves—a modernist moral quest that set Thoreau by Walden Pond, drew Huck Finn's raft down the Mississippi, and did not end with Holden Caulfield contemplating a patch of rye or John Rechy hustling his way across the American nightscape. "Identity politics"—the bête noire of cultural traditionalists, the supposed source of all that divides us—is actually as American as apple pie, canonized in our literature and lived as the national mythos.[11]

Modern identity politics is "self-making" in an era when "man"— implicitly straight, white, property-owning man—is not the only one who makes himself. What is new about today's dialectic of identity and identification, then, is that it now encompasses groups, interests, and ideas that were traditionally excluded. And what is unprecedented is just how problematic the quest for an authentic self becomes in the wake of feminism, gay liberation, and other social movements that challenge received notions of masculinity and femininity, of "proper" ways of inhabiting this or that body, and the power relations these conventions structure.

"Be Yourself" is still the quintessentially American catchphrase—the message of every self-help book and the gospel according to Hollywood— but everyone now knows what slippery work that is and how much ruse and artifice go into every staging of self. In the ensuing cacophony, where moral imperative meets fissioning identities on a many-sided battlefield, those durable categories that once grounded the search for meaning—Man,

Woman—have been thrown to the winds of history, contingency, ambiguity. The call "Be a man" still rings, still finds its echo in the civil rights claim "I am a man"—but it is no longer clear exactly what this means, or even whether men should aspire to be it. "You Make Me Feel like a Natural Woman" still gets airplay—but what once seemed like a transparently earnest yearning for spiritual communion now sounds more and more like a camp put-on, destined from the beginning to serve as a TV ad for hair dye. In an era appropriately presided over by a succession of transvestic cultural heroes—movie characters who can't decide which sex they're opposite, trickster pop artists who go both ways on the gender divide—everyone is compelled by rapidly changing circumstances to participate in an ongoing quest for self-definition. Today that quest is, unavoidably, a perpetual crisis over the true nature of sex.

FLASH: SCIENTISTS CLAIM TO DECIPHER SECRET OF LIFE

Not coincidentally, while identities everywhere are caught up in epistemic flux and volatility, American culture is at the same time awash with highly publicized (and publicity-sensitive) studies asserting a genetic source, a hormonal cause, and/or a hardwired, gendered brain as the basis for all manner of human traits and practices—from individual preferences and aesthetic judgments to cultural meanings and social structures. Hallmark's TV ad alludes, playfully, to a familiar litany of claims about gender, genetics, and genitivity—to notions given an increasingly wide hearing, in popular culture as in the culture of science: Men give gifts; women receive (and appraise) them.[12] Men are competitive, women cooperative. Men are direct creatures whose acts go straight to the object; women's thoughts and deeds are immersed in webs of social ties, emotional bonds, and complex interpersonal calculations. Men are from Mars, women from Venus. So goes a series of claims—snapshots, cartoons—about how men and women really are, in their deepest, most abiding nature. Tales of stability and permanence whispered in the maelstrom. Old bromides decked out in scientific drag.

What is most obvious about these naturalistic and naturalizing representations is that they are so emphatic on matters about which recent history has been so equivocal. They conjure nature as a stopgap boundary, a hastily drawn edge to limit and hem in ceaseless introspection and analytical regress over Man, Woman, Love, and the nature of human affinities. Incantations of symmetry, at once mythic and mathematical, they summon nature as a golden mean, an exemplary center, a principle of balance. In the resulting scientific models (as in the modern folklore from which they draw sustenance), it is the mutual lust of man and woman that insures and perpetuates this oh so balanced complementarity, the conjoining of opposites. It is thus *desire* that anchors the "I" of identity, even as everything else is swallowed by storm: a desire to recognize, to identify, to know the true nature of men and women, a desire to know the truth of desire. A desire to beat a hasty exit from the world of history and struggle and becoming.

In the long shadow of the Human Genome Project, with its dream of deciphering once and for all the Book of Life—thus, in the geneticist vulgate, settling with mathematical precision both the question of identity and the question of true human nature—the one thing slips into the other. Nature, as legible as a greeting card, becomes desire, then the "nature" of desire, which flows into an unrequited longing for nature as an uncomplicated condition of authenticity.[13]

SEXUAL SCIENCE AND ITS DOUBLE-EDGED CLAIMS

It was in the wake of neuroanatomist Simon LeVay's 1991 "gay brain" study—the first of a recent series of scientific studies purporting to document an organic, hormonal, or genetic basis for homosexuality—that I began collecting newspaper articles, magazine essays, and scientific reports on what I would later gloss the " 'nature' of desire."[14] At first, I was primarily interested in depictions of same-sex desire as a genetic tendency or organic condition and in the all too explicable media and political attention these claims received. My interest in the subject was at once personal, political, and professional.

As a gay man, I was—and remain—skeptical of attempts to understand homosexual desire by reference to an underlying organic condition supposedly affecting a distinct minority of the human race. Claims of this order are, of course, nothing new. Homosexual rights advocates have leaned on innatist paradigms since the mid-nineteenth century, when gay rights pioneer Karl Ulrichs first surmised that same-sex desire was a congenital condition. Ulrichs claimed that male same-sex desire expresses "psychic hermaphroditism," the carnal and emotional longings of a "female psyche" trapped in a "male body." Working with a few simple assumptions about the dynamics of erotic attraction, Ulrichs elaborated a general typology of sexual instincts and sexual beings. There was nothing nefarious about Ulrichs's courageous and humane publications, which expressed the author's heartfelt conviction that he and others like him had been "born different."

Ulrichs was optimistic that if it could be shown that men who were attracted to men were not "choosing" their feelings—that if sexuality were thus situated in the domain of nature, rather than that of religion—then dispassionate science would dispel prejudice and superstition against same-sex desire.

Subsequent developments reveal the naiveté of this optimism. The emerging field of psychiatry took up the theme of the homosexual body, held to be physiologically distinct from other bodies. Indeed, the notion became central to early sexological concerns. But where Ulrichs had seen benign variation, Richard von Krafft-Ebing and a host of others saw pathological deviation. Drawing on Spencerian notions of biological evolution and racial regress, Krafft-Ebing depicted "contrary sexual feeling" as a degenerative disorder of the central nervous system. This theme of sexual "degeneracy" pervaded the texts of early psychiatry and medicine. It loomed large in Nazi pseudoscience, where homosexual bodies were among those marked "unfit." It informed decades of sham "therapies" and violent "treatments"—just as it haunts the discourses of sexual science today. Modern variations on the theme include homosexuality as hormonal imbalance, homosexuality as fetal stress syndrome, homosexuality as a genetic disorder associated with fetal wastage, and so on.[15]

As my thumbnail sketch ought to indicate, biological explanations of same-sex desire have a long and problematic history, not the least problem being that they promote a dangerously medicalized view of human sexuality—a conception of sexual orientation that still resonates to the ideas of nineteenth-century scientific racism, eugenics, and phrenology. But it is not just that these explanations represent bad politics. Like their nineteenth-century antecedents, they represent bad science (and especially bad *social* science), as well.

THE CULTURE OF NATURE

Since the Enlightenment, the natural sciences have been recognized as the disciplines that ought to guide our understanding of the world around us and how it works. As such, the natural sciences enjoy enormous cultural prestige. Their findings are the object of considerable interest by non-scientists, whose curiosity about new scientific discoveries on subjects ranging from particle physics to global ecology is served—and stoked—by a variety of electronic and print media, including cable TV channels, popular magazines, and weekly newspaper supplements. Interest is particularly keen in the biological sciences, especially as these touch on how the human body works or fails to work, on medical treatments and their efficacy, and, perhaps especially, on human evolution and genetics.

Fields like evolutionary biology and genetics have always been subject to ideological influences—to the sway of received opinions and untested beliefs about the nature of men, women, and desire. The proximity of their subject matter to folkloric ideas about human nature and heredity has also long sustained the correlative flow of less-than-scientific ideas in the culture at large. The resulting two-way traffic between science and popular culture has perhaps never been more intensive than it is today, and much of this circulation happens in the serious public sphere. Just ask any regular reader of the *New York Times* or *Time* what biological science is good for, and chances are s/he'll tell you that it helps us understand the evolution and biology of human nature. It tells us how men and women got to be the way they (supposedly) are. At best, it might give us some leverage over the genes and hormones that control our lives.

In popular culture today, when it comes to questions about "human nature," biology is almost always taken to be sociobiology, a set of claims organized around the assumption that biology is destiny for humans, that genetic predispositions determine (or ought to determine) our behavior toward others and our institutional forms. Sociobiology and its offshoot, evolutionary psychology, which explains our attitudes and behaviors in terms of natural selection and sexual selection, are thus reductive, and unapologetically so: they reduce culture to nature, pure and simple—and to a very simple conception of nature, at that.

Sociobiological conceits have come to organize an extensive body of writing in popular science and in science journalism. Opinion leaders are so confident in the truth of the naturalistic origins stories they relate that they seldom even name "sociobiology" as such. Rather, they take its assumptions for granted. Entertainers, journalists, and scientists alike invoke its style of argumentation, its reduction of social actions to genetic causes, and especially its pat claims about what the real man does and the natural woman wants. Indeed, so do scientists, particularly when explaining and legitimizing their work in public pronouncements. Assumptions of a sociobiological order undergird some of the urgency and much of the hype associated with the Human Genome Project—no small quarter of scientific practice today. Scientists working in many fields invoke sociobiological stories about human origins in their models of the biology of sexual differences. Neural and hormonal correlations with anatomical sex—no matter how tentatively documented or poorly understood—are instantly interpreted by many scientists in terms of speculative evolutionary stories.

A 1999 episode of the Fox TV show *Ally McBeal* provides a window onto the resulting entanglement of scientific claims with common sense in popular culture. A fictional lawyer, Richard Fish, summarizes his arguments to the jury in terms that all too accurately distill claims widely aired in the serious public sphere as scientific fact. "When we talk about men and women as Mars and Venus, it's because on some issues the two genders really are from different worlds," Fish begins. "And when it comes to sex, basic anthropology tells you [that] the female species [sic] looks for one mate while the male innately looks to spread his seed with as many mates as possible—to propagate the long-term survival of the species itself. Even a gay anthropologist will tell you so. Men and women are guided by a different missile."

The misidentification of anthropology with evolutionary psychology (never mind the unfounded claim about what a gay anthropologist might say) is symptomatic of just how confused questions related to sex, science, culture, and human nature have become in the public sphere. It wasn't ever thus. As Micaela di Leonardo reminds readers, when *Time* magazine ran its front-page story on the then-new field of sociobiology in 1977, it gave extensive space to rebuttals by prominent anthropologists Marshall Sahlins and Marvin Harris as well as Harvard biologists Stephen Jay Gould and Richard Lewontin.[16] The year before that, in a polemical treatise that distilled a devastating anthropological critique of sociobiology, Marshall Sahlins had written of the "good possibility that [sociobiology] will soon disappear as a science, only to be preserved in renewed popular convictions of the naturalness of our cultural dispositions."[17] Sahlins, it turns out, was half right. Sociobiology has become common sense. But by that very virtue, it has by no means disappeared from the scientific scene. The interchange between common sense and sociobiology, as between folklore and science, is a continuous and multifaceted one.

Throughout the 1990s and into the twenty-first century, science journalists at *Time, Newsweek,* and the *New York Times* have essentially served as cheerleaders for evolutionary psychology's pat proclamations on the "nature" of gender and sexuality. The very different reception for reductivist paradigms in today's marketplace of ideas corresponds to no new scientific discoveries that would clinch the case for "nature" over "nurture" on any question of social import. On this point, the facts remain much what they were in 1984, when Richard Lewontin, Steven Rose, and Leon J. Kamin wrote: "Up to the present time no one has been able to relate any aspect of human social behavior to any particular gene or set of genes. . . . Thus, all statements about the genetic basis of human social traits are necessarily purely speculative."[18] Indeed, it is almost embarrassing to point out that sociobiology, which purports to derive human nature from simple genetic models, transmutes into evolutionary psychology, which attributes the origins of human nature to supposed norms of hominid behavior over eons of evolutionary time, *precisely* as an intellectual retreat from the former's strong but ultimately undemonstrable empirical claims.[19]

What has contributed in part to the ubiquity of sociobiological discourses today is the declining visibility of credible "public intellectuals" steeped in the social science traditions, and the consequent erosion of what C. Wright Mills used to call "the sociological imagination" in public affairs. Sociobiological claims once faced systematic, factual rebukes from anthropologists, sociologists, and even biologists—critiques that were informed and substantial but that could still be put in educated laypeople's terms—whereas today's biological reductivist claims face decidedly less decipherable responses from a generation of scholars weaned on the arcane vocabulary of deconstruction and poststructuralism. Advances on the intellectual front have thus drawn heavily against loans on the political front. The resulting dumbed-down public space has been largely ceded to those who can serve up bite-sized homilies and comforting quotes.

The enthusiastic reception for reductivist ideas today no doubt reflects both a changing sociopolitical climate and changes—some subtle, some dramatic—in the staging of claims about human norms and human nature. Reductivism's adjustments, ameliorations, and "softening" over the years have contributed to its present public respectability while facilitating its accessibility—as has a steady stream of sympathetic media reports. Its "positive" spin, crafted through evolutionary psychology's tales about how men got to be risk-taking but protective, how women got to be nurturing but savvy, and thus how the two ought to get together and cooperate (copulate), has ultimately proved more palatable and more marketable than the first drafts of the sociobiological argument, which largely set out to explain and rationalize the "nasty" side of human behavior, most frequently in misogynist terms. (Everybody likes a happy ending, it would seem, and evolutionary psychology—sociobiology lite—thrives on the art of spin.)

Identity politics—the quintessentially modern justification of political action and social redress by appeal to supposedly deep-seated, essential identities—also provides fertile ground for bioreductivism. (And it ought to be clear that "identity politics" is by no means the exclusive strategy of excluded, oppressed, or oppositional groups: women, gays, racial minorities. It is also the means whereby dominant groups define themselves and pursue their interests.) Real-life political columnists pitching the supposed interests of misunderstood straight men invoke evolutionary scenarios much like the one drawn by fictional lawyer Richard Fish to justify his client's actions. Activists steeped in cultural feminism or gay identity politics rush to selectively embrace those parts of biological determinism that serve their purpose—the idea that women are inherently nurturant, or the notion that homosexuality is an inborn condition. More than anything, today's reductivism offers to stabilize identity in the *points de capiton* of biology—that is, it purports to secure stability and certitude in an era when nothing much seems anchored about either identity or biology.[20]

The subsequent pages are critical of the reduction of identity and social relations to biological determinants. In what follows, I shall take my pokes at a number of currently fashionable bioreductivisms: evolutionary just-so stories, neurohormonal essentialisms, and "genomania" in general, the explanation of all manner of happenings and mishaps in terms of genes. But far from being a reflexive science basher, I'm the first to give science its due. I take it as axiomatic that most of us who live in the developed world enjoy modern science's considerable benefits: in a cultural tradition that values free inquiry and reasoned explanation, no less than in elevated living standards and extended life expectancies. I also have no doubt that means and modes of scientific knowledge have advanced dramatically in recent years.[21] It seems that each week brings reports of new discoveries about the nature of the universe, the development of solar systems, the origins of life on earth, and the evolution of human beings. The sci-fi enthusiast in me reads a headline from 2001 with adolescent wonderment: "Scientists Bring Light to Full Stop, Hold It, Then Send It on Its Way."[22] I marvel in particular at recent breakthroughs in the fields of genetics, stem-cell research, and therapeutic cloning, with their implications for how we might some day live—for instance, the discovery of genetic correlates for a number of hereditary diseases, the first apparently successful application of gene therapy to a medical disorder in human beings, the subsequent identification of a genetic mutation that roughly doubles the life span of fruit flies, the use of embryonic stem cells derived from cloned mice to grow healthy organs or organ tissue, the use of gene therapy to restore sight to blind dogs . . .[23] Still, I cannot help but note that many of the more consequential breakthroughs in medicine have occurred in less technical fields. I think here of the development of simple but highly effective physical therapy techniques for the aggressive treatment and rehabilitation of stroke victims.[24] And I remain skeptical of the aura acquired by big science in the 1990s. If it seems incontrovertible that scientists working in many fields have greatly refined their methods for investigating certain questions about the *natural* world, it also seems clear that recent advances in science have fed a kind of generalized rage for bioreductive explanations for questions about the *social* world.

The problem is that the latter mania does not follow from the former success stories. Good as it is at solving certain kinds of problems, science cannot answer every question (a point "the official bad boy of science journalism" John Horgan has been at pains to make).[25] Advances in genetics notwithstanding, not everything has a genetic explanation.[26] And not every proposition advanced in the name of science is actually scientific. It's when those claiming the authority of science give vent to unwarranted reductivisms or when scientists themselves turn from controlled experimentation with well-defined natural phenomena to unchecked (and uncheckable) claims about social life that I become critical—for it is precisely on these grounds that science mutates into something most unscientific: biological folklore, fables contrived out of prejudice, and related forms of pretend knowledge. Not coincidentally, much—though by no means all—of what critics treat as the dark side of the Enlightenment springs from this unscientific practice of science: medical misogyny, scientific racism, and homophobic biology.[27]

Thus, this book proceeds on two tacks. First, because innatist claims about human sexual orientation have never stood up to careful investigation, the perpetual attachment of scientists to such conceits suggests an ideological fixation, not a legitimate scientific interest. In the following chapters, I try to untangle various permutations of this fixation, to trace the meandering course of an ideology against the historical backdrop of changing sexual formations. Second, because such double-edged notions continue to exercise a strong appeal among advocates for gay/lesbian rights—because a naturalistic and naturalizing ideology seems to give stability and uniformity to something as phenomenologically volatile as *desire*—it is necessary to examine critically the deeply felt logical deduction that first becomes visible around the question of same-sex attraction sometime in the mid-nineteenth century: "What desires I do not will or control belong to my inborn nature, my biology."

SPECIOUS CLAIMS AND LOGICAL FLIP-FLOPS

To a practitioner of cultural studies, as to an anthropologist trained on the analysis of cultural variation and historical change, nothing could seem more naive than the notion that sexual preference is somehow indelibly written on the body or "coded" in one's genes. But this is precisely where matters become complicated—where mythic conceptions of nature sprawl in a kinked and languorous way through contested representations and changing practices, in the culture of science, no less than in popular culture.

Take, for example, the phrase that hangs over nineteenth-century models of homosexuality: "contrary sexual feeling." Although Karl Westphal coined the term in 1869, the concept was implicit in Ulrichs's earlier ideas, first published in 1864. The notion that one's sexual feeling might be "contrary" sustains a set of key oppositions and relations that appear in other terms from the period—"sexual inversion," "homosexuality," "intermediary sexual type"—that were appropriated by Krafft-Ebing in 1877.[28] In inaugurating this chain of associations, Ulrichs took it for granted that same-sex desire was "contrary" to the ebb and flow of *natural* attractions: Men were drawn to women and women to men by force of a mysterious "animal magnetism" whereby opposites attract. That very notion, in turn, marks an earlier cultural revolution. As historian Thomas Laqueur shows, the phrase "opposites attract" was actually a novel idea at the beginning of the eighteenth century. And as gay studies researcher Hubert Kennedy notes, by Ulrichs's day, this aphorism already seemed to occupy the status of a timeless universal: Ulrichs conjectured about same-sex desire in a milieu in which attraction to the "opposite" sex had been naturalized.[29] Ulrichs resolved the apparent inconsonance of same-sex desire by simply asserting that inside the body of the man who desires other men is the mind of a woman—for whom attraction to men is proper and consonant. He thus brought what Christian prejudice deemed "unnatural" lust within the purview of nature by folding it under the rule of heterosexual symmetry. Same-sex desire thus turned out really to have been heterosexual desire all along—in disguise, perhaps, or confused by the outward signs of the body.

A delightful series of logical involutions happen when any of this is taken literally—meanders and flip-flops that used to give my smart-aleck friends and me late-night giggles during college sleepovers. For example, a gay man in love with a straight man expresses psychic heterosexuality, but only so long as he desires to play the "female" role. Reciprocal male homosexual desire would actually seem to be a special case of lesbian psychic desire, since it involves the minds of two women. Contrariwise, lesbians are "really" gay men trapped in women's bodies. Like an excellent transvestic gag, nothing is quite what it appears to be. However, the man who wishes to penetrate another man remains something of a puzzle under Ulrichs's associations— unless one conjectures that he is really a "straight" man attracted to the "woman" inside the desired gay man.

The logic of this position was no better a hundred years ago than it is to-day. Still, it is amazing how often one encounters similar propositions, en-tanglements of vernacular gender with common-sense nature in essential-izing representations of sexuality. In myriad contemporary discourses, gays, lesbians, and transsexuals continue to be represented as individuals whose gendered brains are somehow out of sync with their bodies. Such a preconception has settled uneasily into the received wisdom. It guides every search for a "gay brain," a "gay gene," or "gay neurohormonal" pat-terns. (One can hardly understate the naive literalism of present-day sci-ence on these matters: scientists still look for the supposed anatomical at-tributes of the opposite sex embedded somewhere in the invert's brain or nervous system.) And this notion now enjoys a second, third, and even fourth life in political discourses. It is by appeal to such conceits that Aaron Hans, a Washington, D.C.–based transgender activist, reflects on his un-comfortable life as a girl: "I didn't *think* I was a boy, I *knew* I was a boy." Hans elaborates: "You look at pictures of me—I actually have great pictures of me in drag—and I literally look like a little boy that's been put in a dress."[30] Far, far be it from me to cast doubt on anyone's sense of discom-fort with the ascribed gender roles. Nor would I question anyone's sense that sexual identity is a deeply seated aspect of who they are. But testimonies of this sort and appeals to the self-evidence of perception beg the obvious question: Just what is a little boy or girl *supposed* to look like? The photo-graph that accompanies Hans's interview shows a somewhat robust girl. Is this to say that (real) girls are necessarily delicate and (real) boys athletic? (If so, virtually all of my nieces are "really" boys, since not a one of them is delicate or unpresupposing.)

There is indeed something compelling about such intensely felt and oft-invoked experiences—"I knew I was gay all along"; "I felt like a girl"—but that compulsion belongs to the realm of outer culture, not inner nature. That is, if "inappropriate" acts, feelings, body types, or desires seem to throw us into the bodies or minds of other genders, it is because acts, feelings, and so on are associated with gender by dint of the same all-enveloping cultural logic that gives us pink blankets (or caps, or crib cards, or I.D. bracelets) for girls and blue for boys in maternity ward cribs. When we diverge one way or another from those totalizing associations, we feel—we really feel, in the depths of our being—"different." Therein lies the basis for an existential opposition to the established order of gendered associations. But therein also lies the perpetual trap: Every essentialist claim about the "nature" of same-sex desire in turn refers to and reinforces suppositions about the "nature" of "real" men and women (from whom the invert differs), about the "naturalness" of their mutual attraction (demonstrated nowhere so much as in the invert's inversion), about the scope of their acts, feelings, body types, and so on (again, marked off by the deviation of the deviant).[31] Aping the worst elements of gender/sexual conservatism, every such proposition takes culturally constituted meanings—the correlative associations of masculinity and femininity, active and passive, blue and pink—as "natural facts." In a twist as ironic as the winding of a double helix that goes first this way, then that, the search for gay identity gradually finds its closure in the normalcy of the norm as a natural law.

In the end, I am not convinced of the basic suppositions here. I doubt that most men are unfamiliar with the sentiment given poetic form by Pablo Neruda: "It happens that I become tired of being a man." Even psychiatrists who treat "gender dysphoria"—a slick term for rebellion against conventional gender roles—admit that at least 50 percent of children at some point exhibit signs of mixed or crossed gender identity or express a desire to be the "opposite" sex.[32] A century after Freud and fifty years after Kinsey, the rectitude of the heterosexual is as doubtful as the monosexuality of the homosexual. In the throes of the twentieth and twenty-first centuries' ongoing cultural revolutions, it should scarcely be a matter of great controversy to say that desire, like identity, is compound, complex, and volatile. You can't even convince me that feelings of homosexual desire are in any sense the exclusive domain of a distinct minority. One has only to read novels, watch movies, or talk to one's straight friends to realize that everyone agonizes over queer feelings of "difference." One has only to watch the evening news or read the daily papers to see that everyone, gay and straight, participates in a modern identity crisis.

WHY I DON'T BELIEVE IN THE GAY GENE

My perspective on these matters is informed by serious, careful study, but I have little doubt that my relentlessly constructionist view of sexuality— the notion that identities are socially "made," not biologically "given"—is also informed by the idiosyncrasies of my own coming-out experiences.

When at the age of eighteen I announced that I was gay to friends, parents, and fiancée (yes, sadly, it *was* complicated), my interlocutors were, without exception, flustered, upset, and tearful. With all the teenage earnestness I could muster (and in what at that moment must have constituted something of an emotional ambush), I pressed a point: "You're not going to tell me that *you've* never had sexual feelings for someone of the same sex."

From man, woman, and child, the reply was something like "Well of course I've had these feelings. Everyone has. But you don't actually *act* on those feelings. . . . "

Of course, I *did* act on them. (So did a lot of my straight friends, albeit with equivocation or alibi: "I was *so* drunk last night . . . " "I don't normally do this kind of thing . . . ") I came away from these conversations convinced that it was only this "acting"—and the proclamations I insisted on making about it—that separated my queer self from those who stayed on the straight and narrow. I also came away from these experiences convinced that any difference between the inner yearnings of gays and straights was one of degree, rather than of kind: a difference constrained and elaborated by the need to make and maintain a difference.

This sense of things was reinforced when, over the months that followed, feelings of occasional sexual attraction to members of the *other* sex gave rise to panicked thoughts and yet another round of identity crisis: "Maybe I'm not homosexual after all . . ." "Perhaps I've made a terrible mistake . . ." That, as they say, is another story. But the complexity of my own experience is the lived, phenomenological basis for why I don't believe in the existence of a gay gene.

IDENTITY, ALTERITY, AND NATURE

It might be objected that I am giving a one-sided critique of biological determinism, that non-innatist conceptions of sexuality—for example, Christian moralism, conservative Freudianism, Stalinist puritanism—are equally problematic and have been equally implicated in the oppression of gays and lesbians. Undoubtedly, this is so. But what imbues these approaches with an intolerance of same-sex desire is not their "constructionist" bent, but their reliance on a heterocentric and moralizing preconception of nature: at base, the idea that some desires and practices are more "natural" (and thus more "wholesome," more desirable) than others.

Mindful of this complexity, I had originally set out to compose a tight essay on the concatenations of popular-cultural conceptions of nature in scientific representations of homosexuality, only to find that in tracing the meandering tracks of this subject matter, I perpetually encountered much the same problem described by Anne Fausto-Sterling in her early critique of Simon LeVay's "gay brain" study: To write about homosexuality is to write about heterosexuality, and vice versa.[33] But it is also to write about so much more than desire, sex, or even love. Because vernacular understandings of *gender* infuse scientific models of *sexuality*—now, no less than in the nine-teenth century—the "science" of homosexuality invariably rests on, refers to, and reinforces a broader set of cultural conceptions: notions of what a real man is and what a natural woman ought to be, understandings of what a man does and how a woman feels, ideas about reproduction and its role in human life: thoughts, in short, in which "identity" refers to *otherness.* By means of such gossamer threads we are all caught up, and endlessly so, in each other's fables (and in fables about each other).

Thus the mad desire for studies that "prove" men and women really *are* different—in the hypothalamus, in the corpus callosum, in their basic na-tures. Thus the speed with which every real and alleged anatomical differ-ence between men and women gets caught up in the familiar stereotypes and speculations. So, too, the immediate transposition of every putative gen-der difference to the question of sexual orientation: Gay men are believed to have something of the woman about them, and lesbians are understood, essentially, as manly women. In the sexual imaginary, theories about what it means to desire "the same" diacritically embrace claims about the nature of the opposition between "opposites"—claims that invariably turn out to be "folk beliefs." Neuroanatomist William Byne summarizes the current state of these shabby endeavors with the following flat-footed but devas-tating generalization: "Recent years have witnessed numerous attempts to demonstrate that the brains of homosexuals exhibit characteristics that are typical of the opposite sex. In some cases, these attempts have come decades after persuasive evidence suggested that the brain characteristic in question does not differ between the sexes in humans."[34] And with the dissemina-tion of scientific studies through mass media—always the study that as-serts an essential difference between men and women, straights and queers, never the study that falsifies the premise or fails to replicate a find-ing of difference—the circle is closed. What began as preconception and went untested in research returns to its source—common sense, collective prejudice—as a "new discovery."

Science—which has seldom been very scientific about the study of sex—thus plays an important role in the varied imaginings of contemporary sex-ual culture. Its attempts at reducing desire to a thinglike object of study give the illusion of stability and permanence to relations and practices that are ambiguous, contested, and in flux. Its authority as a source of supposedly detached and impartial inquiry lends credence to the most transparently par-tisan claims about human nature. Its attempts at defining an ahistorical, acul-tural, bell-shaped curve of "normal" desire obviously enough serve func-tions of legitimation and social control—as do myriad struggles over what to include in the norm and what to exclude. But such struggles, because they are struggles, are open-ended. They participate in the dialectical tug and tow of history—in *changing* ideas about manhood, womanhood, and desire. So immersed in the politics of representation, the study, no less than what it studies, ought to proceed judiciously—with a sideways glance and a look over the shoulder. And here, I mean to enclose the present study within the give-and-take of cultural introspection and social reflexivity. A lot, after all, rests on how we think about and what we make of nature.

THE NATURE WARS

In the summer of 1998, when organizations of the Christian Right sponsored a series of ads in national newspapers promoting religious "cures" for homosexuality, they were careful to avoid using the word "unnatural." For good reason. The open invocation of "un-nature" would have raised the specter of its corollary, "nature," which in modern discourse belongs to the province of science, not religion. Since the view being promulgated was anything but scientific (or, at any rate, because the discourse trafficked in a defunct science of generations past), these carefully scripted ads instead evoked the conventional, euphemistic stand-in for nature: a longing for purity and wholeness, supposedly felt by ex-gays while they were immersed in "the homosexual lifestyle." Of course, one would have to have been culturally illiterate not to read the unwritten word, just as one would have to be scrupulously inattentive to the religious Right's real practice to swallow the first part of the oft-invoked slogan "Love the sinner, hate the sin."

In this context, then, consider the rhetorical wrangling over nature in the exchange between Urvashi Vaid of the National Gay and Lesbian Task Force and Stephen Black, "ex-gay" preacher of First Stone Ministries, on ABC's *Good Morning America*. At pains to say that he was not simply repressing or denying homosexual desires, Black argued, in essence, that Christ restored him to his true nature—his authentic self—by liberating him from acquired habits of artifice and vice caused by childhood trauma and gender confusion: "As I began having a relationship with Christ, he began to deal with the issues of my heart. . . . I kind of naturally progressed into being able to be free from having same-sex attraction." Black thus echoed the message delivered a month earlier in full-page newspaper ads: "I'm living proof that Truth can set you free," headlined Anne Paulk, "wife, mother, former lesbian."[35] The premise is that gays are inhibited from realizing their own "true" instincts. The conclusion is that they can "go straight" without repression, overt violence, or brainwashing: the outcome of a natural healing process induced by Christ's love, which dis/re-places same-sex love.[36]

Vaid began appropriately by countering the fundamentalists' unscientific claims at their very root: "Homosexuality is not an illness, it's not a mental illness, it's not a medical condition, it's not an affliction. To say that homosexuality is an illness . . . is as nonsensical as saying that heterosexuality is an illness." But then she went further, underscoring in quasi-religious terms not the strength of desire but the truth—and fixity—of identity: "Sexual identity is deep-seated, it's fundamental. It's God-given, if you believe in God." Vaid concluded with a claim about the beneficence and plenitude of nature: "And that's part of our nature. Some of us have same-sex sexual orientation, some of us are heterosexually oriented. . . . It's healthy, natural, and normal . . . for human beings to have sexual identities and to have a sexuality."[37]

It might seem like nitpicking to sift through Vaid's wording. Her performance on the occasion was, as always, effective. In a personal communication, she later added: "It is hard to negotiate these 'discussions' when your whole legitimacy as a human being is being questioned."[38] If my rehearsal of nineteenth-century confusions indicates my discomfort with Vaid's refuge in nature, with the rhetoric of science, and with the implication that people's sexual feelings are immutably "hardwired," I can offer no easy solution to the problems posed by the recurring engagement in which her arguments participate. The long-standing demand, made by religious conservatives, distraught parents, and liberal helping professions alike, is but this: *Change your unnatural desires.* Time and again, the response is given: *I can't change them—They're part of my nature.* Would it be as convincing to own one's sexuality in a voluntarist fashion, to say, simply, "No, I *won't* change them—I'm as queer as I want to be"?[39] Or would it be as effective to take to the airwaves with a post-Kinsey, sex-radical, constructionist position— to argue the legitimacy of one's feelings without recourse to the depths of an inner nature?

In "Queering the State," Lisa Duggan humorously imagines a panel of queer theorists on ABC's *Nightline,* attempting to communicate esoteric theory to a presumably perplexed mass audience.[40] Her point is well taken. Academic work often marshals mysterious jargon in the pursuit of fine points. As work in gay/lesbian studies has become more responsive to academic demands, it has become more recondite—and sometimes less relevant to real social and political struggles outside the conference circuit. Still, the public that put *Seinfeld* at the top of the Nielsens for much of the nineties and followed President Bill Clinton's careful delineations of what might count as "sex" in official depositions probably has a greater capacity for nuance than Duggan's conjecture might allow. I am convinced that sophisticated constructionist positions can be put into simple, accessible language—as with the exemplary antiracist ads that appeared on the sides of buses in 1998: "Race is a fiction, racism is real." Consider this book, then, to be a gambit about the intelligibility and relevance of constructionist arguments.

ANOTHER KIND OF CONTENTION

"Desire and identity are inherently ambiguous," a different kind of contention might begin. "Some of us are more or less exclusively homosexual for most of our lives, many more are exclusively heterosexual," the argument might continue, rightly acknowledging the salient facts. "But sometimes even straight men find themselves infatuated with their best friends and—as any veteran of feminist consciousness-raising can tell you—women who think of themselves as heterosexual sometimes discover a lesbian potential they didn't know was there. It's not unheard of for gay men to fall for women, or lesbians to sleep with men."

Now for the theory: "Freud believed that all human beings have a bisexual potential. Research by Alfred Kinsey, Laud Humphreys, and others suggests that a lot of people *act* on that potential at some point in their lives.[41] Anthropological studies of other cultures have shown that human sexual practices are remarkably varied—that there's more than one way to organize the institutions of family, kinship, and sexual life. Some societies even *require* every male to engage in same-sex relations for extended periods of time. What all of this means is that nothing in 'human nature' gives us a heterosexual norm and a homosexual minority. Sexuality is largely what we make of it."

Then, a dash of social context to make sense of how we "make" sexuality: "In modern America, people are very much in the process of making new things out of sex and sexuality. All around us, relationships are in flux: gender roles are changing, sexual practices are changing, basic institutions like marriage, family, and kinship are changing, all at a dizzying speed. None of this means that people 'choose' their sexuality the way a person might choose a pair of socks. But in fact, many individuals do change over time."

Segue to the argument: "So much variation, experimentation, and change makes some people very nervous: they come up with absolutist claims about an unchanging nature, or, they fall back on the premodern idea of divine law as the last recourse on these matters. But 'nature' explains nothing here.

And nobody really knows very much about why people have the feelings they have." Then, cut to chase: "None of this is an illness or a disease. None of this means that the end of the world is at hand. There's nothing wrong with any way that people can express love, make community, or find consensual pleasure. What's wrong is trying to make people feel sick or evil or perverted about things that are just part of being human. What's wrong—and dangerous—is trying to narrow the range of pleasures people find in our wondrously human bodies."[42]

My brief narrative here keeps to positions I argue throughout this book: first, that representations predicated on a narrow, unitary, or fixed conception of identity work at the expense of human freedom; and second, that reductive arguments from nature will ultimately fail us—because such arguments are conservative precisely where they should be radical, because they cannot grasp the logic whereby identities are really improvised, and because they ignore the volatile history in which present-day struggles are so deeply and inextricably immersed. The current frenzied compulsion to fix everything to an immutable nature—on the one side, to anchor the faintest nuance of desire in the structure of the genome, on the other, to measure every lust against the dictates of God's Law, as it was revealed to Bronze Age sheepherders—is itself symptomatic of just how much, how quickly, and how fundamentally the tectonic plates beneath basic institutions are shifting. Any model of identity unable to embrace its own freedom while grappling with its own historicity will not likely be up to the task of helping to build more livable institutions or providing insight in the struggle against intolerance, for it is precisely on this point that the foes of change fixate: on the cultural anxieties aggravated by the rapid pace of recent changes—on a sense of uneasiness with modernity's blurred genders, sexual transformations, and personal liberations.[43]

In an essay that laments the conservative movement's growing obsession with homosexuality, gay journalist and author Andrew Sullivan—himself a Tory Catholic—gives a glimpse of those anxieties and their inner logic. His description of a 1997 gathering of the intellectual Right, where movement notables plotted the terms of an ongoing crusade against modernity, is revealing:

> Bill Kristol . . . gave the concluding address at a Washington conservative conference dedicated, as its brochure put it, to exposing homosexuality as "the disease that it is." Kristol shared the podium with a variety of clergy members and therapists who advocated a spiritual and psychoanalytic "cure" for homosexuals. One speaker, a priest, described homosexuality as "a way of life that is marked by compulsion, loneli-

ness, depression and disease," comprising a "history-limiting horizon of a sterile worldview divorced from the promise and peril of successive generations." Another speaker decried legal contraception and abortion as the "homosexualization of heterosexual sex," and bemoaned that nonprocreative trends among white Europeans was leading to "race death."[44]

Across the United States, such familiar appeals to natural law and divine law sustain political opposition to civil rights for gays and lesbians. Here is a brief sample of town-hall statements in opposition to Vermont's gay civil union bill: " 'This is a sad, dark day for the state of Vermont, and God help us all,' one opponent said. Another warned, 'What we are trying to do is against the law of nature and against God's law.' A third called the civil unions bill 'social rape' and a sign of 'moral rot,' and a fourth said gays 'choose to engage in unnatural and unhealthy acts.' "[45]

At the outset of the new millennium, then—as at the end of the nineteenth century—we still have the homosexual scapegoat, whose very existence serves as a demonstration of "moral decay." Once again, we have scenarios of "sexual degeneracy": the commingling of sex, gender, and race in panicked disquisitions on the "nature" of desire.

NATURE, THE POSTMODERN CONDITION

In high culture and low, in matters academic and vernacular, and in science no less than in superstition, wherever the truth of the true, the reality of the real, or the authenticity of the self are at stake, partisan positions invariably present themselves as dispassionate claims about the nature of nature.

Thus, in a *New York Times Magazine* essay, a historian attacks "the joyless militancy of American feminism" for "questioning the naturalness of the family" and "upsetting the psychologically delicate ritual of seduction."[46] An anthropologist prefaces his academic text with a broadside against the "suffocating hold" of social constructionism and the "dreadfully passive view of nature it upholds"—echoes of Camille Paglia's charge that gender feminism lacks an appreciation of the "wildness" of nature and sex.[47] Meanwhile, a persistent variant of cultural feminism rhapsodizes over Woman's timeless nature, supposedly embodied in goddess myths, coded in the X chromosome, and expressed in nurturing practices of the eternal Mother.[48] But the nature venerated in certain misanthropic variants of the Deep Ecology movement more faithfully resembles the vengeful old shrew of sixties TV-advertising fame who castigates errant humanity with calamity and plague: "It's not nice to fool Mother Nature." And in the austere pages of the conservative *National Review,* an essayist invokes neo-Darwinian evolutionary stories about sexual difference, social hierarchy, and "natural inequality" to endorse the study of nature: "We may fairly conclude that a Darwinian politics is a largely conservative politics."[49] In fact, one is tempted to conclude, with Jeffrey Weeks, that "there are . . . as many natures as there are conflicting values."[50]

Plainly, now is not the first time nature has appeared in so many contested shapes and volatile forms. Conceptions of nature that encompass both the "nature" of desire and a longing for unadulterated purity go back, at the very least, to the Platonic Christianity of Saint Paul. And certainly, there is nothing new about the conceptual somersault whereby the heady euphoria of Rousseauian nature gives a sharp and gut-wrenching turn to the lawlike structures and dictatorial imperatives of a master design. This is not just the history of sexology from Ulrichs to Krafft-Ebing, it is the larger trajectory of Enlightenment science and philosophy as chronicled by critical theory since the Frankfurt School.[51] But if we might trace the similarities between, say, nineteenth-century conceptions of "psychic hermaphroditism" and contemporary claims about an intersexed hypothalamus—or perhaps between Cesare Lombroso's anthropometrics and certain tendencies today to assert that criminality is an innate, genetic condition—it is good to recall that although history repeats itself, it never quite repeats itself *exactly.* Hence, Marx's pithy old joke: events play out "the first time as tragedy, the second as farce."[52] Old concepts get a new spin in today's genomania. The discourses on nature, however hauntingly familiar, take on a special saliency for us in the midst of ongoing social changes and culture wars. We thus live nature in our own special ways.

In *The Condition of Postmodernity,* David Harvey depicts contemporary culture caught in the cross-currents of global capital flows, technological innovations, and rapid social changes. His snapshot provides a useful context for the themes introduced here. On the one hand, "the more flexible motion of capital emphasizes the new, the fleeting, the ephemeral, the fugitive, and the contingent in modern life." On the other hand, this experience of flux and change coexists with—indeed, it *feeds*—authoritarian nostalgia and a longing for certainty: "It is also at such times of fragmentation and economic insecurity that the desire for stable values leads to a heightened emphasis upon the authority of basic institutions—the family, religion, the state."[53] In Harvey's survey, then, the "postmodern condition" is neither the one thing nor the other, but rather the way the two come together: secular cosmopolitanism with religious neofundamentalism, nativist movements in the midst of unfettered free markets, obsessions with identity alongside the end of clear-cut identifications. It is also, in the terms of this book, the way genetic folklore comes together with religious ideology and the way modern political economy takes up contrary, contending representations of sex.

THIS BOOK

This introduction has attempted to trace the issues at stake in the book. A first immersion in the text as a whole, it suggests how innatist ideas about homosexuality lean on essentialist conceptions of gender, and it suggests something of how these representations flare at moments of acute conflict and uncertainty about sex and the body.[54]

The first section, "Origins Stories," surveys some of the basic intellectual moves of sociobiology and evolutionary psychology—today's creation myths—with special attention to their brisk, uncritical traffic in conventional understandings of family, gender, and sexuality. Like the other parts of the book, this section may be read either as an extended essay on its subject or as a series of separate but related chapters.

The chapters in "Adam and Eve Do the Wild Thing" take up and intensify the themes broached in "Origins Stories." They critique pervasive conceptions of a "natural design" of sex, the body, and identity as these notions have been promulgated both in scientific models and in their social dissemination. Expressly examining the heterosexualization of nature and the naturalization of heterosexuality in scientific models from Darwin to sociobiology, this section also outlines a broad historical and philosophical approach to questions about nature and science. It treats in general terms how scientists, as social actors, are located within the knowledge they craft and the representations they make.

"Venus and Mars at the Fin de Siècle" follows the circulation of meanings between science, politics, and the mass media. It draws on popular science journalism and TV entertainment to trace the changing logic of contemporary stories about men's and women's inner natures: from neo-eugenicist theories about the biology of beauty to retellings of the story of masculine and feminine natures in light of new social realities, and from male revanchist musings on the marvels of testosterone to confused claims about what the laws of nature say about the institution of marriage.

It is clear that the scientists who speculate about the "biological basis" for supposedly universal sex norms work in sublime ignorance of the methodological procedures and empirical findings of *social* science on gender, sexuality, and kinship. Revisiting basic lessons from Anthropology 101, "Varieties of Human Nature" attempts to show just what's wrong with simplistic universal claims about gender relations, sexual mores, and family forms and with misguided attempts to find the roots of labile meanings and ephemeral institutions in an unchanging biological substrate. I claim no great originality in these or other arguments. Rather, I am attempting to articulate, in an accessible language, what most cultural anthropologists or sociologists of culture (especially those who work in gender/sexuality studies) find harrowingly uninformed in evolutionary psychology's much-disseminated claims and shockingly inappropriate about its methods.

The absent presence in all these tales of gender and generation is, of course, that foil to the normal body, the queer body: the body whose desires deviate from the supposedly natural design. "Permutations on the 'Nature' of Desire" directly addresses innatist theories of homosexuality—the "gay gene" and "gay brain" studies—as these questionable scientific ideas have been communicated in serious journalism and as these concepts plodded their way through a changing social context over the course of the 1990s. It is in this notion of the homosexual body, I show, that innatist conceptions of gender both are generated and reach their logical conclusion or limit.

It would not be enough simply to refute, like a good debating partner, the basic arguments of sociobiology, evolutionary psychology, and the heteronormalizing "science" they embody. To combat such dangerous nonsense, it is also necessary to show how these ideas get a purchase—or how they fail to establish themselves—in wider social discourses, to show how these representations participate in struggles over cultural institutions and social resources. I thus regularly tack back to the broader world of sexual, institutional, and economic transitions, transitions about which much of the "science" of sex seems to be in denial.

By treating these ubiquitous tales of origin and design in their present historical context and by following their transmission through the mass media, I hope to suggest that the dominant function of these representations is politically conservative, yet illuminate the contradictory roles they sometimes actually play. Even stories about Adam and Eve, after all, are subject to different interpretations and may be spun to serve up various morals. It is to such quandaries that "The Ends of Nature" attends. This concluding section returns to the agenda stated here and developed in the more historically pitched asides on social transitions. That is, it treats the ongoing entanglement of stories about gender with stories about sexuality—of contending representations with contested practices—and hazards a few guesses about where the social currents we are moving with may ultimately go, or better yet, where our peripatetic desires might take us.

IRONIES AND EQUIVOCATIONS

This, then, is a book about how political interests are variously staged, reflected, refracted, and contested in a public culture where certainty jostles with uncertainty over fundamental premises—an attempt at writing *en cours* about a moment whose characteristic icons include both Hallmark homilies on natural sex roles and stunts by RuPaul, drag queen supermodel of the world. Its intellectual setting is a milieu in which gay activists and academics debate the emergence of "postfeminist" and "postgay" sexual cultures, while news journals and lifestyle magazines explore the emergence of a "poststraight" sensibility.[55] It is not quite a science studies book, nor is it exactly a media criticism book. Rather, the following chapters track the dialectical flow of contrary claims about the "nature" of men, women, and desire: from popular culture to the culture of science and back again through mass media outlets.

They also attempt to follow the not quite predictable swirls and doubling-backs that belong to history in play. Who, for instance, could have foreseen that David P. Barash, a psychologist famous for his genetic reductionist caricatures of human nature, would op-ed against the idea that DNA equals Destiny?[56] Who would have imagined that Natalie Angier, whose science journalism in the *New York Times* provides an apparently endless supply of grist for this book's critical mill, would abruptly reverse the course of her essentializing arguments to publish a thorough, thoughtful refutation of evolutionary psychology's central idea, the notion that men and women evolved to be radically different from each other?[57] And who would have guessed that socialist-feminist Barbara Ehrenreich, noted for constructionist histories of various aspects of modern life, would find fresh appeal in speculative fictions about the effects of hormones on human nature?[58]

It is undoubtedly one of the consummate ironies of our era that attempts at supposedly "queering" science—in order to explain and affirm gay identity—are serving, yet again, to consolidate an astonishingly *heteronormative* conception of human nature: Where homosexuality is understood as the biological exception, it necessarily follows that heterosexuality is the natural default mode.[59] No less ironic is the not so secret complicity between the theological conception of nature and its supposed nemesis in science—a relationship occasionally blared out from news-

Figure 2. RuPaul, Supermodel of the Universe. Photos by
Albert Sanchez for World of Wonder Management/Rhino
Records Media Relations. Courtesy of Photofest, New York, N.Y.

stands, where a picture is worth a thousand words and a sound bite is worth
a dozen books. *Life* magazine's April 1998 cover story, illustrated by twisting blue and yellow billiard-ball strands of DNA, trumpets much of the
zeitgeist of "serious" mass culture in the nineties: "Were You Born That
Way? Personality, Temperament, Even Life Choices. New Studies Show
It's Mostly in Your Genes."[60] Inside, the text suggests that shyness, thrill-
seeking, obesity, aggression, homosexuality, and addiction all have a genetic basis.[61] The same month, in the *Atlantic Monthly*'s lead essay, sociobiologist Edward O. Wilson purports to document "the biological basis

of morality."[62] The *Atlantic*'s cover illustration shows Adam and Eve standing atop stone tablets—the Ten Commandments.

THE TROUBLE WITH NATURE

Although signifiers slide and ironies abound, sociobiology, evolutionary psychology, and related forms of bioreductivism nonetheless revolve around a certain center of gravity. And although it can be "strategically" manipulated by the cultural Left, essentialism is by no means equidistant from all forms of power. With its reification of social acts into natural facts, bioreductive essentialism is far better equipped for conservative, even reactionary undertakings than for progressive ones.

The trouble with nature is that its arguments operate entirely on heteronormativity's strategic playing field: its invocation as the model for how men and women ought to act, its employment in identity work, its use in community building and in political contestations, all leave standing, and even reinforce, the grounding binaries of the heteronormative system. In the dialectic of identity and identification, strategic essentialism in one quarter ratifies and reinforces countervailing reifications in other quarters. The one thing is beholden to the other, and any innatist paradigm can tip over at any moment into outright racism, misogyny, and homophobia. It would be better to reject the naturalized regime of heteronormativity in its totality, no longer to ask ourselves questions about the proper domain for male and female, masculinity and femininity, gay and straight, to be finished with the idea of normal bodies once and for all. This might seem a utopian wish, but many cultural happenings at the moment give expression to precisely this queer desire for a very different kind of nature. Not even straight people really want to live in a straitjacket anymore.

I opened this introduction with the remark that it is becoming difficult to parody the claims of "science" on the "nature" of desire. Parody, perhaps, would be beside the point—but who could resist the temptation to use tools like irony, sarcasm, and humor in the face of such unwittingly seriocomic representations? A perusal of current newspapers finds journalists and politicians invoking a dubious "science" while scientists gossip like village rubes about complexities of the human heart. National gay/lesbian organizations struggle to claim full rights of citizenship under the slogan "I can't help the way I am," while religious zealots strive to recreate the moralizing science of a century past. Everyone marches under the banner of nature, and I find myself being held aloft as Exhibit A in arguments that will undoubtedly shape my reception on the street, my treatment at the hands of authorities, my means of redress—in short, the course of my life. A good sense of irony, at minimum, would seem necessary to maintain the proper critical distance on questions so immediate and intimate.[63]

Comedy notwithstanding, I hope I have suggested something of the seriousness of the problem at hand, the depth of its repercussions, its global scope. Questions about the meanings of nature and the "nature" of desire are not the preoccupations of a vague "politics of signification." They get directly at the crux of institutional arrangements: who's in and who's out of legitimate family, kinship, and other systems of affinity and social support. They also get at political-economic power in the modern world, where values material no less than valuations ideal are distributed according to sex and one's proper or improper performance of it.

Such questions matter, most obviously, to gays and lesbians. What one believes about the nature of "male" and "female," masculinity and femininity, identity and alterity, will affect how one thinks about homosexuality. That is, what one makes of and how one explains same-sex desire will have clear enough practical implications *for us*. But these questions matter for heterosexuals, too—a great deal, as it turns out. For each relationship is correlative, reversible: Every claim about the supposedly "biological" nature of homosexuality has the effect of naturalizing sex roles. And every discussion of homosexuality as an "exception" to the rule points to a *rule* for how it is that "normal" men and women are supposed to be and act.[64] In the regime of nature, as in the prisonhouse of language, those on both sides of the law are subject to its tyranny.

Origins Stories

A need is considered the cause of the origin: in truth, it is often merely an effect of what did originate.

FRIEDRICH NIETZSCHE, *The Gay Science*

1   In the Beginning, Nature

Begin at the beginning: with the natural world, and its perception.[1] "Nature," it seems, "is there from the first day."[2] Lucien Herr's aphorism expresses a partial and provisional truth: We do not "posit" the world around us. We encounter it. So much is self-evident.

Still, men and women have encountered wondrously varied things in the world, seeing a thousand different constellations in the night sky, animating nature with the most contradictory designs. One tribe's Big Dipper is another's Great Caribou. There is a sea, and (like the rest of the natural world) it is really there and has been from time immemorial, independent of any human intentions, but all good reasoning suggests that it was experienced differently by the medieval fishermen who populated it with monsters than by the modern romantic vacationing on a cruise liner. The sun holds us, too, in its regard, but whether we know it as God's first light or as a burning ball of hydrogen, we cannot quite see it the same way as the Aztecs or the Norse did.

Not even the most universal occurrences give rise to any clear consensus about the "nature" of nature or what might constitute its "firstness." All peoples have contemplated the mysteries of life and death, but whatever their divergent findings, most cultures have located these events outside the body, in relation to "the wider cosmos: planets, stars, mountains, rivers, spirits and ancestors, gods and demons, the heavens and the underworld."[3] More than likely, it will seem self-evident to the reader that life is a property *of* the body: a force contained within the envelope of skin that, for modern Western culture, marks the fixed boundaries of an individual being.

More than likely, it will also seem self-evident that nature (as in "human nature") is somehow allied with the body (as against the mind, and all that is conscious and volitional). But this is scarcely what all humans everywhere have thought about how a body inhabits a world. As Carol MacCormack and Marilyn Strathern meticulously show in their classic collection of essays *Nature, Culture, and Gender,* not every culture asserts a continuous line of demarcation between nature and culture, mind and body, and the other familiar dualisms. Indeed, not every culture gives place to the singular and unitary concept, "nature." And even in modern, Western cultures, nature's place is by no means stationary or unchanging.[4]

THE NATURAL BODY

In making claims about nature, modern Western people often point to the self-evidence of the body, its firstness, and the ground it undeniably provides for both thought and activity. But in their own bodies, no less than in the meaning of life or the shape of nature, people have discerned the most variegated forms and functions. The anatomical charts used by Chinese acupuncturists trace structures unseen in Western biomedicine. *Santeros* in New York City sometimes insist that there is a hole in the top of the head, an aperture opened up by Yoruba orishas during spirit-possession rituals. More than one culture has projected its social preoccupations as fantastic body parts: the "wandering uterus" of early modern medicine, for instance, or that magical substance testosterone, mythologized in so much twentieth-century folk medicine.

Physical bodies, like the material world that encloses them, really exist. But by the very nature of their existence, nothing is actually self-evident about what will be *seen* as self-evident in the nature of the body. And often, what seems so self-evident about the experience of the body belongs to history, not nature. Thomas Laqueur's detailed historiography illustrates that this is perhaps especially true where sex is concerned.

For thousands of years, the genitalia of men and women were each construed as variations on an essentially "phallic" theme—the man's parts turning outward, the woman's turning inward. But in eighteenth-century medicine, an astonishing metamorphosis occurred. Male and female bodies took shape as complementary antitheses, each radically different from the other, the one concave where the other was convex. Men and women became "opposites"—and not just that, but "opposites" held to "attract" each other.[5] As counterintuitive a scheme as any ever devised, true heterosexu-

Figure 3.   The Uterus of
Vesalius (circa 1543). From *The
Illustrations from the Works of
Andreas Vesalius of Brussels*, by
J. B. DeC. M. Saunders and Charles D.
O'Malley, Dover Publications, Inc.

ality thus was conceived (if not quite yet "born")—and we, its heirs, are at pains to see bodies differently.

First and second principles, then, for a social history of nature: Nature, even in its "firstness," is in no small part what we *make* of it—if we make anything of it at all. And inasmuch as knowing and doing are linked, what we *see* in nature, in the world, and in human bodies is very much caught up in questions of a social and political order—that is, in what we *want* to see.

HETEROSEXUAL BY DESIGN

Start with creation, go on to procreation: Today, the perception of bodies is nowhere more saturated with cultural meanings than around the question of sexual practices, where some are still seen as "natural" and others as "unnatural." And nowhere is the discourse on nature more resonant with practical, political implications than in those modern origins stories, models of biological evolution—our way of thinking about what it is that seems to confront us, ready-made, with a design of its own from the depths of the flesh. This problematic "nature," as visualized in a questionable "science," is endlessly cited, invoked, and contested in struggles over the sense of the body and the meanings of sex, as the following chapters show in some detail. It supposedly tells us what the real man is really like, what the natural woman truly wants—and how the two of them ought to get together, according to the logic of a natural plan.

Martha McCaughey summarizes the matter: "Anyone questioning the natural and therefore privileged status of heterosexuality today is likely to meet up with an evolutionary narrative: 'After all, how could the human species have survived without heterosexuality?'"[6] In similar terms, Michael Warner underscores the unselfconscious projection of heterosexual normalcy in pervasive narratives about evolutionary origins and human nature: "Het culture thinks of itself as the elemental form of human association, as the very model of intergender relations, as the indivisible basis of all community, and as the means of reproduction without which society wouldn't exist."[7]

According to the prevailing conception, whose assumptions are rarely stated aloud, then, heterosexuality is "there" from the beginning. Said to follow from the biological givens of male and female bodies, heterosexuality has a logic, a purpose—what philosophers call "ontological priority." A voice nags, like some incessant church bore: Human existence, in its very conception, owes a debt to heterosexual beneficence, and supposedly it is thanks to this original and originary heterosexuality that we are all here today. (Thanks, Mom and Dad!) That is to say: Heterosexuality, like nature, is "there from the first day." Or rather, heterosexuality *is* the first day—the very principle of origin, creation, and generation. By extension, heterosex—that is, reproductive sex: penis-in-vagina-to-the-point-of-ejaculation sex—is "real" sex, manifestly revealed in the design of the genitalia. (That's what sex is *for*, isn't it?) Everything else is derivative, secondary, artificial, or tainted, and by comparison to heroic heterosexual intercourse, the way homosexuals and lesbians invest their desires seems wasteful, frivolous, selfish. Consequently, according to a thousand scientific accountings, the very attributes that allegedly define human nature—a tendency for men to compete and think, for women to network and feel—all supposedly emanate from that magisterially self-evident concave-to-convex scheme that seems so undeniably clear, yet took humankind such a long time to discern in human forms.

Penis to vagina as form to function: Bodies, nature, life itself, are conceived as heterosexual by design.

Third, fourth, and fifth principles for a social history of nature: "Scientific practice," writes Donna Haraway, "is above all a story-telling practice." "Stories," she adds, "are means to ways of living."[8]

And stories about origins are inevitably stories about destiny. Or as Edward Said puts it: "A 'beginning' is designated in order to indicate, clarify, or define a *later* time, place, or action. In short, the designation of a beginning generally involves the designation of a consequent *intention*."[9]

ORIGIN AS DESTINY

So what's a fag to say, when speech about the nature of desire has been pre-scripted by such an exclusionary code? And how's a queer to feel, when all her organs of perception are caught up in a dense social process of visualization, imagination, and contestation? And how's a pervert to know his own true nature—or is it unnature?—without immolating himself in a knowledge that burns, without negating the real lessons his peripatetic desires compose?

You and I have been forewarned—and on the authority of science, no less: It would be folly to flout this nature, so full of implications not only for gay and straight, male and female, but also for deviation and norm, and for distinctions of every kind. We are thus invited to pick our place in nature as either *variations on* or *deformations of* a heterosexual design. Or perhaps we might choose to claim our own good and separate nature, apart from the rest of creation—a break in the Great Chain of Being.[10] Such are the politics of nature at the beginning of the third Christian millennium: so many unsatisfying options. So many demands to conform to one's own true nature, however that nature is conceived. So much wrangling over the question of what is "essential" and what is "supplementary."

These dilemmas have salience in part because desire, like nature, always seems to be there ahead of us, to come to us unbeckoned, to impress its will upon us, independent of any intention we might have toward it. These questions have urgency because claims about the nature of sex inform decision making in scientific, medical, educational, legal, and other institutional settings.

Sixth principle (corresponding to the sixth and last day of creation): Origins stories invariably script plots for how, in returning to our true nature, we might emulate models that were social to begin with. The return to origins is really a return to history.

PROSPECTUS

Start with nature, go on to sex, conclude with politics. Or rather, let us begin, if not at beginnings proper, then at least with the idea of a beginning: with the making of men's and women's bodies; with the modern (pro)creation myth that posits bland, conformist, heterosexual necessity as the precondition for human existence. Let us commence with the "straight" body and "normal" desires—with a neo-Darwinian Adam and Eve, not Adam and Steve, whose sex lives occupy the center of the naturalist vulgate, whose binary desires form the Procrustean bed of heterosexual metaphysics. And in following the kinked and circuitous course of desire through changing claims about nature, let us remain alert to the wondrous diversity of meanings human cultures have actually made of nature, bodies, and desires—the better to ask: How might science be remade? What might nature become? How might we design to live in a nature so cultured? What might become of us, and what might we aspire to become, on this unexceptional planet?

2   The Normal Body

"The most plausible explanation for our relative hairlessness—or indeed, nakedness—is that it results from sexual selection in ritual courtship."[1] So begins an inventive recent survey of our evolutionary history, a modern morality play in scientific garb.[2]

Consider how coherent are the contours of normal bodies sculpted by natural selection in this story—how seamlessly science and ideology come together in this closely woven tale of how things got to be the way they are:

> The upright stance reveals the full beauty of one's own and another's primary and secondary sexual organs, and it enables hunters and gatherers to carry the fruits of their labors home (and perhaps then to remember who gets which share)—for which, moreover, it helps us to have a home to carry things home to, thus a ritually charged place, and a kinship system that determines the rules of distribution. The upright stance also enables parents to carry in their arms babies who are helpless because they require a much longer infancy period than the young of other species, a long infancy demanded by the need to inculcate children in the complexities of tribal ritual. The upright stance, moreover, undoubtedly contributed to changing the mating position from mounting to face-to-face, thus encouraging that extraordinary mutual gaze which is the delight of lovers and the fundamental warrant of the equality of the sexes—an equality that was absolutely essential if the human traits of intelligence, communication and imagination were to be preferred and thus reinforced.[3]

From heterosexual "courtship ritual" Frederick Turner thus deduces anatomical "nakedness" and upright walking to unfold further the various complex facets of human culture: collective labor and social sharing; ritual, kinship and tribe; timeless practices of home and hearth . . . Indeed, in Turner's remarkable account, so full of whimsy and conjecture, straight walking *is* heterorectitude.

The author purports to have deduced all this from a dispassionate gaze at the contours of the human body. But in this tale of origins, we have encountered not denuded humanity—an essentialist notion Clyde Kluckhohn used to give as "man in the raw"—but rather a distinctly modern, Western, even American creature, fully clothed in the fashion of a certain day and age.[4] Turner's arguments "work"—his zany scenarios seem plausible, even convincing—only if they are abetted by the silent labor of the reader, a reader of a certain disposition, who fits familiar form to functional norm, thus converting common sense into analysis, fancy into fact, what conceivably might have been into what indubitably is, and historical contingency into universal biology. Readers such as these supply the *need* that sustains the argument by filling in the gaps—readers who, more than anything, desire that their own longings, their own experiences, be seen (both by others and by themselves) as "natural," "normal," and "healthy." For the reader willing to project himself or herself into Turner's text, all that is uncertain, equivocal, and contested in modern life is turned into a set of ready platitudes. And what awaits the reader who refuses to give silent complicity is a certain kind of existential blackmail: S/he is cast outside the charmed circle of nature.[5]

Consider the political work of the reader in sustaining the text, the movement of his or her desires within Turner's arguments. The reader knows— as our foraging hominid and early human forebears likely could not have known—what a "home" is: a settled and secure place, returned to daily, invested with sentiment and meaning.[6] (Or is it a divided place, rent by conflicts and held together for convenience' sake—imagined otherwise only in the reveries of a misplaced nostalgia?) Modern readers are prone to believe, as their predecessors of only a few years ago did not—and as Turner's own sociobiological antecedents emphatically did not—in that "fundamental warrant of sexual equality" here lauded, it must be noted, not for its tonic political implications but for its eugenic effects, its improvement of the breed. The relatively liberated reader might even think that "the beauty of the genitalia" could serve as a model for beauty more generally. But the very notion of genital beauty has been vigorously contested, even denied, in many cultures—our own included. (Freud, who understood aesthetic enthusiasm as a sublimation of sexual feeling, nonetheless felt compelled to note that "the genitals themselves, the sight of which is always exciting, are nevertheless hardly ever judged to be beautiful"—a riddle that prompted John Updike to title his review of a museum exhibit of Egon Schiele's paintings with a question, not a conclusion: "Can Genitals Be Beautiful?"[7]) Finally, the reader knows, or thinks s/he knows, or at any rate claims to know, what "sex" is, or ought to be, or ought to be seen as. But is it really so clear that the human mating position *has* changed from "mounting" to "face-to-face"? An ample heterosexual lore celebrates the joys of fucking "doggy-style," not to mention the yin-and-yang delirium of that numbered act, "soixante-neuf." The "missionary position"—icon of normal, real sex—was so named not because missionaries discovered face-to-face intercourse with the man on top to be the universally practiced norm of coitus—far from it—but because it seemed to be the peculiar and singular predilection of colonizing Europeans.[8] Since so many natives in so many lands required instruction on how truly human beings might "properly" have "normal" sex—and since the colonizing Europeans were anything but sexual egalitarians—it is unclear how face-to-face positioning could provide the evolutionary ground either for human equality or for universal rules for how an upright creature ought to act.

CAUSE AND AFFECT, ADAM AND EVE

A scientific narrative that weds common sense to conjecture and need to explanation might well draw inspiration from mythology.[9] Framing his flat reductions as explorations of a reflexive, biocultural dynamism, Turner expressly mines the "intuitive truths" of the Book of Genesis:

> Many of our creation myths show an intuitive grasp of the strange
> process by which the cultural tail came to wag the biological dog.
> The story of the clothing of Adam and Eve, where (the awareness of)
> nakedness is the result of shame, which is in turn the result of self-
> knowledge, expresses one aspect of it. Again in Genesis, Eve's punish-
> ment for her acquisition of knowledge, that she must suffer in child-
> birth, expresses the fact that one aspect of our viviparous species, whose
> cranium at birth is large relative to its size at maturity, is the capacity
> of the female pelvis to allow the passage of a large skull—hence also
> the beauty for the male of the female's wide hips. The larger (and to
> the heterosexual male, attractive) breasts of the human female, and her
> dependency upon a protective male during lactation—also referred to
> in Genesis—are likewise the sign of a nurturing power that can deal
> with a long infant dependency, and thus produce human beings of
> intelligence, wisdom, and aesthetic subtlety. Infants without protecting
> fathers must enter adulthood earlier and cannot be fully instructed in
> tribal ritual; they thus need smaller brains, and smaller hipped and
> breasted mothers to bear them.[10]

Figure 4. Diorama of *Australopithecus afarensis,* also
known as "Lucy," and mate. Courtesy Department of
Library Services, American Museum of Natural History,
New York, N.Y. Negative Number: 2A 21119.

In fact, here—as in much of that emergent field of ex post facto stories,
"evolutionary psychology"—the moralizing tale has come to wag the evolutionary dog. Nature grants a grudging equality between the sexes, but
only in the context of a cozy, heterosexual-pair-based family, and with much
gratuitous finger wagging at welfare mothers who refuse to depend on a
male breadwinner. In the absence of any real evidence—for this family portrait is nowhere preserved in any archaeological record—an "ought" is not
so much "read" as "written" into the human anatomy: the heterosexual
imperative. And not just a supposedly natural imperative for heterosexuality, but for a *certain kind* of heterosexual relationship, stable monogamous unions with a clear division of labor. With all the trappings of scientific authority—but in exactly the same mode as the biblical narrator, who
sets out to justify sexual inequality and human unhappiness in general—
Turner tells readers that in glimpsing the time of origins, they are also bearing witness to the truth of the design of their bodies.

And thereby hangs a tale—a tale of norms, both stated and implicit. Under the sign of normal desire, Woman joins Man at last, as Eve to Adam:

Figure 5.   Neanderthal burial diorama. Photo © Chip Clark, Museum of Natural History, Smithsonian Institution.

coevolutionary coheroes, different but equal, and joined by mutual need—the woman's need perhaps being a little greater than the man's, since it is she who needs protection (from other men, I dare say).

MEANWHILE, THE DEGENERATES GET WHAT THEY HAVE COMING TO THEM

But on what script is this costarring drama of nature cast, dreamed up out of what author's reveries? Who will play what roles? Who will get to play hero, and who will be consigned to act the foil? Who will get speaking parts, and who will be silenced in this modern-day morality play? What readers, which spectators, will be included within that "circle of the we," the "charmed circle" of normalcy?

Meditating on the grim logic of sacrifice, Turner makes his implications quite explicit. Evolution is the story of human progress and degenerate "throwbacks," of species hygiene and castaway "defectives." In this saga, "fragile virtues" and "moral growth" are purchased by "a dark and terrible element" in human affairs: a violent selection *against* "the dullards, whiners, liars, blowhards, hoarders, spendthrifts, thieves, cheats and weaklings." This "terrible and most shameful selection process," which winnows out the sick, the weak, and the criminal, "has not yet ceased, and we had better face up to that fact."[11] Francis Galton—the father of eugenics, whose thinking shaped that distinctly modern concept, "the normal body"— would be pleased by this postcard from the future.[12]

HETERO HIERARCHICUS

Likewise adjuring us to "face up to facts," Berkeley sociobiologist Vincent Sarich used to explain the development of male homosexuality to his classes by conjuring the image of that boys' game, King of the Mountain, with its agonal play of push and shove, tug and tow. King of the Mountain, like the sociobiological sexual economy, is the ultimate zero-sum game. Only so many boys can stand on the hill: one at the top, others taking up jealously guarded posts along the slopes in descending order. The bigger, stronger, more aggressive boys dominate the summit; the smaller, weaker, less aggressive boys are driven to the fringes and valleys.

He-men on top, sissy boys at the bottom: How very like so many other (supposedly natural) phenomena—primate dominance hierarchies, business competition, bell curves . . . Child's play is Domination for Beginners. And what an instructive game—from an insuperably blinkered point of view— so evocative of "real life" winners and losers. The higher ranking, successful, dominant males assume positions at the top of the hierarchy, take the women for themselves, and procreate. The damaged, dominated, leftover men make do with each other's solace, with homosexuality. In the order of nature, as in the order of society, to the winners go the spoils. To the losers, consolation prizes.

CONSTANT CRAVINGS

Not all such tales about the time of origins are mean and ugly. Here is a different sort of story about evolution and the nature of desire, a thoroughly modern parable about cravings, obsessions, and the human design.

> If something feels so good that we tend to do it to excess, you can bet there's a plausible evolutionary explanation for the strength of that desire. We love fat and sweets: fat because it's high in calories and it behooved our chronically undernourished forebears to seek fatty foods,

and sweets because fruit was another ancestral staple and sweetness signals that fruit is ripe and thus digestible. We love shopping, and logically so: we're planners and hoarders, and by stocking up on available goods now, we can weather lean times tomorrow. Who would have guessed that our basic appetites would be turned against us by the invention of refrigerators, golden arches and credit cards?[13]

New York Times science writer Natalie Angier tells a damn good story, in the just-so tradition—a narrative apparently innocent of the homophobia vented by Vincent Sarich. Still, one wonders how innocuous this cheery tale could really be, given its structural similarities to Frederick Turner's chilly, eugenicist musings—in fact, given the origins and legacies of the sweet-tooth story itself. Although Angier's discussion is less pernicious than the usual biomythology, it borrows heavily from the kind of reasoning developed by Robert Wright in *The Moral Animal,* a pivotal text in the transition from sociobiology to evolutionary psychology. Wright invoked the now endlessly recycled "sweet-tooth" fable as a parallel to the problem of male sexual jealousy, both supposedly serving as examples of basic human biological instincts:

> The classic example of an adaptation that has outlived its usefulness is the sweet tooth. Our fondness for sweets was designed for an environment in which fruit existed but candy didn't. Now that a sweet tooth can bring obesity, people try to control their cravings, and sometimes they succeed. But their methods are usually roundabout, and few people find them easy; the basic sense that sweetness feels good is almost unalterable. . . . Similarly, the basic impulse toward jealousy is very hard to erase.[14]

Angier's short illustrations, like Wright's chatty arguments, are instructive, especially in the move from "plausible explanation" to positive accounting, from what might have been ("fruit was another ancestral staple, and sweetness signals that fruit is ripe and thus digestible") to what supposedly is ("Similarly . . . jealousy is very hard to erase"). Angier does not quite claim that there is a gene for Big Mac attacks or for credit-card abuse. Rather, she claims that "our basic appetites" give birth to fast-food chains and credit cards—and that our organic cravings, like boundless founts, are always overflowing their cultural expressions. By the same just-so logic that prompts certain stand-up comedians to extend primitive and supposedly "male" hunting tools imaginatively into TV remote control devices, Angier telescopes the modern practices of an undifferentiated "we" through the sociobiological optics of evolution, genetics, and desire.

Narratives like these provide a series of hooks and attention grabbers that

go straight to the reader's (supposed) habits. There we are, you and I, yet
again—either as we really are or as the author would have us be—spread
out in the smallest detail across the tableau of biological history, exactly as
we need become. More than that, it is our very *yearnings* that we see, already
there before us, emerging from primordial mists at the dawn of the first day.
And if there is something unspeakably vulgar and immediate about the conception of "need" that Angier and Wright communicate—need being, quintessentially, organic need, hunger—their tales have the compensatory benefit
of a certain ironic twist: Modern, Western, affluent humans today confront
the excess of their own success, awkwardly equipped for the modernity that
flows from primordial needs and desires. Although our nature has its reasons, it can also go awry. We are trapped in bodies from an alien time.

The sweet-tooth story conveys the gist of many current arguments about
human origins and human nature. Such stories reason backward from certain "common-sense" propositions to hypothetical evolutionary scenarios.
But invariably such reasoning fails all tests of logic and evidence. Is it really true that human evolutionary time was shaped by chronic shortages?
Little in the archaeological evidence would indicate that. Quite the contrary,
the best evidence suggests that low-density foraging societies enjoyed relatively high standards of living (as gauged by nutritional standards, height,
dentition, susceptibility to infectious diseases, leisure time, and other such
measures), absolutely higher than those of agricultural producers in most
state-level societies.[15] And were our ancestors really acquisitive pack rats?
Not likely, since foraging requires constant movement, and stocks of goods
would only impede mobility.[16] Could they even have been "planners"? By
all indications, planning beyond a few days' time becomes feasible only with
settled life and agriculture—well after the emergence of *Homo sapiens* and
far too late to register the effects of natural selection. And is it even the case
that everyone has a "hardwired" taste for fats and sugars?

A Nigerian acquaintance I met at a Middle Eastern takeout in a gay neighborhood provides an equally plausible (and equally informed) narrative in
our impromptu discussion of food likes and dislikes: "You Americans eat
too many sweets. There's too much sugar in the diet here. Your parents give
you sweets when you're young, so you get used to eating that way. In Nige-
ria, our parents don't give us so many sweets, so we don't crave them. I don't
like them at all."[17]

What is wrong with Angier's apparently innocuous tale, and others like
it, is that it purports to know more than what anyone could really know
about cravings, desires, and needs. It gives the writer, and the reader, a fake

sense of mastery, a feeling that s/he understands, hence controls, all those strivings that seem to come to us unbeckoned and with wills of their own. Apart from any overt political claims, such forms of pretend knowledge invariably give a reduced and impoverished picture of human existence. They also promote a general dumbing-down of discourse, a sense that answers to complex questions are as easy as a quick trip to the shopping mall. These tales only *appear* to be innocent of politics. It can never be innocent to seal up human capacities within a few familiar motifs, or to foster faux analyses and clever stupidity.

PSYCHOLOGIZING EVOLUTION

Such anecdotal narratives about human nature sprawl across a vast terrain today. They characterize the primary methods of so-called evolutionary psychology, an offspring of 1970s sociobiology whose musings now dominate discussions of evolution in the public sphere and have almost come to define "popular science" in many venues.[18] Like Turner's just-so story, or Sarich's fable, or Angier's parable, the narratives loosely grouped as evolutionary psychology attempt to explain the emergence of supposedly universal human characteristics—especially putative differences between the sexes— as the outcome of a long evolutionary selection process in which men (presumably) hunted and women (theoretically) gathered, in which families were (supposedly) formed in certain ways, and through which certain dispositions (are said to have) thus acquired a genetic, hormonal, and neural cast. . . . Evolutionary psychologists breezily refer to this mythic time and place as "EEA—the environment of evolutionary adaptation. Or, more memorably: the 'ancestral environment.'"[19]

As though everything impinging on the evolution of human beings could be compressed into one definable space and time to yield (voilà!) a legible, definitive blueprint of human nature. Oh, how I'd like to take a walkabout in that ancestral environment (which I imagine to be a kind of neo-Darwinian theme park), since it is in this virtual space that heterosexuals claim to see themselves reflected (if only as two-dimensional, Disneyland automatons). And since this fabled land, the very vortex of origins, is also the inner sanctum of heterosexuality—a space in which every queer is told s/he'll find nothing but a howling nothingness if s/he dares to gaze there—I'll take it as a magical dare to stroll about the premises, to gauge the distance between hard fact and fairy tale, between legitimate science and ideological conjecture.

Figure 6.   Nicole Eisenman,
Betty Gets It, 1992. Ink on
paper, 14 inches × 11 inches.
Courtesy of Jack Tilton/
Anna Kustera Gallery,
New York, N.Y.

Obviously, what one gets out of a spaciotemporal concept like EEA depends a great deal on what one puts into it. That is, everything depends on how one marks epochs and posts boundaries. In fact, it is difficult to say just which human traits evolved on the African savanna and which ones evolved in the forests of Asia. And contrary to several generations of just-so stories about human origins, recent fossil evidence suggests that upright walking developed in the hominid line not on the dry, golden savanna, so our ancestors could see above the tall grasses, as the old story goes, but in lush tropical forests, where this trait's "adaptiveness" is far from clear.[20] But even if we assent to the questionable notion of an "ancestral environment"—and even if we pretend that the mythic savanna was all one undifferentiated and invariant environment, without forest fringes here, dry areas there, millennial droughts, and many species' migrations over time—we still cannot logically conclude very much about the way foraging strategies, human sociality, and kinship practices coevolved. That is because all the "data" turn out to be little more than conjecture, a point highlighted by a generation of feminist physical anthropology and archaeology. We know relatively little about how our evolutionary antecedents labored and even less about whether or how they divided their labors and tools according to age and gender. We know considerably less about what any of this might have *meant.*

And since meanings leave few and indirect traces in the archaeological record, we know almost nothing about how affinities were made, how kin relations were structured, and how reproduction was organized in what were presumably varied and variable hominid and human bands.[21]

In fact, there is no reason to believe that these matters were substantially less variable in the human and prehuman past than in the present. On this point, the lessons of anthropology are clear. Among present-day foragers, there are almost as many divisions of labor, kinship practices, and systems of social organization as there are foraging societies. The ethnographic record describes technologically simple societies where consumption is based on immediate returns, others where value is stored in various delayed-return forms; societies with a high degree of sexual equality, others with a high degree of sexual stratification; societies that involve bilateral kinship, so-called monogamous marriage (meaning only one husband, one wife, *not* one sex partner for life), and nuclear families, still others that involve complex lineage systems, polygamous marriage, and various living arrangements.[22] "Human nature," it would seem, actually admits the most variegated of practices—even for those societies supposedly nearest the baseline of human origins stories. There is no reason to suppose that this splendid variety is not the real plotline of a plausible story about human evolution.

TALES OF THE MODERN SUBJECT

Notwithstanding the paucity of any compelling evidence, narratives of normal order and good form sound out above the collapse of empires and the siege of white male privilege. Manifestly, such accountings for biology, beauty, and normal desire are narratives less *about* than *of* twentieth-century anxieties—especially race, class, gender, and sexual anxieties.

When Lionel Tiger and Robin Fox describe sexual and generational politics in their influential book *The Imperial Animal,* it ought to be clear enough that they are sketching a social picture counter to the upheavals of 1971: a tidy, well-ordered world where dominant and dominated keep within proper bounds, where men are more powerful than women, but both strive to get along, and where youth accept what their elders teach.[23] The authors are telling us less what *is* than what in their view *ought* to be.[24] What sociobiologist Edward O. Wilson files as regrettable "facts" under the heading of "human nature"—for example, "We tend to fear deeply the actions of strangers and to solve conflict by aggression," and "Men, it appears, would rather believe than know"—are actually interpretations expressing the author's orientation toward social and historical events around him.[25] When Robert Wright and others claim that "sexual jealousy" is hardwired into the human male, they argue nothing of great scientific rigor; they simply contest modern demands for sexual freedom with one of the most privileged arguments available in Western society: "It's against human nature." When Natalie Angier contemplates the supposedly ubiquitous sweet tooth or the allegedly irresistible urge to stock and hoard, she reflects not on the biology of cravings, but on the obsessions of a postmodern suburban consumer culture where soccer moms express quintessentially contemporary desires by binge shopping and fighting the battle of the bulge. And when anthropologist Helen Fisher asserts in a recent book that women's biological disposition tends to hamper their rise to the top of hierarchies—but also makes them especially benevolent rulers once they get there—she expresses little more than her own anxieties and hopes about the place of women in a changing social world.[26]

These accounts of a supposedly universal human nature are actually tales of the modern subject and his or her subjectivity as lived in a very particular social context. In these stories, we encounter our own familiar selves in familiar situations: the play of schoolyard bullies, the violence of the urban jungle, the heroism and villainy of the marketplace, the battle and the truce of the sexes, and every crisis of gender and sexuality. But as is appropriate for myth, the logic is allusive: All the terms have been transposed. We view really existing persons not as they are, but as they might idealize themselves—or as they might fear themselves becoming. In this mirror world, social forces have become forces of nature, and persons have become genes, while genes are likewise personified. And everything is set in the archetypal time of evolution.

The need for security in the midst of upheaval becomes a comforting tale of home and hearth. Or alternately—and at the same time—the harshness of the social world becomes the storm of nature. Natural history becomes a straight line connecting us (or at least some of us) to ourselves.

3 The Human Design

In one of my more demented moments (possessed, perhaps, by the edgy ghost of Walter Benjamin), I had imagined composing a documentary text consisting solely of origins stories, quoted directly and unaccompanied by any commentary or critique. Such an experimental text would include scientific-sounding folktales about the natural design of men's and women's psychologies, biological fables about queer deviations from the basic floor plan of human nature, and of course, a sampling of cautionary tales—those evolutionary stories whose narrators, in purporting to describe the basic blueprint of human nature, also claim to see the future, to know in advance the outcome of today's social struggles because they know the limits of what is humanly possible. . . . Properly arranged, one story would undermine or contradict another, so that, read in sequence, the net effect would be that of a spontaneous deconstruction.

I have not followed this mad scheme, partly because a handful of undergraduates on whom I experimented with an early draft all but shattered my faith in deconstructionist shtick about how certain kinds of texts "deconstruct themselves." Some readers were so delighted by evolutionary psychology's story-telling techniques that they tended to lose sight of how those texts were inconsistent or contradictory, even when they offered diametrically opposed claims about human nature.

With good reason: above all else, evolutionary tales about human nature make for good entertainment. Like all good drama, origins stories provide writer and reader, teller and listener, the means for a certain kind of uninhibited acting out: the occasion for phantasmic dances of desire and anxiety, for the imaginary breaking of taboos around sex and violence, for the expression of thoughts that might otherwise go unsaid. Even I want to play along, but I stop short. For among the guilty pleasures these tales typically indulge is the ritual murder of the scapegoat. And it is my own self I in-variably see in the role of the sacrificial offering.

So rather than give these narratives any uncontested space, I will try to in-terrupt their fun and games—to stumble them up on the very grounds of their staging, in "scientific facts." And rather than entertain speculation about what the time of origins might have looked like—if only we had time machines and could visit it—I will underscore how these not altogether con-sistent stories both reflect on and participate in current events and histori-cal happenings. At stake in this exercise is what makes a story a good story—what causes it to convince journalists and lay readers, not to mention the community of scientists. What is also at issue is these narratives' claim to science—and what kind of science, what kinds of narratives, might justifiably treat of the human condition.

Whatever their defects as science, sociobiology, evolutionary psychol-ogy, and related forms of biomythology provide a cache of irresistible mo-tifs for telling stories about how we (or rather, some of us) are who and what we (supposedly) are. That is, these representations give identity to subjects whose definitions are plural, contested, and in flux. Like the antecedent form of biological determinism, Social Darwinism, these latter-day stories about nature shape public perception and lend the authority of "science" to the everyday chatter of ideology. But because these stories join history at a par-ticular juncture, they take up old motifs in new ways. They give the stamp of "nature" to those often complex, sometimes contrary, and always chang-ing "structures of feeling" and "whisperings from within" that define the political moment.[1]

HARDWIRED

The teaser for an *Esquire* article by Michael Segell, "The Second Coming of the Alpha Male," articulates much of the current mood and also shows how questionable scientific ideas are taken up in journalistic and pop-cultural accounts:[2] "After thirty million years of evolution, three thousand years of civilization, and thirty years of feminism, it's time to put the *man* back into *manhood*. By balancing traditional male virtues—aggressiveness, courage, power—with new emotional insight, you can unleash the big dog within."[3]

Segell's piece instructs men on how to "reclaim" their wounded mas-culinity in an age of empowered women. It begins with a tale of buddy identification and role modeling so transparently homoerotic that at first I took the whole thing as a very camp put-on—an impression fed by the unsurpassed silliness of its pictorial collage of alpha males, an iconography that includes General Dwight D. Eisenhower, prizefighter and convicted rapist Mike Tyson, the cartoon detective Dick Tracy, samurai actor Toshiro Mifune, bodybuilder-turned-actor Arnold Schwarzenegger, and Bolshevik revolutionary V. I. Lenin, chin thrust defiantly forward.

Despite such flourishes of Generation X cutesiness—including an obnoxiously silly cover loudly announcing "THE BIG DOG GETS THE GIRL"—Segell's article appears to be earnest about its subject matter. It expresses sentiments aired widely in both academic and popular culture: a desperate yearning to get back in touch with one's own true self, and (mirroring cultural feminism's claims to a universal sisterhood) a raging desire for global fraternity. Evoking the style of Robert Bly's articulation of men's-movement archetypes, Segell's piece is strewn with quotes from Margaret Mead, allusions to the research of Gilbert Herdt, and gossipy parables from David Gilmore—assorted ethnographic trinkets offered as therapeutic counsel for the bewildered modern man, displaced, but not quite removed from his primal nature. This New Age tribal mélange is rigged up to support familiar universals about essential masculinity, male bravado, manly risk taking, male bonding, and dominance hierarchies. The piece is also littered with declamatory pseudoscientific palaver like the following sample: "The feminist critique of modern masculinity has exhausted itself," "The fact is that men who are effective in the world are the products of hardwired instincts," and "Of course, some women are endowed with thrill-seeking traits that exceed some men's . . . but many women's aversion to risk and the single-minded pursuit of status may stifle their rise to the top in a Darwinian business world."[4]

BORN TO BE BAD

Segell's disquisition on how to "reclaim" the "alpha male within"—all those "good things, engraved deep in biology"—while rejecting the nastier legacies of manhood—male domination, violence, and self-destruction—brings to mind Oscar Wilde's quip: "To be natural is such a very difficult pose to keep up." But in its swaggering ambit between cultures and across evolutionary history, the idea that alpha-male high jinks and bad-boy antics are *especially* "hardwired" into *certain* male bodies brushes up against another vigorously contested border in that danger zone of sex and power. Even cleaned-up and made presentable, these celebrated, emulated "alpha males" bear a more than passing resemblance to "problem males"—those hyperactive lads, disruptive boys, misbehaving male youths, and violence-prone men whose impulsive behaviors fall increasingly under the aegis of a neo-Victorian biomedicine. Nature here turns menacing, and a mesh of new, disciplinary powers embraces the hapless lads caught in its grip. On the dark side of this brave new world of biomedicine, "antisocial," "attention-deficient," and "aggressive" behaviors are labeled according to specious psychometric "norms" tailored to the schoolroom. Teachers and counselors arbitrarily apply categories reminiscent of phrenology and criminal anthropology to mark off "tomorrow's predators," "future recidivists," and "affectionless personalities." Increasingly, such "undesirable" traits are attributed to organic, even genetic, causes—and consequently, those boys who exhibit *too much* "boyishness" are "treated," drugged, surveilled, and administered, as Philomena Mariani shows in her article "Law and Order Science."[5]

In this context, the lapse of Frederick Goodwin, former director of the National Institute of Mental Health, made while addressing the American Psychiatric Association on the goals and logic of the NIMH's "Violence Initiative," speaks volumes on how pseudoscholarly conceits on gender, sexuality, and race filter into mass culture—and conversely, how pop-cultural conceits animate scholarly research designs:

> If you look, for example, at male monkeys, especially in the wild, roughly half of them survive to adulthood. The other half die by violence. That is the natural way of it for males, to knock each other off and, in fact, there are some interesting evolutionary implications of that because the same hyperaggressive monkeys who kill each other are also hypersexual, so they copulate more and therefore they reproduce more to offset the fact that half of them are dying.
>
> Now, one could say that if some of the loss of social structure in this society, and particularly within the high impact inner-city areas, has removed some of the civilizing evolutionary things that we have built up . . . maybe it isn't just [a] careless use of the word when people call certain areas of certain cities *jungles,* that we may have gone back to what might be more natural, without all of the social controls that we have imposed upon ourselves as a civilization over thousands of years in our own evolution.[6]

Subtly, but with spectacular vulgarity, the sociobiological argument from inescapable, irresistible nature has twisted around until, coming full circle, it replicates the Victorian perspective: nature as something to be overcome, rather than emulated.[7] Drives as forces to be harnessed and sublimated by civilization. Manhood as a balancing act on a slippery slope. But if Goodwin's remarks recapitulate all the key ideas of nineteenth-century scientific racism—the notion that some men are, unhealthily, more "natural," more given to a brute animal nature, than others, the image of a culture gone wild—his picture of the urban jungle also contrasts with other, contending uses of similar conceits.

MEAN GENES

Consider the very different implications forwarded in Richard Morin's *Washington Post* column, "Unconventional Wisdom," which offers "new facts and hot stats from the social sciences."

> Here's one big reason why prejudice is so hard to eradicate: It may be hardwired into our genes.
>
> Psychologist Harold Fishbein of the University of Cincinnati does not suggest that there's a racism gene that makes whites hate blacks or one that provokes Bosnian Serbs to despise Croatians.
>
> But he does maintain that these and all other bitter flavors of prejudice are the direct result of "a genetic predisposition now gone awry."
>
> His argument goes something like this: Until relatively recently, humans organized themselves into small, genetically closed tribes who struggled in life-and-death competition with other neighboring clans for roots, berries, the tenderest mastodons, etc.
>
> Over time, this fear and loathing of "out groups" became wired into our genetic code.
>
> One problem: "We evolved in a tribal environment," Fishbein said. "But we no longer live in tribal environments"—yet we still retain this "genetic predisposition" to favor our own kind and disfavor others.
>
> Okay, but in the absence of other competing tribes, and if there isn't a racism gene, how do we know whom to hate?[8]
>
> Political leaders, the media and other authority figures tell us, Fishbein said. . . . And inevitably, we listen, he said, because humankind's allegiance to authority figures is itself gene-based, a claim first made by the renowned geneticist C. H. Waddington.
>
> Fishbein isn't a pop sociobiologist with a gift for the glib. His new book, *Peer Prejudice and Discrimination,* recently received the Eleanor Maccoby Book Award from the American Psychological Association. He has been invited to discuss the genetic basis of prejudice in an address at the APA convention next summer in Chicago.
>
> Fishbein says his ideas often provoke pessimism from other scholars: Why not just surrender to our bad selves, since we can't write prejudice out of our genetic code?
>
> "We should do quite the opposite," he says. "We know who the en-

emy is, and what the mechanisms of prejudice and discrimination are. Authority figures have to take the leadership role to turn it around. It won't be easy. It's an ongoing battle against ourselves."[9]

TRIP-WIRED

Whatever Fishbein's stated intention, the idea that human beings have a natural disposition to hate outsiders gives rise to a certain resignation in the face of prejudice, as evinced by Andrew Sullivan's speculative natural history of how hatred became hardwired in the human animal. "Hate is everywhere," Sullivan begins, somewhat unconvincingly. He goes on to generalize about generalization, as though hatred were just a special case of over-generalization: "Human beings generalize all the time, ahead of time, about everyone and everything." Sullivan immediately moves from this abstract generalization to the speculative case: "A large part of it may even be hardwired." Having thus formed a working hypothesis, he frames an imaginary evolutionary sequence as though it constituted evidence for his argument: "At some point in our evolution, being able to know beforehand who was friend or foe was not merely a matter of philosophical reflection. It was a matter of survival." Sullivan claims that science (and the true, dispassionate picture of nature it gives) authorizes such assertions. But the evolutionary record reveals no such thing, and this is, of course, the same move made by Frederick Turner and Natalie Angier in the previous chapter—indeed, the same move made by everyone else working in this genre: It is the move from "might" to "is," the conjuring of an ancient "innate drive" to explain a contemporary social institution or a cultural fact.[10]

Sullivan goes on to argue, perplexingly, that those who mobilize politically against intolerance are, in their way, even more hateful than those who commit hate crimes. He concludes, on the basis of no real evidence and a less than rigorous logic, that it is best to do nothing about the politics of hate.

4 Our Animals, Our Selves

Today's sociobiology and evolutionary psychology present an amalgam of different parables about human nature, a collection of tall tales that owe nothing to any empirical evidence and everything to projection, ideology, and outright fantasy. Vernacular beliefs about sex, "blood," temperament, drives, and heredity do not taint these "scientific" models from the outside. They constitute them from the inside, from their moment of inception, through untested beliefs about what a man is, what a woman is, and how each ought to act; in unexamined feelings about sex and bodies, and especially in preconceptions about what defines sex and what bodies are for; through myriad culture-bound understandings of what is normal and abnormal, healthy and pathological, fit and unfit. One of the principal genres of just-so stories about human origins deals with narratives based not on how we ourselves behave, but on how our relatives—whether near or distant—in the animal kingdom behave. Therein lies another, all-too-familiar dimension of biomythic discourse.[1] Mother Nature is made to speak her meanings to us through the refractions of a particular social imaginary, the bestiary.[2]

THE BESTIARY

The role played by animals in stories about human nature is most significant. Since all of the evolutionary just-so stories about pair bonding, family formation, and so on turn out to be based on nothing more than dark conjecture or wish-fulfillment—and since the empirical evidence from cultural anthropology gives no support to pat parables on timeless human nature— examples from the animal kingdom become pivotal to the credibility of the analysis and to its status as "science." Other species undeniably exist. They can be observed today, their behaviors reported and tabulated.

But the citation of animal behaviors as empirical evidence on questions of human social action is problematic in many ways, beginning with the problem of commensurability: Is animal communication really comparable with human language? Are animal behaviors comparable with transformative social acts? Sociobiologists and evolutionary psychologists sidestep this basic problem by selectively invoking examples from the animal kingdom to match examples of (and claims about) human actions. Animal "aggression" is held to be manifestly comparable to human violence. Animal reproduction is claimed to be similar to human courtship, and the similarity is held to be self-evident. Avian "pecking orders" and primate dominance hierarchies are said to be commensurate with social stratification. And so on.

Consider the associative logic that flickers along the great chain of being in naturalistic and naturalizing accounts: Humans are like certain birds—"monogamous pair bonding"—but also like bees—"social hierarchy." Like baboons, humans naturally exhibit male dominance. Like pack animals, humans have a "communal nature"—but the individual can also act as a "lone wolf." However, humans are certainly not like those species of worms and insects that mate at random, without regard to sex. Nor are we like the emu, that giant bird whose females lay but whose males hatch the eggs, in something of a role-reversal of the usual logic attributed to sexual selection. And we are decidedly not like those peaceful, bisexual, matriarchal practitioners of free love, the bonobo chimpanzees, who often appear as heroines of a kind of sociobiology in reverse.[3]

Selective associations of this sort provide material for much of the endearing patter of popular science journalism, but they are also surprisingly common in "serious" scholarship on the "nature" of sex, where they are invariably conjured to clinch arguments.[4] Such ad hoc associations—decontextualized traits and atomized behaviors selectively drawn from a scattering of species—have nothing to do with sound biological principles, plausible evolutionary sequences, or degrees of phylogenetic relatedness among species. The function they serve is strictly rhetorical: They give the appearance of "empirical data" to ideological whimsy. Readers are thus invited to see themselves, to pick and choose their own attributes from assorted creatures in the familiar parade of nature. The resulting tales *sound* like science, as Jonathan Marks notes. They've got animals and animal behaviors, "experiments and genes." But they constitute a "science of metaphorical, not of biological connection," a "nonscience."[5]

When instructive examples of animal behaviors are drawn from a more disciplined and more principled sample, the results are not much better. Stephen Jay Gould once quipped of a *Science* article whose authors had constructed sociobiological models of panhandling in terms of inclusive fitness, reciprocal altruism, and food sharing among chimpanzees and baboons, but without any (direct) reference to American big-city realities of race, class, and gender: "I would have accepted [the article] as satire if it had appeared verbatim in the *National Lampoon*."[6] Gould's badinage underscores important points. Naively and without distinction, reductivist arguments shuttle between facts of nature and facts of culture, between selectively invoked animal behaviors and unrelated human practices, between culturally motivated intentional actions and culturally constructed unintended effects, and between "higher-order" symbolic activity and biological reflexes, taking them all as the evolutionarily adaptive expressions of genetic predispositions. We are still in the realm of a questionable associative logic.

The resulting analysis is logical and consistent, but in the same way as what Claude Lévi-Strauss calls "the totemic illusion": "The animal world is . . . thought of in terms of the human world."[7] Kin group A is said to be like species X, subpopulation B is held to resemble phylum Y, and gender C is fancied to be similar to organism Z. Not unlike what Gould calls a basic "category mistake," this associative reasoning proceeds in a feedback loop.[8] First, concepts, relations, and activities characteristic only of humans—"society," "bonding," "hierarchy," "slavery," "rape," "harem"—are used to describe animal behaviors. Then, the very behaviors so described—anthropomorphized—are invoked to shed light on human social practices. Nothing escapes, nothing exceeds, this closed circle of reification and fetishism. The bestiary is us, reflected, refracted, and objectified. Scientistic thinking about human nature attributes human motivations, desires, and characteristics to nonhuman animals, including our unknowable hominid forebears, while at the same time denying the special complexity and historical specificity of culturally mediated practices, only to read those very actions back into human social forms as . . . "nature." Animal kingdom? More likely a magical kingdom.

"MONKEY JUST THE SAME AS YOU"

When scientists say that we human beings are "like" one species and "unlike" another, they make a certain sense of the human world by recourse to a social reading of the animal world. They thus take nature as mirror and screen, finding there social reflection and ideological projection in equal measure. On this fundamentally misguided count, the difference between the free-associating journalist and the focused scientist is anything but self-evident. The scholars of sociobiology do not appeal to sounder evidence or more rigorous logic than do the lay people. Nor do the "vulgarizers" draw conclusions qualitatively different from the scientists who invoke evolutionary psychology to background their research. Given the methodological naiveté of this reductive structure, given the perpetual recourse to arguments on the "hidden hand" of genetic competition, given sociobiology's conflation of analogies (traits or behaviors in different species that resemble each other in function but have different evolutionary histories) with homologies (similar characteristics in different species that share a common evolutionary development)—indeed, given its basic blurring of analogous analytical terms with the very different referents they designate (e.g., human social stratification as compared with "hierarchies" in bee and ant "societies")—then one could construct an explanation of anything and everything according to this scientific-sounding language, which, not coincidentally, tells us what we intuitively "feel" to be true—all points developed in detail by Marshall Sahlins in his trenchant, incisive *Use and Abuse of Biology*, a book as timely now as it was in the 1970s.

But such confusions did not spring, autochthonous, from the seminal works of Edward O. Wilson, Robert Trivers, Lionel Tiger, and Robin Fox, as Donna Haraway shows in her history of twentieth-century primatology.[9] The sociobiologists' reification of the heterosexual pair, the nuclear family, patriarchal forms, and social inequality—no less than their strategic misreadings of evidence from the animal kingdom—drew on the legacies of 1950s and 1960s primate studies and, to a lesser degree, on the coetaneous ethnographic studies of foraging societies they influenced.[10] It is precisely here, in the "primary data," in the "observation" of "facts" about wild animals and wild men, that vernacular narrative structures and social self-interest are so entangled with scholarly findings as to be almost inseparable.

As Susan Sperling economically puts it, "Studies of monkeys and apes have never been just about monkeys and apes."[11] The physical anthropologists who studied baboon behaviors in the savanna understood male-headed troops as the primate norm and took these baboon troops as the model for prehuman hominids and early human patterns of group behavior. In the 1950s and 1960s, baboons appeared a particularly instructive species, not for any "objective" or "value-neutral" reason, but because of what physical anthropologists already believed (or took for granted) about the nature of human sexual relations and social groups. Drawing on the adaptationist logic of structural-functionalism, scholars projected a certain distorted image of modern institutions—marriage, family, hierarchy, and male domination—as broad, transspecies universals, thus refining a certain enduring image: alpha man, the hunter and protector, beside his little woman of the hearth.

This view is deficient in many ways, as Sperling goes on to show. Phylogenetically, we are far more closely related to chimpanzees (and bonobos) than to baboons, and subsequent studies have put under question a number of the "observations" first recorded about the baboons themselves. Observers, it would seem, tend to describe the behaviors of female apes in stereotypically "feminine" terms, those of males as "masculine," no matter what the actual behaviors in question. Moreover, it is difficult indeed to generalize about primates. The different species of our nearest primate relatives—chimpanzees, bonobos, gorillas, and orangutans—each exhibit distinctive ways of bonding, banding, and reproducing. Differences increase when monkeys are added to the great apes.

To make matters yet more complicated, each species behaves and groups differently depending on the environment it occupies (e.g., open savannah versus dense forest) and upon other conditions (e.g., low versus high population density). For example, the Indian gray langur monkeys studied by Susan Sperling live all over the subcontinent in many different ecological settings. In low-density, forested areas, they form multifemale, multimale groups with "fairly relaxed dominance relations." In high-density areas, they tend to form troops consisting of one adult male and multiple females, groups characterized by high levels of intermale competition (e.g., male raids on nearby troops, fast turnover in the resident male position, etc.).[12] Baboons, likewise, exhibit a range of "social" behaviors. In the open-country savanna, baboons live in male-headed troops. In the forest, baboon groups cluster, without clear dominance hierarchies, around a female and her offspring.[13] Which is a more "natural" setting? A meaningless question, since gray langurs and baboons "naturally" live in different habitats over wide territories.

Some studies suggest that, contrary to both early ideas and to socially informed common sense about big bullies, primate "dominance hierarchies" tend to reduce, not enhance, violence. More important, it is now known that these dominance hierarchies are not all one single hierarchy, stable over long periods of time. Recent studies distinguish, among others, hierarchies for grooming, for eating, for aggression, and for sex. A chimp, for example, might be superordinate to a peer in the grooming hierarchy, but subordinate to him in aggression. Not only do these various hierarchies change over time, but reviews of the data on many primate species find little correlation between dominance and reproductive success—supposedly the big Darwinian payoff driving such behaviors—indeed, the very linchpin of every sociobiological association.[14]

THE OTHER PRIMATES

Meanwhile, that mainstay of mid-twentieth-century social anthropology and precarious buttress for so much nonsense today, "hunting-and-gathering studies," had by the late 1980s entered a crisis from which not even basic concepts could be rescued. Since in the chapters that follow I've generally cast anthropology and anthropologists as heroes and heroines, counter to bioreductivist foils, I should explain.

Once upon a time, mid-twentieth-century cultural anthropologists studying contemporaneous foraging peoples depicted their craft as a practice of "ethnoarchaeology," their subjects as "living fossils."[15] That is, in observing present-day foragers, ethnographers claimed to be viewing the human and hominid past, to be reconstructing plausible accounts of human evolution, and to be getting at a picture of "human nature" uncontaminated by artifice and modernity. But that is Jean-Jacques Rousseau talking, not Franz Boas. (Boas, often referred to as the parthenogenetic "father" of modern anthropology, expressly rejected the notion that present-day "primitives" provide a window on the human past. All cultures, he reasoned, are equally "old" in that they and their antecedent cultures have been improvising solutions to human problems for the same lengths of time.)

Even so, the relationship between bioreductive fables and ethnographic descriptions is a slippery one. For all their deficiencies, these mid-twentieth-century ethnographies informed sociobiology and evolutionary psychology more in terms of form—the idea of a primitive human nature associated with a time of origins—than in terms of content. Little in the ethnographic data on hunting-and-gathering societies actually supported sociobiological turns on alpha man the aggressive hunter, dominant over his little domestic woman of the savanna. Quite the contrary, most ethnographers reported a high degree of sexual equality in the foraging societies they studied.[16] No doubt the basic pairing of "male hunting" with "female gathering" supported many a heteronormalizing notion about sexual complementarity, but, truth be told, very little in the way of systematic sexual research was ever undertaken among present-day foragers—just a few pages of testimony from that late classic of its genre, *Nisa*, the occasional aside . . . [17] At any rate, it was already apparent by the early 1960s that a sharply demarcated sexual division of labor was by no means universal in foraging societies—

that *Man the Hunter* and "Woman the Gatherer" (to cite the titles of two representative studies) simply did not stand as synecdoche for human relations in technologically simple societies. Women sometimes hunt. Men often gather.[18] Neither would seem bound to the other *of necessity*. And at the same time, abundant evidence already suggested that "pair bonding for life" is a relatively rare occurrence in many present-day foraging societies, some of whom have divorce rates comparable to those in contemporary U.S. society.[19]

Most of the field findings from foraging studies, such as they are, would seem to weigh against the classic sociobiological notions. But matters are even more problematic than a simple tallying of basic "facts" might suggest.

First, it is by no means clear that studies of hunting-and-gathering societies are even remotely relevant to the question of early human and hominid subsistence in the evolutionary past. Some physical anthropologists, primatologists, and biologists believe that collective, organized hunting was a relatively late development in human history and that our late hominid and early human ancestors were more likely to have been opportunistic scavengers than hunters and gatherers.[20]

Second, although evolutionary psychologists loosely correlate the practices of contemporary foragers to practices *surmised* to have occurred in the EEA—the Environment of Evolutionary Adaptation—the modern foragers whose practices supposedly shed light on evolutionary antiquity have almost all been reduced to the most marginal environments on the planet: the Kalahari desert, the Arctic tundra, or remote portions of rainforests. These are not, by any stretch, the environment in which human antecedents are likely to have evolved.

Worse yet, under inspection, not a single "primitive isolate" in the ethnographic annals proves to be either "primitive" or "isolated." All such societies participate in elaborate multicultural trading networks. All now live within the frontiers of modern state powers, and most have done so since long before the anthropologists arrived to study them. All have played bit parts—subject peoples, soldiers, day laborers—in the histories of local colonialism and world empires. Nor can anything resembling an unbroken continuity with prehistoric social practices or cultural meanings be firmly established for any foraging people—some of whom have horticultural, even proletarian, pasts, and any number of whom would seem more logically cast in the role of "refugees," "outcast peoples," or "subordinate populations" than as "noble savages."

Napoleon Chagnon's famous texts and films depicted the Yanomami as latter-day wild men: as an instructive example of how "primitive" human beings once "naturally" lived. What the anthropologist actually studied more than three decades ago was a group of twentieth-century subsistence horticulturists: latecomers to the Amazon whose territory was being encroached by miners, loggers, farmers, and ranchers, and whose villages were subject to lethal racist attacks by the same.[21] Restudies have cast that consummate hunting-and-gathering society of the Kalahari, the Ju/'hoansi (also known as !Kung), not as living relics of the distant past, but as a dispossessed subject people reduced to abject poverty in an ethnic caste system.[22] And with the scandals surrounding the "discovery" of the Tasaday—in which journalists and researchers appear to have been duped by the Marcos regime into "documenting" the existence of a "Stone Age" people living in the Philippine mountains—the field of hunting-and-gathering studies (and with it the now-romanticizing, now-denigrating mythos of anthropological primitivism) succumbed to its own contradictions.[23]

Foraging societies are undoubtedly different from horticultural, agricultural, or industrial societies. Their study is indispensable to a non-ethnocentric record of human ways of life.[24] But only by means of a profound indifference to standards of evidence and good logic could people immersed in the history of states and empires be depicted as people "without history," their lives situated beyond the scope of regional trade and global commerce, their cultures imagined to be as distinct as billiard balls.[25] And only by means of the most outlandish framing devices could anything of great detail be claimed as secure knowledge of foraging antiquity. Modern foragers, no less than the author or reader of this text, inhabit a world shaped by global capitalism and modern nation-states.

The ethnographic stock figures in whose name broad generalizations about human nature were drawn come perilously close to being a hoax. At best, what remains of the field of foraging studies is—as Richard Lee has put it—a way of looking at scattered forms of *modern*, even postmodern resistance to social stratification, forms that flare up at the edge of empires, in deserts or tundras or rainforests, wherever states are weak. These forms are as contemporary as any other.[26] At worst, what remains of foraging studies after its encounter with both history and modernity is little more than mythology: hunters and gatherers as stand-ins for Adam and Eve.

SOCIAL (F)ACTS

Observation is an embodied, social, and collective act. Undoubtedly, we *are* "like" certain animals in some ways and "unlike" certain animals in other ways—but everything depends on which traits we choose to foreground and how we wish to define the characteristics in question. Undoubtedly, bands of Khoisan people forage in the Kalahari (or did so until very recently), but how we see their labors will depend on how we frame them and how we put them into narratives of history (or its denial). That is to say, what we see as an objective "fact" will depend on a large number of subjective, constitutive, social *acts*. Even what counts as a phenomenon worth studying, as a result worth reporting, as a similarity or a difference—these are not self-evident in the nature of the world. They depend on what perspective one takes—or refuses to take.

The constant reading-off of instances from the animal kingdom in sociobiological narratives serves to convince the reader that s/he is in the presence of honest facts. The rhetorical presentation of contemporary "primitives" serves up supposedly convincing lessons about our own supposed past. But every such "fact" is *made*, a duplicity properly recorded in the very etymology of the term. As Paul Rabinow, Donna Haraway, and Bruno Latour have all, in different contexts, pointed out, "fact" and "factory" both derive from the common Latin root, *factum*, meaning "something made," "something fashioned," "something done."[27]

The basic facts about hunters and gatherers remain open to question. One could mobilize contradictory evidence from widely differing foraging societies to show that a sexual division of labor is or isn't practiced, that lifelong monogamous pair bonding is or isn't practiced, that men and women are or aren't equal, that homosexuality is or isn't condoned. The only thing we cannot reasonably conclude is that any of this has anything to do with the time of human origins, with humanity in the raw, with the basic human blueprint.

So, too, with primate behaviors. Because our simian and primate relatives exhibit such a wide range of behaviors, one could selectively mobilize real instances to support virtually any claim about "human nature." And because so many primate behaviors are ambiguous, one could attribute to the evidence any number of contradictory readings. Unchallenged, mid-twentieth-century scholars tended to select and interpret those behaviors and species that made sense to them as relatively affluent, mostly white, putatively heterosexual men living during the Cold War era. As Donna Haraway shows, they tended to compose, through stories about apes, a certain selective portrait of themselves.[28] Basic ideas about the human design—about the origins of kinship, the meanings of sex, the social division of labor—were spun in the warp and woof of heteronormativity. A generation later, sociobiology incorporated these pictures into its narratives. Nineteen-fifties ideology became 1970s conservative revanchism and 1990s evolutionary whimsy and nostalgia.

Embedded in each story, then, we find another story: an intricate narrative architecture of fact and fantasy, discovery and projection, science and morality, a design, derived from social experiences and projects of the past, that is also part of our cultural design today: knowledge as a form of not knowing; vision as a practice of not seeing. The "making" of "nature."

Adam and Eve Do the Wild Thing

The Science of Desire,
the Selfish Gene, and Other Modern Fables

He who does not know how to put his will into things at least puts
a *meaning* into them: that is, he believes there is a will in them
already.

FRIEDRICH NIETZSCHE, Twilight of the Idols

5   The Science Question
Cultural Preoccupations and Social Struggles

The form of criticism whose arguments I have been rehearsing—the constructionist approach to science questions—has lately come under considerable fire. "Social constructionism" views scientific models as narratives that (like other stories) are influenced by the prevailing cultural ideas, stories whose representations, in turn, contribute to the architecture of society. This approach has lately gained a reputation for issuing frivolous claims—asserting, for instance, that our physical bodies are "posited" much as one might posit notions in a philosophical exercise, or that reality is "made up" much as one might write a work of fiction. Alan Sokal—the prankster physicist whose hoax paper consisting mostly of gibberish was unwittingly published by the editors of *Social Text* in their "Science Wars" issue—thus hectors his opponents with all the bravura of a petty realist, inviting constructionists to take a flying leap: "Anyone who believes that the laws of physics are mere social conventions is invited to try transgressing those conventions from the windows of my apartment. (I live on the twenty-first floor.)"[1]

Undoubtedly, the time was ripe for an exposé, and Sokal's table turning enjoyed considerable celebrity in the mass media. The hapless *Social Text* editors did not understand the mangled physics or nonsensical mathematics in Sokal's prank paper, but they published it anyway because it made gestures toward the kind of politics they lauded. Pesky constructionists were thus hoisted by their (our) own petards, found guilty of a *Left* version of the same ideological bias they (we) usually attribute to the scientific establishment on the cultural Right: a willingness to see what they believe rather than to believe what they see. Turnabout is fair play, by all accounts, and Sokal unquestionably scored major debating points against his adversaries in critical science studies. Still, it is worth noting what role Sokal's cause

71

célèbre played in media venues. The physicist's full-throated cry for "scientific objectivism" generated front-page stories in newspapers that, at that
very moment, were trumpeting claims (later quietly retracted) that scientists had discovered a "gene" for "thrill seeking," thus giving genetics a
magical aura more appropriate for palm reading than for science.[2] In the
same newspapers, science journalists were circulating fantastic (and counterfactual) tales about the evolutionary origins of human nature, giving
the imprimatur of "objectivism" and scientific authority to everyday sexual stereotypes.

Critics like Sokal are not entirely wrong when they suggest that a current of social constructionism has succumbed to the temptations of magical thinking: the belief in a dream world where discourse alone makes
things happen, where intellection encounters no resistance from material
obstacles. Jacques Derrida's choice aphorism, featured on the first page of
Judith Butler's *Bodies That Matter*—"There is no nature, only the effects
of nature: denaturalization or naturalization"—would seem to confirm
Sokal's worst charges.[3] And a common apostrophization of the constructionist perspective—"nature is to culture as the constituted is to the constituting"[4]—already verges on idealism, if "culture" is understood as an
ideal schema that "acts" (constitutes, constructs), and if material life, labor, social relations—in a word: history—are thereby evacuated from the
process of construction.

But the properly enunciated constructionist claim is more subtle than
the idealist propositions sometimes staged in its name. Constructionism asserts, not that the facts of gravity are at our whim, but that in transforming the motions of terrestrial objects into "natural laws," Newton participated in the elaboration of a mechanistic model of the world—a view that
carried social implications no less than technological applications. Constructionism says, not that atoms or DNA are mere figments of an ideological imagination, but that the image of the atom—individual, discrete,
autonomous—resonates in complex ways with the preoccupations of an in-
dividualistic civilization and that when biologists refer to DNA as the "master molecule," they plant ideology squarely at the center of scientific description. Constructionism maintains, not that men and women, gays and
straights, are spectral fictions, but that these and other distinctions of the
human world are socially elaborated, not naturally given. Constructionists,
in their right mind, do not deny the inevitable pull of certain life events:
physical growth, metabolism, sleep, illness, death. They simply state that,
for human beings, what is most important about those questions is their
meaning—and that what so often appears self-evident and timeless about

these meanings belongs to history, not nature. A properly thought-out constructionism does not deny the materiality of physical things, but it does suggest that the objectivity of objects is itself the product of a certain highly subjective work. It argues that what marks the object as such is countless unmarked decisions about what to foreground and what to background, what to hold constant and what to see as variable. . . . A defensible constructionism does not hold that the natural world is simply "made up" or "posited." Rather, it stresses that how we *think* about natural beginnings, what we *want* from the world, what perspectives we consciously *take* or unconsciously *refuse* all contribute to what we perceive in and *make* of nature. Constructionism thus attempts to trace the flow of history through the congealed, naturalized objects of common sense—and thereby to give us a little more critical leverage over the facts whose facticity we usually take for granted.[5]

CONSTRUCTING NORMALCY

Consider for a moment the flow of history in and around a scientific distinction that is now all but ubiquitous—that schema of all classificatory schemes, "the norm." Whatever premodern peoples made of good and evil, rule and infraction, desirability and undesirability, health and infirmity, it never quite occurred to them to amalgamate these concepts under a classification of human phenomena as "normal" and "abnormal." Until the nineteenth century, "normal" was primarily a geometrical term. Doubtless, as Ian Hacking points out, an evaluative concept lurked in the background of this geometrical terminology. An orthogonal or normal line "is a 'right' angle, a good one."[6] And doubtless, concepts like "the golden mean," "moral rectitude," even "straightness" (no pun intended) had been linked, since the Greeks, to a faith in geometry as truth and to a highly mathematized understanding of "nature" as instructive ideal. Still, the term "normal" itself had meant no more than "standing at right angles; perpendicular."[7] Whatever thoughts and practices its nearest conceptual relatives had admitted all seem different from those motivated by the modern notion of normalcy. For example, since Aristotle defined the best human characteristics as *extremities* (not averages), his conception of the mean admits no calibration and is thus "radically different from that of a century that defines degree of intelligence by a Normal distribution with a mean scaled at 100."[8] The "golden mean" of yore is not the modern "norm."

Although myriad developments in the eighteenth-century culture of science gave presentiment to the emergence of such concepts, there was as yet no such thing as a "normal" person or an "abnormal" desire. It was in the 1820s, with the French doctor François-Joseph-Victor Broussais, that the task of modern medicine was first clearly defined: the identification of excitations that cause tissues and organs to "deviate" from their "normal" state into an "abnormal," pathological, or diseased state.[9] After Broussais, the conceptual distinction between "normal" and "abnormal" underwent a rapid development and a broad dissemination, as Georges Canguilhem shows.[10] Belgian astronomer Adolphe Quetelet contributed the notion of the *homme type*, or "average man"—and, more important, put the concept of the norm on a statistical footing, charting social phenomena (crime rates) on a normal (or bell-shaped) curve.[11] With Francis Galton—Darwin's cousin and the founder of the "science" of eugenics—these statistical principles were applied to anthropometrics, and "normal" became a kind of body, a type of mind, and an inner condition.[12]

What occasioned Galton's influential distinction was the idea that human progress requires selective breeding.[13] That notion, of course, was prompted by racist beliefs complicit with the practice of colonialism and by correspondingly racialized conceptions of social stratification and social order. In the background of these ideas about "race" and character were traditional English folk beliefs about "blood" and "heredity." The rest, as they say, is history—a history that links anthropometrics and scientific racism to the rise of modern medicine, psychiatry, and sexology; discourses on "nature" to institutions of "normalization"; and folklore to science and science back again to everyday life. Henceforth, "normal" became the measure, all at once, of the ideal, natural, healthy human being. "Abnormal" became the pathological deviation from desirable rectitude. And after a process of institutional routinization, everyone ever after has been prompted to ask the peculiar question, "Am I normal?"[14]

I trust that my highly abbreviated example suggests something of the stakes involved in constructionist accountings. In good Foucauldian form, the statistical application of geometrical principles to human populations ("normalization") has reshaped the world both as it is inhabited and as it is perceived by human beings, who were themselves transformed on the grid of the bell-shaped curve.[15] It is not that Broussais or Quetelet or Galton created human norms ex nihilo, or even that the idea of the norm is an empty or useless category. Rather, the concept of the norm brought together preexisting ideas in a novel constellation—a constellation that proved *socially* useful and *materially* productive. It is thus that the idea of normality got caught up in the momentum of history, conveyed along by powerful currents, while at the same time speeding along other changes, some intended, others unforeseen. Today, at the end of a long and complicated process, we call "normal" what is deemed good, healthy, or desirable (not to say "natural"). It now seems self-evident that a healthy heart functioning as we might hope is "normal." We discern in the world "normal" and "abnormal" weather conditions, just as it might seem that men's and women's bodies, when "normal," are "opposites" that attract each other. Or, alternatively, we claim that same-sex attractions, too, are normal, by which we mean natural, healthy, and good.

If my example gives rise to some modern wonderment about how people expressed themselves in the face of usual and unusual events before the era of normalcy, it might also provide the occasion for stating the constructionist claim in its full thickness and resonance. In the course of their everyday practices, people elaborate ideas about the world. These ideas do not fall from the sky. On the one hand, they refract and creatively reconstitute the preexisting notions people have already had; on the other hand, they reflect the practical orientations and intentionalities of social actors toward a socialized world. Subject to the push and shove of contending social interests, some of these ideas find a purchase on men's and women's thoughts and actions, while others do not. And so begins the second part of a dialectical process of social construction. Arbitrary distinctions, conceived in the womb of culture, and ideal schema, forged in the crucible of history, are socially articulated in and through institutional practice. Routinized as ritual, congealed in social relations, sedimented in law, stamped in products of labor, and habituated in the body, they thereby become part of the warp and woof of everyday experience and practice—they thus somehow impose themselves in that oblique social space where the self-assurance of common sense meets the self-evidence of the senses.

It is this oblique space—this "blind spot"—that constitutes the real object of every constructionist inquiry. It is here that history is buried, layers deep, in habit, where social acts are taken as natural facts and ephemeral constructs pass themselves off as eternal verities.[16]

THE SITUATION OF KNOWLEDGE

The field of critical science studies, boiled down to its basic proposition, says no more than this: Science, which attempts to know nature in a methodical and objective way, can no more escape the tug of culture than can any other form of reflection. Because scientists are human, perceiving subjects—and because perception is impossible without some perspectival standpoint, some point of view, some complex interplay among intention, idea, practice, and desire—the most elemental practice of science is saturated with politics, and from many different angles.[17] These points are perhaps best made by Donna Haraway, whose magisterial *Primate Visions* traces the tracks of twentieth-century politics and social preoccupations across the history of primatology and physical anthropology: "natural sciences, like human sciences, are inextricably *within* the processes that give them birth."[18] Such a robustly constructionist framework also provides the scaffolding for Stephen Jay Gould's critical studies of the history of scientific measurement: "Science, since people must do it, is a socially embedded activity. It progresses by hunch, vision, and intuition. Much of its change through time does not record a closer approach to absolute truth, but the alteration of cultural contexts that influence it so strongly."[19] Margaret Wertheim, writing in *The Sciences*, underscores the basic point: "The current bitterness engendered by the so-called science wars has obscured the fact that postmodernism expresses an essentially reasonable insight: all knowledge is derived within a particular cultural framework and will therefore reflect aspects of that culture."[20] All knowledge, in short, is "situated."[21]

The eye of the scientist, like that of everyone else, is a *trained* eye that has *learned* to see. The act of looking, and what is sought, affects what nature discloses. Even the "objectivity" of the scientific object is the result of a certain highly subjective work—social elaborations, cultural framings, and perspectival distancing. . . . All the more so, because scientific observation is almost never "simple observation" of some object, but first a series of interventions designed to mark, cut, extract, foreground, or isolate—to "objectify"—whatever is under study. For example, Anne Fausto-Sterling shows that anatomical parts such as the corpus callosum or the hypothalamus, as studied by biologists, are *not* "naturally occurring" parts of the anatomy, but hypothetical constructs. What biologists actually study are cross-sectioned, haphazardly measured, and arbitrarily divided objects that are partly "real" and partly "invented."[22]

Necessarily, then, presupposition affects the way scientists shape their objects of study. More than one scientist has noted how faith in the order and coherence of the prevailing scientific paradigm sometimes all but precludes evidence to the contrary. "You don't look for what you don't believe in," explains Albert Goodyear, an archaeologist, reflecting on how archaeologists collect evidence about the human past based on what they (rightly or wrongly) already believe to be the case.[23] Confident that the Americas had been exclusively settled by slow-wending, foot-propelled migrations from across the Bering Strait, archaeologists thus seldom looked for evidence of archaic settlement by coastal boat routes in southern South America or in the southeastern United States, and they tended to discount findings contrary to the prevailing model until the slow, happenstance accumulation of evidence allowed no alternative to a rethinking of basic assumptions.

Stephen Jay Gould gives the following pointed example of how presupposition can blinker the practice of science and even shape what "counts" as a "finding": "A former student of mine recently completed a study proving that color patterns of certain clam shells did not have the adaptive significance usually claimed. A leading journal rejected her paper with the comment: 'Why would you want to publish such nonresults?'"[24] "Nonresults," in this case, were in fact results that had negative implications for the ultra-adaptationist variant of Darwinism prevailing in the culture of science, a variant that views every aspect of physiology and behavior as a positive "adaptation" to the pressures of natural selection.[25]

In cases like these, the validity of an approach or the significance of a finding relies on what Bruno Latour describes as the "communicative community" of scientists.[26] Scientists, presumably, are experts. Their opinions, informed by a body of evidence and experimentation, logically ought to have a certain weight. But in a surprising number of instances, scientific opinion cites and relies on nothing more substantial than social convention or prevailing sentiment—history buried in habit, convention taken for nature.

Newton thus clung to a misguided belief in the mysterious substance "ether"—not because his physics needed a medium in which to operate, but because he believed that ether could provide the continuous medium through which God's will might be expressed in the universe. Linnaeus devised the term "Mammalia" (literally, "of the breast") not because lactation is the only (or even the best) criterion for distinguishing this class of animals but because he participated in the obsessions of his day and age, a moment when "doctors and politicians had begun to extol the virtues of mother's milk" as the natural sign of women's role in domestic (as opposed to public) affairs.[27] "Race" seemed a salient biological concept for so long not because it economically captured phenotypic variation (it did not), but because a colonial history framed the way people, including scientists, perceived and thought about human bodies.[28] And as long as racist and misogynist views held sway in the society that contextured scientific communication, there was never any shortage of scientists who were willing to construe every real (and very many imaginary) anatomical difference(s) between African and European, or between men and women, as a demonstration of natural inequality.[29]

The dispositions and perceptions of scientists, like those of other human subjects, are shaped on the lathe of history—by the rise and fall of empires, by the heave and shove of classes, and by the give and take of various ideological struggles. As a result (and the protestations of some scientists notwithstanding), the question of where "science" ends and "ideology" begins is a difficult, open-ended one. This is not to say that science (or even perception) is always "politics by other means" (although manifestly it sometimes is). Nor is it to claim, globally, that the "idea" exercises a fascistic tyranny over thought and perception (although sometimes it indeed seems this way). It is to say, rather, that the framing of questions—which shapes what can be seen or known, which forecloses what other inquiries might disclose—always occurs within an ideological horizon. It is to say, moreover, that socially conditioned "common sense" affects inquiries in myriad ways—beginning with the definition of a "proper" problem, including the formation of a "plausible" hypothesis, and not excluding what might be counted as "evidence." It is to say that systems of scientific classification inevitably carry traces of systems of social classification—for insofar as scientific classifications are useful systems, they turn on the practical interests of the socialized beings who use them to see and understand the world. It is to say that perceptual practice is both empowered and befuddled by political habit and by the complex body of acquired dispositions that gives us intelligible objects to begin with.

When the object in question belongs to the realm of astronomy or physics—areas theoretically far removed from the heave and shove of social interests—the work of social intentionality shows up in varied and layered ways: in the use of humorous or clever names to designate mysterious entities, in characterizations of things that verge on personifications, in the attribution of human foibles to physical objects, in the uncanny resemblance of certain natural laws to the balanced formulae of social rules. . . . But when nature is "human nature," cultural precept is often taken for natural percept outright.

THE DIALECTIC OF NATURE

Marx and Engels telescoped the urgency of these issues when they commented on the irony that although Darwin—whose work the two socialists admired—had "dealt the metaphysical conception of Nature the heaviest blow," he still unselfconsciously projected onto the screen of nature a no less fantastic picture: "It is remarkable that Darwin recognizes among beasts and plants his English society with its division of labor, competition, opening up of new markets, 'inventions,' and the Malthusian 'struggle for existence.' It is Hobbes's 'bellum omnium contra omnes,' and one is reminded of Hegel's *Phenomenology*, where civil society is described as a 'spiritual animal kingdom,' while in Darwin the animal kingdom figures as civil society."[30]

Marx does not claim that natural selection is but one narrative among other origins stories—an argument best left to creationists and religious fundamentalists. What Marx indicates is the give and take of a two-sided process.[31] On the one hand Darwin, for all his insight, could not help but draw—sometimes consciously, sometimes unconsciously—on the social language, cultural vantages, and political habits available in a bourgeois civilization. As a result, he could both *see* and *not see*. When he gazed upon nature, he caught glimpses of himself, his own English society and habits. On the other hand, when Darwin's ideas about natural selection returned to the society that gave them birth, the mist-enveloped time of evolution became a mythic time. Darwinian insights mutated into vernacular notions of individual agency, economic competition, business practices, and social progress.[32] The idea that what is most "natural" about human nature belongs to the hominid past became caught in the coils of a time-bound understanding: an all too familiar world in which cultured nature mirrors naturalized culture.

6 Sexual Selection

Eager, Aggressive Boy Meets Coy, Choosy Girl

Today's origins stories, some of which we've sampled in the first part of this book, owe not to the best but to the worst part of Darwin's thinking. At the origins of the origins story, and even before the coining of the term "heterosexual" or the widespread modern usage of the word "normal," Charles Darwin's theory of sexual selection laid the groundwork for much of what would subsequently come to be understood as "natural" about desire.

"The males are almost always the wooers," he argues, reading off a page from Victorian romantic conventions.

> The female . . . with the rarest exception, is less eager than the male. . . . [S]he is coy, and may often be seen endeavoring for a long time to escape from the male. Every one who has attended to the habits of animals will be able to call to mind instances of this kind. . . . The exertion of some choice on the part of the female seems almost as general a law as the eagerness of the male.[1]

Having drafted a "general law" for the animal kingdom, Darwin gives a predictably Victorian picture of the female of his own species, her condition defined by motherhood: "Woman seems to differ from man in mental disposition, chiefly in her greater tenderness and less selfishness. . . . Woman, owing to her maternal instincts, displays these qualities towards her infants in an eminent degree; therefore it is likely that she would often extend them toward her fellow-creatures."[2] And where the woman is described in terms of nurturance, the man, contrariwise, is defined by competition: "Man is the rival of other men; he delights in competition, and this leads to ambition which passes too easily into selfishness. These latter qualities seem to be his natural and unfortunate birthright."[3] Happily, the woman is there to civi-lize the savage male.

Such is the Darwin that has come down to us as sociobiology and evolu-tionary psychology. Such passages simultaneously project English social conventions as natural laws and rebound to naturalize those very conven-tions. They lay the groundwork for the subsequent attribution of not just heterosexual desire to timeless nature, but also a definite kind of man, a cer-tain kind of woman, and specific patterns of sexual attraction and seduction: the eager, wooing male, the coy, choosy female . . . the draw of opposites, the ritual of the chase.[4]

Darwin presents his portraits of male nature and female nurture as though they were unobjectionable, hence timeless. The author's character-izations thus lean on the consent of the reader: "Every one who has attended to the habits of animals will be able to call to mind instances of this kind." But in fact, such commonsensical notions were of recent vintage in Dar-win's day: They trace the course of political, economic, and institutional changes then under way in Northern Europe and the United States. Dar-win's description of universal maleness conspicuously highlights emergent masculine conventions of the liberal marketplace, with its competition and rivalry between ambitious individual men. At the same time, the weakly sexed and pacific creature that Darwin describes as the universal woman evokes his own era's "angel of the house," whose role it was to tame men's passions and to maintain propriety. Far from constituting a timeless or uni-versal consensus on woman's nature, this idealized view of femininity ac-tually took shape in the wake of the democratic and egalitarian revolutions in America and France. As Carol Groneman explains, the picture of the ten-der, selfless woman reconciled the promise of democratic equality with the reality of sexual inequality: It served to justify the fact that women could not vote and lacked equal property rights.[5]

Contrary to Darwin's depictions of male desire and female choosiness, it was in fact *women* who had been regarded as the more highly sexed crea-tures in classical antiquity and throughout the Middle Ages. Elizabeth Ab-bott recounts a telling myth: Zeus and Hera ask Tiresias to settle an argu-ment over who gets more pleasure from sexual intercourse, male or female. Hermaphroditic Tiresias promptly replies that women do—claiming, more-over, that women receive nine-tenths of the pleasure of sex and men a mere tithe.[6] In the same vein, Londa Schiebinger notes that "for Aristotle, writ-ing long before the historical rise of the 'passionless female,' mares were sexually wanton, said to 'go a-horsing' to satisfy their unbridled appetites. If not impregnated by a stallion, these dissolute females would be fertilized by the wind."[7] Contrary to Darwin's associations of motherhood with general "tenderness" and nonaggression, an ample folkloric tradition going back to Aristotle, Aelian, and Pliny dwells on the special ferociousness of she-wolves, she-bears, lionesses, and female animals in general when pregnant and after parturition.[8]

FITNESS

It is useful to mark the departure of passages like these from the wider thrust of Darwin's work—and to mark the contrast between Darwin's general approach and those of other nineteenth-century naturalists. Nowhere is the revolutionary impact of Darwin's ideas more pronounced than around the concept of "fitness." A brief comparison between Darwin's ideas and those of two other prominent thinkers of his day will suffice to show the stark contrast between naturalistic fables and natural science.

In the early nineteenth century, William Paley's *Natural Theology* had served up lessons from comparative anatomy as homilies on divine design. For example, since they served an underground existence, the mole's tiny eyes and palmated feet were offered as proof of nature's benevolent design, hence of God's existence.[9] By contrast, Darwin's arguments on natural selection show that the "fit" between physiognomy and environment results not from an intelligent design but from eons of incremental adjustments. His arguments also suggest that evolution devises less than perfect meetings of form and function.

And whereas Herbert Spencer, one of the grand social philosophers of the nineteenth century, imagined "survival of the fittest" as an absolute, universal, and progressive principle—with evolution as a series of stages through which all species "advance" along a single gradient to "higher forms"—Darwin argued only that "fitness" is a measure of adaptation to *local* conditions at a *given moment*. Even a relatively subtle change in environmental conditions can render what was supremely fit one epoch "unfit" the next—as witness the veritable avalanche of extinctions recorded in the evolutionary record. For Darwin, then, as opposed to Spencer, "fitness" is an entirely relative concept.[10]

The Origin of Species occasionally lapses into culturally conditioned phrases and notions—existence as a "struggle" or "competition," a "utilitarian doctrine," notions of "progress," and so forth—but the main argument of the text is properly read as turning on contingencies, accidents, and happenstance adaptations. Indeed, Darwin initially rejected the use of the term "evolution," since its pre-Darwinian understanding implied a linear development from a simple or lower condition through stages to a higher, more complex, or more advanced state. In contrast to earlier ideas, the theory of natural selection provides a distinctly nonteleological approach to evolution, an account in which organisms change or become extinct not because things were predestined to come out a certain way, but through the inadvertent proddings of local selective pressures on chance creatures. Since nothing in Darwin's theory can vouchsafe the triumph of the complex over the simple—and because "fitness" is always a precarious, makeshift arrangement—Darwin's ideas, at their core, cannot sustain concepts like "progress," "advancement," or "perfectability." As Stephen Jay Gould summarizes, natural selection is a radical idea because of "its power to dethrone some of the deepest and most traditional comforts of Western thought," namely, the theological notion of "nature's benevolence, order, and good design" and the notion that "nature has meaningful directions"— indeed, that it serves up lessons of any sort.[11] In short, Darwin's key ideas about fitness and adaptation in *The Origin of Species* are incompatible with just-so stories about anatomical design and with moralistic parables about the meanings of nature.

This is not the case with Darwin's arguments on sexual selection. In the chapters of *The Descent of Man* called "Selection in Relation to Sex" (and especially in "Sexual Selection in Relation to Man"), the language of Victorian convention—ideas about sex, race, and progress—all but overwhelms the scientifically serious hypothesis that, for certain species under certain conditions, it is "possible for the two sexes to be modified through natural selection in relation to different habits of life . . . or for one sex to be modified in relation to the other sex."[12]

NATURAL SELECTION, SEXUAL SELECTION

The theory of sexual selection supplements that of natural selection in the following manner: Natural selection "is about living long enough to reproduce," whereas sexual selection "is about convincing others to mate with you."[13] Specifically, the theory of sexual selection hinges on the idea that there can be intraspecies competition (especially competition between male members of a species) as well as interspecies competition. Darwin gives two main variants of this competition: the struggle, by way of tooth, horn, and claw, between individuals (usually males) "to drive away or kill their rivals," and the struggle (also usually between males), by means of brilliance, dance, ritual, or song, to "excite or charm" members of the opposite sex.[14]

At its most convincing and useful, the theory of sexual selection treats certain anomalies not altogether explicable under the theory of natural selection. The peacock's spectacular coloration and weighty tail, for example, appear to be "unfit" under the logic of natural selection. These traits would seem to leave their bearer more visible to foxes and might even impede his escape from predators. Such extravagant plumage might make good biological sense, however, if it somehow singled out the peacock to the peahen—that is, if excellent markings conferred upon their bearer some reproductive advantage over other, less visibly marked, peacocks.

The hypothesis itself—the idea that reproductive pressures, in some species, have differentially "selected" for modifications to the sexes' anatomies and behaviors—is a serious one, obviously useful in explaining traits like the male salmon's hooked jaw, the pheasant's plumage, or the deer buck's antlers. The demonstration of a "general law," however, presupposes a scrupulous, difficult practice of "bracketing," the separation of received wisdom and local prejudice from cautious, careful accountings of what male and female animals (not to say men and women) really *do* under various circumstances. That's a difficult phenomenological feat under the best of circumstances. Almost nowhere has this caution been evident among biologists wishing to advance the argument or evidence for sexual selection. Indeed, the influence of ideology is evident from the first premise. The theory itself attempts to explain what Darwin viewed as a general tendency for males to be not only larger, stronger, and more prominently marked, but also more active, more pugnacious, and more intelligent than females—a general tendency, that is, for "male excellence" and, correlatively, for female mediocrity.[15] Much of what Darwin counts as the evidence and how he describes it proceeds from, rather than leads to, this conclusion.

In developing a narrative about sex, dimorphism, and selection, Darwin (no less than latter-day Darwinists) attributes practices of great variety to the rule of this singular "general law." Despite the author's framing, however, it is difficult to see how "male eagerness" and "female coyness" manifest themselves in the mating behaviors of certain spiders:

> The male is generally much smaller than the female, sometimes to
> an extraordinary degree, and he is forced to be extremely cautious
> in making his advances, as the female often carries her coyness to a
> dangerous pitch. De Greer saw a male that "in the midst of his prepara-
> tory caresses was seized by the object of his attentions, enveloped by

her in a web and then devoured, a sight which, as he adds, filled him with horror and indignation."

Darwin dismisses examples that clearly diverge from the general rule in the same the way that he treats inconvenient information about sea worms: "These animals are often beautifully coloured, but as the sexes do not differ in this respect, we are but little concerned with them." Or worse yet, the exception is cited as though it manifested the general rule: "In almost every great class a few anomalous cases occur, where there has been an almost complete transposition of the characters *proper* to the two sexes; the females assuming the characters which *properly* belong to the males" [emphasis added]. Having noted that some species in every class *invert* (what he believes to be) the general rule, Darwin perplexingly refers in the next sentence to the "surprising uniformity in the laws regulating the differences between the sexes."[16]

The problems with these descriptions are not minor, for it is on the force of such descriptions, misclassifications, and dismissals of evidence that the author invokes the generality of the "general law." Because Darwin presupposes male and female creatures of a certain familiar sort (eager/aggressive, coy/choosy), his descriptions tend to eclipse serious consideration of the varied means by which species actually mate and reproduce. Those means include lack of or minimal sexual dimorphism, highly varied forms of dimorphism (i.e., different kinds of variation by color, shade, size, shape, anatomical feature, or behavior), "inversions" of the "proper" dimorphism, "polyandry" (one female takes multiple male mates), "polygyny" (one male takes multiple female mates), and the like. Although Darwin touches on such variety— the better to enclose it within a monotonous "general rule"—he does not even contemplate what every farm boy (and surely every naturalist) knows about the prevalence of nonprocreative and nonheterosexual sex in the animal kingdom. The consequences of such elisions are manifold, both for the terms in which "science" understands "sex" and for the circulation of scientific ideas in the culture that gave them birth.

First, everything in Darwin's text on sexual selection converges to make reproduction the privileged, unique, and sole aim of sex. This descriptive and analytical strategy inaugurates an impoverished ethological tradition that, to this day, excludes or suppresses from the evidence much of what animals actually do: autosexual behaviors, nonreproductive sexual intercourse between males and females, same-sex behaviors.[17] Second, because it occludes consideration of the diverse ways that sex actually happens and even what might count as sex in the animal kingdom, Darwin's text forecloses the possibility of other approaches to the question—among them, the possibility that sex might serve a variety of different functions in animal life and the possibility that sexual selection might operate by diverse, even contradictory, means.[18] Third, it is by means of these exclusions and suppressions that heterosexuality *of a certain sort* (the pursuit of the coy, choosy female by the eager, aggressive male) comes to be equated with "nature," with the condition of origins, with the first day. And finally, it is this anthropomorphic and heterosexualizing language that gives the "general law" its self-evident applicability to human practices—indeed, its urgency as a moral lesson about the immutable nature of desire.

WARPS AND KINKS
IN THE NARRATIVE ON SEXUAL NATURE

Darwin extrapolates from the animal kingdom to human society: It is presumably the larger, stronger, more competitive men who have better access to women and who thus are favored in sexual selection. This idea would later become an enduring trope in gender culture, from Charles Atlas's famous ads for his "physical fitness" technique to some of sociobiology's darkest ruminations on male competition and sexual domination. But such a notion proves difficult to extricate from Victorian fantasies about primitives, wildness, and raw human nature—fantasies that give rise to passages like the following hallucination of male power and female vanity:

> Man is more powerful in body and mind than woman, and in the savage state he keeps her in a far more abject state of bondage than does the male of any other animal; therefore it is not surprising that he should have gained the power of selection. Women are everywhere conscious of the value of their own beauty; and when they have the means, they take more delight in decorating themselves with all sorts of ornaments than do men.[19]

Note that the above passage purports to describe a *reversal* among human beings of the rule Darwin had sketched for much of the animal kingdom, a widespread tendency for *males* to be conspicuously adorned, brightly colored, or beautifully marked. Arguing analogically, with modern "savages" standing in for "primitive" humans, Darwin is suggesting that at some time in the human past, men used brute force to seize "the power of selection." This passage serves as a critical link in Darwin's argument. It connects human institutions to animal behaviors (that is, social facts to natural facts) by way of evolutionary claims about the lives of imagined "primitive" peoples. But the "facts" Darwin attempts to explain are by no means self-evident, and the argument about human nature chases contradictory objectives.

Darwin seems to believe—erroneously—that in all human cultures, women are the more adorned sex. Worse yet, the picture he gives of "savage societies" draws on descriptions provided by colonial travelers and missionaries, accountings that invariably gave lurid and prejudicial pictures of sexual savagery and primitive violence. The odd result, as Rosemary Jann meticulously shows, is that Darwin's claims about savage sexual depravity work at cross purposes to the general run of his narrative on sexual selection.

Consider the not altogether logical circulation of meanings in Darwin's claims. Having projected Victorian social conventions onto animal behaviors in the theory of sexual selection to begin with—the eager, wooing male, the coy, choosy female—Darwin's narrative returns to the contemporary social scene to suggest that Victorian women exercise some *natural* power of choice in courtship rituals. Darwin is disposed to argue that English-women's coy choosiness "advances" the race in that they "select" the "better" (more "fit") men (and this is somewhat progressive of him, especially by comparison to the ideas of other biologists about female agency—or rather, the lack of it).[20] But the central problem with Darwin's analogy is that among the *animals* he describes, it is the *males* who are the beautiful, adorned, colorful, and alluring creatures. And according to Darwin's arguments, it is "male beauty" that provides the ground for "female choice" in the animal kingdom. This is certainly not what Darwin believes to be the case in human societies, least of all his own. How, then, to reconcile the pattern Darwin discerns in the animal kingdom with what he believes to be the contrary human case, in which it is *women* who are the beautiful, vain, preening creatures? Darwin simply postulates that at some point in the past, men "appropriated" the power of selection, a usurpation that presumably induced women to adorn themselves, so as to make themselves more attractive to choosy men. But of course this scenario and the historical trajectory it suggests throw into jeopardy what Darwin wishes to argue about Victorian women's choosiness: If primitive men long ago usurped the power of selection, then just how did Victorian women get it?

Seemingly aware of the inconsistencies in his arguments, Darwin contrasts the "elevated" state of Victorian women to the abject condition of women in savage societies as both a measure of human progress *and* as a demonstration of the validity of the theory of sexual selection. But this move by no means eliminates the logical inconsistencies in the argument. Indeed, the sexual "degeneracy" of modern savages, analogically linked to the way "primitive" hominids are imagined once to have lived, suggests a warped or perverse turn on the evolutionary narrative: Modern Victorians' gender roles, family forms, and sexual practices are depicted as "more natural" than those of either modern savages or past primitives—the very figures whose existence (at the supposed baseline of the human condition) lends rhetorical credence to the idea of "natural origins" to begin with.[21]

Ultimately, such contorted ideological reveries lead Darwin to pose a confused concluding defense of his theory—a defense that once again calls upon readers to ally themselves, as modern Victorians, with noble animals and against violent savages:

> The main conclusion arrived at in this work, namely, that man is
> descended from some lowly organised form, will, I regret to think,
> be highly distasteful to many. But there can hardly be a doubt that we
> are descended from barbarians. The astonishment which I felt on first
> seeing a party of Fuegians on a wild and broken shore will never be
> forgotten by me, for the reflection at once rushed into my mind—such
> were our ancestors. These men were absolutely naked and bedaubed
> with paint, their long hair was tangled, their mouths frothed with ex-
> citement, and their expression was wild, startled, and distrustful. They
> possessed hardly any arts, and like wild animals lived on what they
> could catch; they had no government, and were merciless to every one
> not of their own small tribe. He who has seen a savage in his native
> land will not feel much shame, if forced to acknowledge that the blood
> of some more humble creature flows in his veins. For my own part, I
> would as soon be descended from that heroic little monkey, who braved
> his dreaded enemy in order to save the life of his keeper, or from that
> old baboon, who descending from the mountains, carried away in tri-
> umph his young comrade from a crowd of astonished dogs—as from
> a savage who delights to torture his enemies, offers up bloody sacrifices,
> practices infanticide without remorse, treats his wives like slaves, knows
> no decency, and is haunted by the grossest superstitions.[22]

Darwin's picture of dark-skinned "savages," his narrative on progress and regress, is a familiar one. Haunted, brooding, and anxious, the depiction of his encounter with natives of Tierra del Fuego conveys astonishment at the sight of "wild" men, then a sense of the "shame" their nakedness might provoke, with intimations of violence, sexual depravity, and wife abuse. Implicitly, modern Victorians, like other animals, are somehow both more "natural" and more "civilized" than "savages," who still somehow provide a model for (human) "nature"—or at least the unpleasant parts of it.

It is this sort of ideological daydream about gender, genitivity, and the nature of desire that leads Darwin to the following conclusion on how to vouchsafe the social good:

> As Mr. Galton has remarked, if the prudent avoid marriage, whilst the reckless marry, the inferior members tend to supplant the better members of society. Man, like every other animal, has no doubt advanced to his present high condition through a struggle for existence consequent on his rapid multiplication; and if he is to advance still higher, it is to be feared that he must remain subject to a severe struggle. Otherwise he would sink into indolence, and the more gifted men would not be more successful in the battle of life than the less gifted. . . . There should be open competition for all men; and the most able should not be prevented by laws or customs from succeeding best and rearing the largest number of offspring.[23]

Darwin here departs from the eugenicists of his day not because he discounts their notions of racial progress and biological regress, but because he believes a laissez-faire marriage market better corresponds to the already eugenic logic of sexual selection and would thus improve the species more efficiently than state regulation.

The point is not to demonize Darwin—far from it, since it is the better part of Darwin's thinking that tends to refute the worse part—but to trace the tracks of social struggles and historical happenings in the never-perfect formation of scientific ideas and to suggest something of how scientific ideas serve in the ongoing construction and reconstruction of social forms.

Representations of gender, sexuality, and race work together in Darwin's meditations on sexual selection—not quite logically, with the one thing leading to the other in a deliberate or considered manner, but in a certain kind of texted, textured, and associative way. Darwin's claims about the general nature of male and female creatures rely on and require the suppression of nonreproductive, nonheterosexual sex acts from the space of nature. These propositions, in turn, cite the "negative example": racist tropes about brown-skinned and black-skinned people's sexual degeneracy. In the flow of associations from one body to another and from one claim to the other, nature oscillates with (and is supported by) its evil twin, unnature, and civilization, discontented though it may be, is sometimes more "natural" than nature itself.

Over the course of the eighteenth and nineteenth centuries, such narratives on wildness, sex, and natural dispositions served as both fantasy of and justification for British colonialism and class rule.[24] The same motifs were invoked by all sides in social struggles over women's rights, race privileges, and the treatment of homosexuality and lesbianism.[25] Wildness was—and remains—the "wild card" of these associations, subject to divided desires and sudden dialectical transformations. It was dreaded by the high Victorians, but by the end of the nineteenth century, white, middle-class men sought out a bit of wildness in initiation cults such as the Boy Scouts and men's groups such as the Moose Lodge as a "cure" for the malaise of an "overly feminized" civilization.[26] And in the coming century, Darwin's naturalization of Victorian sex roles, set within phantasmic tales of progress and regress, civilization and wildness, would disseminate across the culture of science, institutions of higher learning, the medical professions, mass entertainment, dating advice books, fitness guides, and self-help books to become part of the warp and woof of heteronormative cultures on both sides of the Atlantic.

7   The Selfish Gene

The stories Darwin first told about men, women, savagery, and sex were not cut from whole cloth. Like other scientists, Darwin drew on the familiar motifs of his day, on what Victorian gentlemen of a certain class knew went without saying. And like other scientists of his time, Darwin took up these questions in response to the proddings of historical events: in the aftermath of democratic political revolutions, in the midst of ongoing economic upheavals associated with colonialism and capitalist development, and in the context of a consequent restructuring of institutional forms, especially family life and gender relations. The question of nature acquired a special salience in these events—for it was to nature that the era's philosophers, politicians, and social reformers looked for guidance. If women were to be legitimately excluded from full rights of citizenship in the new republics, it would have to be on the basis of what Condorcet had called (so as to discount) a "natural difference" between women and men.[1] And if Africans and other colonized peoples were to be governed without their consent, it would have to be on the basis of what Rousseau called (again, to minimize) "natural inequality."[2] In telling stories about nature (especially human nature), nineteenth-century naturalists contributed to the institutional design and ideological architecture of much that followed. What "went without saying" for men like Darwin became scientific doxa, then official dogma.

The plotline is taken up today, under other circumstances, to other ends. It provides a set of political parables and convenient just-so stories told and retold to men and women in the throes of a crisis associated with a new epoch of social transformations. Flash forward, then, to another era of rapidly changing conventions and to its counterdiscourses of a timeless human nature.

SOCIOBIOLOGY

Conceived in the social turmoil of the 1960s and born of the conservative revanchism of the 1970s, sociobiology fashioned a set of contemporary origins stories in a neo-Darwinian mode. Like earlier applications of evolutionary ideas to social questions, this emergent intellectual movement envisioned human nature as an acultural and ahistorical order where everything is set in motion by natural selection, by mortal combat, by the law of procreate or perish. Unlike the antecedent Social Darwinism, which rose and fell before Mendel's discoveries were widely known, the "new synthesis" of sociology with biology attempted a "gene's-eye view" of human nature—a nature as wild as any Darwinian fantasy, but whose very wildness was construed in terms of high-tech cybernetic coding and design.

It was without the slightest wisp of irony or self-reflection that sociobiologists appealed to this double image, depicting what was supposedly "oldest" in human nature by recourse to metaphors drawn from what was most decidedly recent. For Lionel Tiger and Robin Fox, the "facts of nature" are "as comprehensible as a computer program." "The genetic code *is* a program, and it *is* a way of transmitting a program from generation to generation," they assert, perhaps a bit too emphatically.[3] This genetic "code," and its transmission, was and remains the pivot around which everything in sociobiology has turned.[4] As Edward O. Wilson put it in a gnomic turn of phrase: "the organism is only DNA's way of making more DNA."[5] According to the logic of this genocentric proposition, carnal feelings, sexual desires, and so much else besides unfold—directly, unproblematically—from the genes' consequent "drive" to reproduce themselves.

Echoing motifs developed at length by Richard Dawkins in *The Selfish Gene*,[6] Dean Hamer and Peter Copeland approvingly convey the gist of the sociobiological argument in their book *The Science of Desire*:

> Genes are selfish. They only think [sic] about themselves. For an individual gene, the human body is just a temporary vessel to be used briefly and discarded on the march through time. The gene has only one mission—to endure—and the only way it can continue to exist is if its host multiplies and passes on the genetic information to the next generation. The cold, calculated process of evolution is mercilessly unkind to genes that don't contribute to reproduction, cleansing these genes from the species, causing them to die out quickly.[7]

Nature, according to the authors, is an engineer as ruthless as he is relentless. (I think this "nature" must be rendered male, the punishing father, rather than female, the nurturing mother, in the usual English vernacular.)

Note that Hamer and Copeland do not make the modest Darwinian claim that genes that *actively interfere* with reproduction tend to be eliminated over time. They claim instead that genes that "don't contribute to reproduction" (indeed, genes that do not actually "multiply") are "quickly cleansed." This assertion, typical of sociobiological maximalism, entertains closer affinities to Paley's natural theology or Spencer's "survival of the fittest" than to Darwin's natural or even sexual selection: Every gene, and every trait the genes "express," are said to be selected by nature to enhance the overall reproductive fitness of the human organism.

Hamer and Copeland distill, in micro, the magical, anxious, and compensatory world first concocted by sociobiology, a world where, with little effort, the thinker can think his or her way into the position of a gene. In this mirror world, genes bear an uncanny resemblance to our newest and most lifelike machines—yet at the same time, genes are individuals, with thoughts and missions, embarked on a long march. Evolution, rather than being this march, or its happenstance results, is likewise personified. Readers are entreated to enter into this miniaturized world, to throw themselves into the spirit of a zero-sum game, and to solve various life riddles from the perspective of this individual, personified gene.[8] The science of desire promises to tell us something about our human feelings, but all it really says is that genes burn with a desperate desire to reproduce themselves.

THE NATURAL LAW OF DESIRE
AND THE RETURN TO (HETERO)NORMALCY

As with all science, serious or specious, the arguments of sociobiology are plotted against the coordinates of historical happenings. For all their racism and misogyny, nineteenth-century evolutionists acted "progressively." That is to say, their arguments about biological competition and sexual selection—even their notions of natural hierarchy—captured liberal ideas at their moment of upsurge. Everywhere, scientists told heroic tales of individual initiative, free competition, social progress, and human betterment (Spencer by intent, Darwin in spite of himself). On all fronts, the Social Darwinists mustered arguments against hereditary privilege, religious orthodoxy, and restraint on commerce—a constellation of intellectual engagements that prompted historian Richard Hofstadter to wonder "whether, in the entire history of thought, there was ever a conservatism so progressive as [Social Darwinism]."[9] Indeed, for Herbert Spencer, the forces of competition and conflict assured that human society would move toward an eventual state of utopian equilibrium: "the establishment of the greatest perfection and the most complete happiness."[10]

The sociobiologists, by comparison, have played a most reactionary role in late-twentieth-century culture. Their models return to key concepts of Social Darwinism—competition, conflict, and hierarchy—but they seize the logic of a system in crisis. Sociobiological stories reflect the bad faith of a society convinced of its utility, but no longer sure of its own ultimate goodness.

In the first draft of the sociobiological argument—a draft whose outlines still show, even in subsequent, altered versions of the thought—animal lust founds two hierarchies at the same time and by the same implication: male/female, and straight/deviant (or rather, reproductive success/reproductive failure) . . . that, and more. Logically unfolding the implications of the narrative succinctly modeled above by Hamer and Copeland, fundamentalist sociobiology also claims that all manner of hateful practices and unjust institutions—gender inequalities, racism, class stratification, war, even genocide—are but the phenotypic expressions of our fixed genotypes, ultimately derivative from the same brutish "laws of nature," the same maximalist logic of "genetic competition."

Men thus rape not because they express in an extreme way dispositions that have been solicited and shaped by a misogynist culture, but because men are, by nature, sexually aggressive—and because rape serves as an effective strategy for broadcasting one's genes.[11] Wars occur not because historically constituted nation-states struggle over religious nuances, ideological meanings, or the control of natural resources, but because violence is the nature of man, and war is but an extension of the pursuit of reproductive advantage.[12] Racism exists not because power saturates perceptions of the human body by historical means, but as an extension of even cruder, baser instincts: because we naturally tend to favor our own kin and to despise those with whom we compete to reproduce.[13] Culture, particularly the ugly, unpleasant side of culture, is thus reduced to nature, conceived as the struggle of all against all. Breed or be outbred, kill or be killed out.

It is no mystery what motivated the scripting of such explanations, the formation of such a worldview. Just as Spencerian ideas were "conceived in and dedicated to an age of steel and steam engines, competition, exploitation, and struggle," the renaissance of Social Darwinian perspectives on violence, domination, and reproduction took shape under the sign of Vietnam, in the aftermath of anticolonial wars of national liberation, and in the wake of the civil rights, feminist, and gay liberation movements.[14] The arguments of 1970s sociobiology were at every turn enmeshed in the social struggles of their day. They represent, essentially, attempts to turn back or at least to

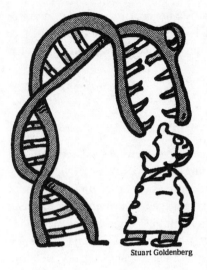

Stuart Goldenberg

Figure 7. DNA as
a devouring monster.
Stuart Goldenberg, *New
York Times* letters section,
July 27, 1993.

stop the wheel of history—or, at any rate, attempts to accomplish this goal magically, in the miniaturized world of the gene.

Like salve for a bad conscience, sociobiology offered intellectual comfort to a confused and sometimes mean-spirited America straining for "normalcy" after the 1960s. The message, repeated over and over again, was basically a modern sermonette on nineteenth-century liberalism, which is to say, twentieth-century conservatism. Life is a cruel struggle, nasty, brutish, and short. We are selfish because of our genes, which are selfish, and our genes are selfish because of "the cold, calculated process of evolution," which everywhere inculcates and rewards selfishness. It is therefore not only futile but most likely counterproductive to try to change social inequities. They serve a purpose, and they are hardwired into our deepest human nature. Even what appears most arbitrary about the circuitry of power flows from an inexorable, immutable logic. That is why there can be no more social progress— no racial equality, no liberation of women and gays, no end to warfare, no eventual state of equilibrium, no happy ending to all this suffering and travail. To yearn for freedom is to flout the law of nature, which lays down limits. Above all else, desire must subserve reproductive utility—for to desire beyond the virtuous heterosexual circle is tantamount to that old eugenics concept, "race suicide."[15] These things are there before us, from the beginning, as natural as men and women, as eternal as families, as transparent as desire, and as inevitable as these very bodies. . . .

And so, with one hierarchy standing in for so many others—hierarchy upon hierarchy, all of them interchangeable—an image of human nature congeals: competition, violence, inequality. The corporeal economy embraces a larger political economy, as the selfish gene gazes longingly into the face of cold, calculating evolution. We are caught, you and I, with no exit, in that great timeless mirror, nature.

A GENE'S-EYE VIEW

A gene's-eye view of life implies a gene's-eye view of need, longing, and desire. A neo-Darwinian view of culture implies a neoliberal view of nature. The one statement says the same thing as the other. But an atomized gene, personified as a selfish individual engaged in a competitive survival mission—such a picture has myriad more implications, few of them ending in civility, much less in democracy, as Hamer and Copeland's unhappy invocation of "cleansing" might suggest. Such a "gene" tells us far more about the state of culture and of the orientation of certain people toward certain events than it could ever reveal about the state of nature.

This is not to argue that evolution is irrelevant to the shape of the human body or the breadth of human capacities. The creative human capacity to make things in and through social labor is undoubtedly as much a part of human biology as it is a part of human culture.[16] Nor is it to say that genetics has no place in the study of human beings. Certain features of human anatomy follow simple, Mendelian rules of inheritance, while certain maladies owe their existence in no small part to out-of-sequence DNA. And *everything* human beings do is "genetic"—in the sense that all manner of practices, even contradictory ones, lie within our genetic potential. But as Stephen Jay Gould has eloquently put it, "potentiality" is not the same thing as "determination."[17] Peace and war, love and hate, greed and generosity, Apollonian reserve and Dionysian abandon all unquestionably lie within human biological potential, but none of them can be said to be "determined" or "caused" by human biology.

What is intellectual folly in sociobiology is the leap from general pronouncements about "biological potentiality" to categorical statements about a "hardwired biological program." What is logically unwarranted is the selective, self-serving, or commonsensical derivation of this particular form or that particular practice—monogamy, nuclear families with breadwinner dads and stay-at-home moms, polygyny, polyandry, male aggression, compulsory heterosexuality, Puritanical morality, playboy exploits, imperialism, capitalist business practices, collectivism, to name but a few contradictory instances—from made-to-fit stories about evolution served up as scientific bromides. What ought to be suspicious, prima facie, is the characterization of nature in terms of a male "wildness" so plainly cultivated in Western culture. And what seems shockingly naive is the attribution of distinctly human cultural characteristics—desires, needs, personalities—to those chains of nucleotides we call "genes."

THE "BIOGRAM" (OR BIOLOGICAL PROGRAM)

Indeed, there is more than a little wrong with the image of the isolated, individuated (never mind personified) gene—the hero of every sociobiological tale and as likely a candidate as any for the logo of the decade.[18] Nor is the idea of genetic "causation" an entirely rigorous one.

As Ruth Hubbard and Elija Wald show in *Exploding the Gene Myth*, the conception of genes embodied in sociobiology, scientific shorthand, and journalistic reporting alike makes for bad molecular biology. "Molecular biologists, as well as the press, use verbs like 'control,' 'program,' or 'determine' when speaking about what genes or DNA do. These are all inappropriate because they assign far too active a role to DNA. The fact is that DNA doesn't 'do' anything; it is a remarkably inert molecule. It just sits there and waits for other molecules to interact with it."

As Hubbard and Wald go on to argue, the "geneticization" and "genetic fetishism" now in vogue—the notion that genes provide a "blueprint" for the organism, the reduction of traits and behaviors to genetic causes, the attribution of motives and volition to DNA—ultimately lose sight of the very organism and organic processes such approaches pretend to describe:

> It is an oversimplification to say that any gene is "the gene for" a
> trait. Each gene simply specifies one of the proteins involved in the
> process. . . . We need to think of DNA, RNA, and proteins as all act-
> ing upon one another, rather than assuming a neat line from DNA to
> protein. . . . When scientists talk about genes "for" this or that mole-
> cule, trait, or disease they are being fanciful. They attribute excessive
> control and power to genes and DNA, rather than seeing them as part
> of the overall functioning of cells and organisms.

In fact, the authors point out, "within the same gene, base sequences can vary a great deal without any change being apparent in the corresponding trait." Although "genes affect our development because they specify the composition of proteins . . . it is more realistic to think of genes as participating in various reactions than as controlling them. . . . In very few cases can a gene legitimately be said to be 'for' any one thing."[19] In other words, genes do not exist in isolation. They do not "act," nor do they "express" a given content in an empty world. They do not carry, in micro, a "blueprint" for the organism, much less for his needs and motivations, or for her thoughts and feelings.

Richard Lewontin critiques the ubiquitous genomania in similar terms: The problem with the "program" metaphor, he explains, is that "it is bad biology. If we had the complete DNA sequence of an organism and unlimited computational power, we could not compute the organism, because the organism does not compute itself from its genes. Any computer that did as poor a job of computation as an organism does from its genetic 'program' would be immediately thrown in the trash and its manufacturer would be sued by the purchaser."[20]

Even if we restrict the question to genetic processes proper, thousands of genes interact with each other in ways that are far too complex to support reductive causal statements about a singular genetic "program." The route from DNA to protein is more involved than the usual shorthand, "DNA makes RNA makes protein," and at the end of this process, proteins fold and thereby affect cellular physiology not simply as a sum of their constituent amino acids, but within specific cellular environments. All of these biochemical processes happen within an overall organic context. At the far end of this system, the organism interacts with its environment— including the other organisms living there. It is not even enough to say that an organism's traits are the outcome of interactions between a genotype and the environment, as though the environment were a fixed variable, because the *timing* of certain interactions—the exposure of the organism to certain conditions at certain times—also affects development, sometimes in dramatic ways. (To give an obvious and extreme example, a child exposed to famine conditions is affected differently than an adult.) To make matters yet more complicated, every organism affects and to some extent creates its own environment. The external "nature" within and against which the organism acts is to some extent a product of its own actions. Finally, accident, chance, and random events—what Richard Lewontin calls "developmental noise"—further affect the organism's development at all levels. Lewontin thus concludes: "even if I knew the genes of a developing organism and the complete sequence of its environments, I [still] could not specify the organism."[21]

In sum: Any living organism is a complex feedback system. There is almost never a singular "cause" for anything. Swarms of dynamic interactions between the organism and its environment over the course of time affect the expression of even the simplest physical trait in complex, unique, and often unpredictable ways—an interactive (one might say "dialectical") relationship evoked by the title of Richard Lewontin's recent book (on whose arguments I have leaned in the last couple of paragraphs), *The Triple Helix: Gene, Organism, and Environment.*[22]

CRACKING THE CODE OF LIFE?

I made these notations in the week following the announcement that the first stage of research in the Human Genome Project had been completed. Newspapers bulged and airwaves crackled with oversimplified, reductive, and misleading claims—the legacy of two and a half decades of sociobiology and especially of its PR-savvy offspring, evolutionary psychology.

"Genetic Code of Human Life Is Cracked by Scientists," reads the banner headline in the *New York Times.* In fact, scientists have sequenced or "mapped" about 90 percent of 3 billion base pairs of nucleic acid—no mean accomplishment—but they have not determined where an estimated 30,000 to 50,000 different genes begin and end in twenty-three lengthy chains of DNA. Much less have they "deciphered" the complex roles played by different, interacting genes in various traits. Talking heads and radio pundits give the impression that a host of life-saving gene therapies and medical treatments are five to ten years away, treatments that will "attack" the singular genetic "source" of maladies like cancer, Alzheimer's disease, or schizophrenia. In fact, very few illnesses appear to be "caused" by a single gene, and the genetic contribution to some "genetic" maladies involves no more than an elevated risk that the person will develop the disease. Scientists do not understand the role of genes in more than a small handful of illnesses, and the medical applications of such knowledge remain largely hypothetical. It is one thing to identify a genetic influence on a disease; it is an altogether different matter to develop and deliver effective medical treatments at the subcellular level or to "correct" genetic "typographic errors."

Against the complexity and equivocation warranted by today's news, what stands out is the unabashed triumph of Digital Darwinism in public discourse: a set of claims all too obviously suited to the era of dot-com capitalism, with its unabashed faith in the cybernetic code and its reveries on grand design. DNA is described by *New York Times* science reporter Nicholas Wade as "the hereditary script, the set of instructions that defines the human organism." In only slightly less deterministic language, Natalie Angier also identifies "us" with our "genes": "the human genome has given rise to the creators of every language uttered, every ballad sung, every Poké-

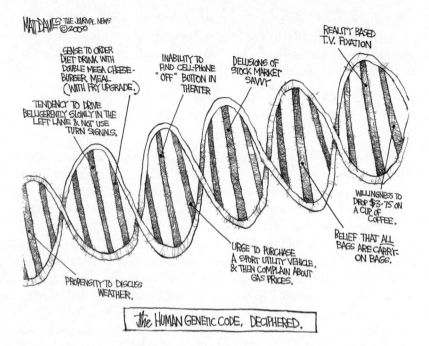

Figure 8. "The Human Genetic Code, Deciphered." Matt Davies, *The Journal News*. Courtesy of Tribune Media Services, Inc. All rights reserved. Reprinted with permission.

mon card traded." Economically combining scientific fetishism with Christian mysticism at a White House ceremony, President Bill Clinton embraced the lowest common denominator of each when he asserted, with frightful hubris, that "we are learning the language in which God created life."[23]

Wade takes up and elaborates on such themes—genomanic, technophilic, mystical, and more than a little Frankensteinian—in a celebratory (and most unscientific) editorial:

> Just as surely as any shrink-wrapped Microsoft disk, the human genome is a program, a set of precisely coded instructions that evolution has written for the design, operation and unfortunately rather limited maintenance of a person. That people should be operated by such a specific script is as amazing as it would be to find that robots have souls. . . .
>
> Over the next few years we should get to read those behavioral instructions. Whatever they are—instincts to slaughter or show mercy, the contexts for love and hatred, the taste for obedience or rebellion—they are the determinants of human nature.

"We have caught the first glimpses of our instruction book, previously known only to God," Dr. Francis Collins, head of the Human Genome Project, said in announcing its decipherment at a White House ceremony last Monday.[24]

Book of Nature, Book of Life—the Human Genome Project is touted as a kind of "owner's manual" for the body, a text that, properly understood, will give direct knowledge of the Creator's mind and intent. A more apt analogy has been suggested by medical anthropologist Margaret Lock. Sequencing the human genome is actually like acquiring all the disparate parts of a Boeing 747—but without any accompanying picture of the plane as a whole, and without any knowledge of aerodynamics.[25]

8 Genomania
and Heterosexual Fetishism

I repeat, in abbreviated form, points that have been made by many others: Sociobiology reduces meaningful social activity to a genetic base. It simultaneously attributes human social characteristics to unconscious (but eerily mindful) chains of nucleotides. The explanations modeled by sociobiology thus shuttle between reductionism on the one hand and reification ("thingification") on the other—a perfect snapshot of the kind of thinking Marx called "fetishistic."[1]

FETISHISM

Marx called a representation "fetishistic" if it attributed human powers, characteristics, and relationships to things. His use of the term drew on an analogy with what nineteenth-century European intellectuals imagined they knew about primitive religion. In "the mist-enveloped regions of the world," as Marx put it, "the productions of the human brain appear as independent beings endowed with life, and entering into relation both with one another and the human race. So it is in the world of commodities with the products of men's hands." Marx employed this (admittedly flawed) analogy (based on a European fantasy of what "primitives" supposedly believed) to underscore the magical powers of commodities and money in capitalist society following the objectification of social labor in labor's products. Commodity fetishism "is a definite social relation between men ... that assumes, in their eyes, the fantastic form of a relation between things." In the resulting mirror world (and by way of a conceptual "table turning"), the more that agency, volition, and consciousness are attributed to "things," the less those qualities appear to inhere in human beings who made them. The more lifelike the object, the more objectlike the laborer.

Notably, Marx concluded his discussion of commodity fetishism by commenting on the classical political economists' "dull and tedious quarrel over the part played by Nature in the formation of [commodity] exchange-value. Since exchange-value is a definite social manner of expressing the amount of labor bestowed upon an object, Nature has no more to do with it, than it has in fixing the course of exchange." Classical political economy, Marx suggested, practices "fetishism" by concealing the role of social labor in setting the price of commodities—that is, by *naturalizing* the commodity.[2] Marx's critique of the science of classical political economy, then, is to be understood as a work of "defetishization," or better still, "denaturalization."[3]

The attribution of human volition and social relations to relatively inert molecules—genes—is, of course, "fetishism" par excellence. The facts of a social order have been translated over into the mirror world of nature, where they appear to exercise intelligent, deliberate control over the world of social beings. The kinds of agency so often depicted as mandated by eternal "laws of nature" are in fact historically contingent practices of a human social order. A definite social relation between people assumes, in their eyes, the fantastic form of a relationship between nucleotides. The fetishism of genes, like that of commodities, reveals something about actually existing social conditions—albeit in a distorted, inverted, fast-frozen form.

It is no accident that references to God, the great fetish creature of Western origins stories, keep popping up in reflections on the miniaturized world of the social gene and in virtually every claim that genes provide the "blueprint" for human nature. Today's genomania articulates the same view of reality that Theodor Adorno dubbed "one part positivism, one part magic." It ought to be clear enough what role laissez-faire business practices play in sociobiological models, which very much embody the perspective of the genetic entrepreneur, anxious to reproduce his chromosomal capital in a dog-eat-dog world.[4] The title of Michael Rothschild's book—*Bionomics: The Inevitability of Capitalism*—says it all.[5] Likewise, it ought to be clear that these models—certainly most of the early sociobiological models—are "phallocentric" in that they are scripted from and within a male point of view. The "selfish" gene engaged in life-or-death "competitive" exploits is almost certainly personified as *male*.[6]

What has been less fully articulated, and what I have been attempting to stress in this chapter and in the preceding ones, is the role "heteronormativity" plays in holding together the diverse narratives of sociobiology and evolutionary psychology—reified stories about competition, rank, and universal norms of human behavior.[7] "Heteronormativity," as I am using the term, implies more than just the normalization or naturalization of heterosexual desire, along with the concomitant casting of homosexuals as deviants. It is that, and more.

HETERONORMATIVITY

Heteronormativity is perhaps best understood as that total system of associations that gives meaning to the paired oppositions "male" and "female," "masculinity" and "femininity," and to the dynamics of sexual attractions— a picture impressively sketched in Gayle Rubin's pioneering model of the sex/gender system. Rubin sees the connection between compulsory heterosexuality and the social division of gender in axiomatic terms: "Gender is a socially imposed division of the sexes. . . . Far from being an expression of natural differences, exclusive gender identity is the suppression of natural similarities. . . . Gender is not only an identification with one sex; it also entails that sexual desire be directed toward the other sex."[8]

Dennis Altman provides a similar analysis of the connection between gender (or "sex roles") and sexuality (the social suppression of nongenital, nonheterosexual desires) in his groundbreaking book *Homosexual: Oppression and Liberation*. Homosexuality, he argues, represents "the most blatant challenge of all to the mores of a society organized around belief in the nuclear family and sharply differentiated gender differences"—a society in which, "as Kate Millet observed, 'sex role is rank.'" Drawing on the Left-Freudians in the vernacular cadences of the early 1970s, Altman elaborates:

> The repression of polymorphous perversity . . . is bound up with the development of very clear-cut concepts of masculine and feminine that dominate consciousness and help maintain male supremacy. . . . Being male and female is, above all, defined in terms of the other: men learn that their masculinity depends on being able to make it with women, women that fulfillment can only be obtained through being bound to a man.[9]

Judith Butler uses the term "heterosexual matrix" to designate a similar concept: "that grid of cultural intelligibility through which bodies, genders, and desires are naturalized."[10] Monique Wittig indicates much the same practice of hetero fetishism when she writes of Western culture and philosophy that "heterosexuality is always already there within all mental categories. It has sneaked into dialectical thought (or thought of differences) as its main category."[11]

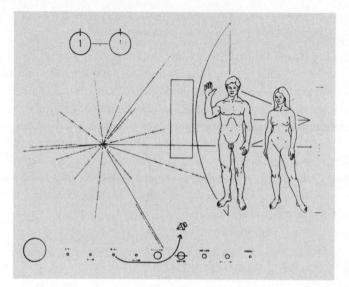

Figure 9.   Plaque for the Pioneer 10 spacecraft, the first human-made object to leave the solar system. Courtesy of NASA.

Michael Warner gives a memorable snapshot of space-age heteronormativity in his discussion of the visual plate attached to NASA's *Pioneer 10*. The plaque depicts a nude male-female pair sent as pictorial emissaries to potential recipients from other solar systems. The picture presupposes that its anticipated viewers will "interpret the outlines, not as individual objects but as generic persons, images for something like 'humanity.'" Warner continues:

> There are other, more damaging, assumptions here as well, for the NASA plates do not carry just any images of persons in their attempt to genericize humankind. They depict—if you share the imaging conventions of postwar U.S. culture—a man and a woman. They are not just sexually different; they are sexual difference itself. They are nude but have no body hair; the woman has no genitals; their heads are neatly coiffed according to the gender norms of middle-class young adults. The man stands square, while the woman leans one hip slightly forward. To a native of the culture that produced it, this bizarre fantasy-image is immediately recognizable not just as two gendered individuals, but as a heterosexual couple (monogamous, one supposes, given the absence of competition), a technological but benign Adam and Eve. It testifies to the depth of the culture's assurance (read: insistence) that humanity and heterosexuality are synonymous. This reminder speeds

to the ends of the universe, announcing to passing stars that earth is not, regardless of what anyone says, a queer planet.[12]

In short, heteronormativity is that coercive gridwork of cultural meanings and institutional practices that defines sexual difference—the manliness of the man, the femininity of the woman. The one thing complements, naturalizes, or justifies the other. Within the regime of heteronormativity, it is the supposed radical alterity of the sexes that implies—demands—their sexual union. Convex to concave, eager-aggressive to coy-choosy, the sexes are "fit," by design, for each other. Heteronormativity (or "heterosexual fetishism") naturalizes—fetishizes—sexual difference as the only legitimate basis for mutual desire to the exclusion of same-sex desire and to the distinct disadvantage of women.[13]

CONSTRUCTING THE HETERONORM

I would not want to go so far as to say that heteronormativity is "always already" there in Western categories, not even in categories of biological thinking. Such a claim would seem to grant the heteronorm more cultural power and historical scope than it actually has. Such an abstract philosophical claim also runs directly counter to sophisticated historical research on the fashioning of modern sexualities and the making of modern sexual identities. Whatever else the Greek inventors of dialectics might have believed about men, women, and nature, they never imagined sexual desire to be the exclusive attraction of a man for a woman.[14]

Nor would I want simply to copy over Marx's admittedly flawed analogy with an exact parallelism of my own: Neither "heterosexuality" nor "fetishism" defines some primitive baseline of human existence. In fact, it is possible to imagine alternatives to the regime of heteronormativity in no small part because human societies past and present have been organized along substantially different lines.

I certainly do not mean to argue that Darwin, who posited the "nature" of desire in the attraction of coy, choosy woman for eager, aggressive man, or Krafft-Ebing, who systematized a psychiatric view of "normal" and "abnormal" sexual desires and beings, or Charles Gilbert Chaddock, an early translator of Krafft-Ebing who brought a German cognate into English as "homo-sexuality," or any other solitary historical figure single-handedly invented heteronormativity.[15] I hope it is clear that the process of social construction I have been evoking is an ongoing, contentious, and continuous one involving the collaboration of a large cast of characters (sometimes unbeknownst to them and sometimes against their will) over a long period of time. What I want to adumbrate is something of the ongoing history of heteronormativity as discerned in gay/lesbian studies from Jeffrey Weeks and Michel Foucault forward.[16] Enlightenment thinkers put certain Christian ideas about "natural" and "unnatural" sex on a "scientific" footing; eighteenth- and nineteenth-century politicians and social reformers invoked the resulting "nature" in propagating institutional reorganizations of gender, sexuality, and family life; Darwin's perceptions, shaped by such a milieu, discerned the rules of a transforming culture yet again in *new* laws of nature; Krafft-Ebing and others drew on Darwin's influence, in turn leaving a wide imprint on all those institutions related to the regulation of sexuality; and so on. . . .

And what I specifically want to mark in this ongoing history of heteronormativity is the role played by theories about the "nature" of desire. From the moment male and female anatomies were conceived as "opposites" that "attract," the histories of modern biology, medicine, and psychiatry were bound up with the history of heteronormativity. The fetishistic naturalization of heterosexuality leaves unmistakable traces across the development of key ideas about evolution in particular: It is the very essence of Darwin's "general law" of sexual selection. The same spectral presence animates selective descriptions of "sex" and narrow, impoverished discussions of what "sex" is "for" in the animal kingdom, and it haunts the resulting ideas about "fitness," "normality," and "degeneracy." These representations and elisions, with their projection of certain human social conventions onto the animal world, bend the theory of sexual selection toward a most un-Darwinian turn: the scientific study of "laws of nature" transmutes, by degrees, into Christian parables on "natural law" and tales of divine design. The result is science at its least scientific—that is to say, its least susceptible to experimental or even just empirical checks—but that is not to say at its least significant or least influential. Everywhere incorporated into the basic architecture of modern social institutions, such ideas about opposition, attraction, and origins have become a "second nature" for twenty-first-century hominids living in the industrial countries of the Northern Hemisphere.

Sociobiology and evolutionary psychology lay hold of and distill this long narrative and institutional tradition, asserting the "second nature" as though it were the first one.[17] In so doing, modern authors rewrite the script of a heterosexual metaphysics: an updated Gospel according to Darwin in the spirit of a high-tech Saint Paul. Copying over this process of transcription and retranscription, I thus rewrite the awkward italicizations of Lionel

Tiger and Robin Fox when they assert, "The genetic code *is* a program, and it *is* a way of transmitting a program from generation to generation."[18] In fact, the genomania currently driving research in the business of science and everywhere trumpeted in fantastic headlines *is* genetic fetishism, but with its reproductivist obsessions and its caricatures of essential gender roles— the one implied or naturalized by the other—it is *also* heteronormativity for the cybernetic age.

A WALK ON THE WILD SIDE:
MORE LESSONS FROM NATURE

Subsequent theory has cartooned the culture-bound images of gender that Darwin first sketched in the theory of sexual selection, as the story line on the "nature" of desire has been taken up by latter-day naturalizing naturalists. Where Darwin naturalized a certain distorted image of prevailing courtship rituals in his day, sociobiologists today geneticize certain gender conventions—not necessarily those of the present moment, perhaps, but at any rate those of the scientists' youth.

Galton had once remarked on the "willy-nilly disposition of the female ... as apparent in the butterfly as in the [hu]man."[19] The thought has been given modern cast by Robert Trivers, who suggests that the different relative investment of males and females in reproduction dictates both the familiar sex roles *and* a natural double standard: Because sperm is cheap (plentiful) while eggs are dear (less plentiful), males and females are said to have very different biological interests in sex and reproduction. Supposedly—since sperm is abundant—it is in the male's genetic self-interest to inseminate as many females as possible. Contrariwise—since her start-up capital investment is greater—it is in the female's supposed interest to prolong the courtship ritual (as a means of insuring the genetic health and physical fitness of her mate) and to use flirtation and sexual rewards to entrap a male, in the process capturing ongoing male paternal support for her offspring.[20] As Edward O. Wilson summarizes it in familiar terms: "It pays for males to be aggressive, hasty, fickle and undiscriminating. In theory, it is more profitable for females to be coy, to hold back until they can identify the male with the best genes."[21]

Terms like "investment," "payment," "profitability"—the familiar and commonsensical language—paper over a real problem with a central sociobiological axiom. As Anne Fausto-Sterling notes in her tour de force review of sociobiological arguments about sexual selection, the differential-investment ratio alleged by sociobiologists is by no means self-evident, and the sociobiologists have provided no proof or demonstration of their basic claim. "How one weighs the energy costs of judicious manufacture of a small number of rather large, well-stocked [egg] cells against the profligate production of millions and millions of tiny ones [sperm] . . . is far from clear."[22] That is to say, what sociobiologists and evolutionary psychologists regard as implicit in the facts of reproduction does not follow from biology, but is entailed in a certain set of cultural presuppositions.

(Such presuppositions are most pronounced in conjecture-based "sciences" like sociobiology and evolutionary psychology, but they also color scientific descriptions of reproduction in standard biology and medicine. For example, biologists typically refer to the spermatozoa as heroically active, agentic, and penetrative and the ovum as passive or inert. In so doing, they both draw on and ratify conventional ideas about gender: the active man, the passive woman. In a pithy essay, Emily Martin shows that such descriptions simply do not square with experimental evidence and notes that one could just as logically wax poetic about the powerful magnetic pull of the ovum, without which the erratic spermatozoa would flounder.)[23]

So goes the familiar shuttle between ideological precept and scientific percept, between how what goes without saying constrains what one sees and how what has to be said figures into claims about nature. And so too goes the tug and tow of a familiar war between the sexes—a contest between bad boys and good girls, between the eager male and the coy female, between manly "salesmanship" and ladylike "sales resistance" (choosiness).[24] The resolution to these familiar conflicts of interest lies in nothing other than—voilà!—the familiar institutional compromise: "The building block of nearly all human societies is the nuclear family," Wilson asserts, repeating a cozy burgher platitude that is not even remotely passable as serious anthropology.[25] It is with the same sense of scientific authority that Richard Dawkins matter-of-factly declares: "The female sex is exploited, and the fundamental evolutionary basis for the exploitation is the fact that eggs are larger than sperms."[26]

Others extend the logic of eager males and reluctant females to its brutally logical implications. Outside the virtuous heterosexual hearth, sex is a risky, violent business, and adventurous women are apt to be punished. In a reprise of Darwin's confused and phantasmic ruminations on the sex life of savages, Randy Thornhill speculates that "in human evolutionary history, larger males were favored because of the increased likelihood of successful rape if they failed to compete successfully for parental resources [i.e., access to females]." David Barash (who, in another passage, sees rape as being so abundant in nature that he describes the pollination of certain flowers as "rape") notes that "perhaps human rapists, in their own criminally misguided way, are doing the best they can to maximize their fitness."[27] Extrapolating from Robert Trivers, Pierre Van Den Berghe and David Barash brood over the topic of sex and violence to suggest that there is something "natural" about severe social sanctions against adulterous women. (How else could genetically self-interested men insure that the progeny they support are really their own?)[28] And with nothing at all in the way of evidence, Robert Wright breezily conjures an imaginary evolutionary trajectory from wild nature to domesticated culture, from rape to romance: "as we evolve from a species whose males forcibly abduct females into a species whose males whisper sweet nothings, the whispering will be governed by the same logic as the abduction—it is a means of manipulating females to male ends, and its form serves this function."[29]

Frans de Waal has sardonically captured the gist of such ideas: "survival of the rapist."[30] A terrible thing it must be, to live a neo-Victorian dream where desire is motivated by desperation and where world-shattering violence is linked by only the slightest displacement to what one loves in one's most sentimental reverie. A bad thing, in particular, for women and queers.

THE MAELSTROM OF NATURE AND THE CRISIS OF HETEROSEXUALITY

Form to function, like penis to vagina (or is it like perpetrator to victim?)— these are snapshots of a heterosexual metaphysics: a divine biology, a postmodern science in Spencer's image. Not always a pretty picture, the image of natural man naturally desiring natural woman is indispensable to naturalizations of the social order in sociobiology and evolutionary psychology. At best, an Adam-and-Eve logic marks every aspect of the theory and is in no small part the pretext for the entire theoretical apparatus: the stable, domestic pair, bonded for life; the creation myth in scientific form; the origins story as role model. . . . At worst, such passages give intellectual pretext to male fantasies about wild men and domesticating women—fantasies now violent and vengeful, now romantic and sentimental. The one side of this binary world supports and reinforces the other: woman and man "fit," *need,* each other (the woman's need being somewhat greater than the man's); pair-bonded monogamous marriage and the nuclear family are necessary *because* men are so wild, so competitive, so violent. . . . In any case, the idea of a hard-wired (hetero)sexual program gives sociobiology both its saliency and its urgency.

Or rather, there is a kind of urgency that impels the entire theoretical trajectory, from the heroic image of the plucky little gene, to nostalgic tales about lifelong pair-bonding, to scarifying stories about the biology of rape. It is the premise, seldom stated outright, that reproduction is a task so difficult, an outcome so improbable, that it poses a dire and ongoing state of emergency for the human species. It is in view of this ongoing crisis that men pursue sex with such zeal while women frantically calculate value: Everyone must strive, against the odds, to replicate his or her chromosomes. Everything must subserve reproductive fitness. That is, everything in nature must give impetus to the heterosexual imperative.

Ironically, this ultra-adaptationist, ultra-reproductivist view of the "nature" of desire takes shape at a moment when even heterosexuals no longer act as though reproduction were the only or even the primary aim of sex. But that, it would seem, is the whole point. All of the institutions connected with sex are in upheaval, ongoing reverberations of a world reshaped by sexual revolution, feminism, and gay liberation, not to mention transformations in livelihood and lifestyle wrought by a capitalist consumer economy. Sociobiologists and evolutionary psychologists translate the terms of this ongoing social and historical crisis into a *biological* state of emergency. The models they give represent "fetish-work" in both the Marxian and the Freudian senses of the term. That is to say, they project the real crisis of the heterosexual nuclear family onto the miniaturized world of the gene—the better to imagine that they understand and have mastered it—the better to enact fake resolutions to real problems.[31]

The resulting "science" does not withstand much scrutiny. It is scarcely necessary to imagine that nature is preprogrammed for heterosexuality because it is scarcely necessary to understand successful reproduction as a statistically improbable event. David F. Greenberg has applied a simple logical test to the ultra-reproductivist premise that underlies certain forms of sociological functionalism, but that is amply manifest in other forms of crude, heteronormalizing reductionism, including sociobiology and evolutionary psychology. He notes that even if every copulation were *random* with regard to sex, the human population would still register absolute increases. As Greenberg goes on to point out, this is exactly how some species of worms behave.[32]

OTHER BIRDS, OTHER BEES

Now, as in the nineteenth century, social events are variously reflected and refracted in scientific theory. Cultural phantasms and social beasts stalk about in the guise of "nature." Now, as then, this picture of nature relies on the suppression of nonreproductive, nongenital, and nonheterosexual acts from the space of nature. It also puts under erasure a not insubstantial part of the animal kingdom: all those creatures who are neither male nor female, or who are simultaneously both, or who are one or the other at different times.[33] The resulting nature is a businesslike place without diversity, plenitude, pleasure—or affection.

Of course, there are other ways of recording animal behaviors, of seeing nature, of thinking about the biology of sex. There always have been. And even within the most domesticated portraits of nature, wildness remains something of a "wild card," subject to nostalgic longing and parasympathetic dread, but also open to sudden, dialectical transformations.

Consider first a bit of earthy folklore, this one told to me by my lover (who recounts it as a tale of his native Puerto Rico). Don Goyito, the old rooster, instructs a young cock on how to be a real *caballero* (gentleman) in barnyard courtship rituals. "You don't just step up to the hen and move on," says Don Goyito in the slow, deep voice of an old rooster. "First, you say, 'con per-miiiiso, se-ñor-a' [With your permission, ma'am]. Then you step up [*pisar*, "step up," but also "step on" or "tread," a common euphemism for intercourse] and when you're finished, you say, 'Graaacias, se-ñooor-a. . . .'" The young cock thanks Don Goyito, then proceeds to implement his lesson—somewhat more rapidly than the old rooster's slow method. "Con permiso, señora Fela—¡Gracias, señora Fela! ¡Con permiso, señora María!—¡Gracias, señora María! . . . Con permiso, Don Goyito—¡Gracias, Don Goyito!"

Or ponder the following colorful vignette from the pages of *National Geographic:* a bit of postmodern folklore on gender bending and the "wildness" of sex in nature. "Many fish and other aquatic creatures can change their sex, but peppermint shrimps are the life of the party," quips the author, John Eliot. These shrimps "begin life as males. Then most change to a female phase—with a twist. 'These "females" retain their male ducts, produce sperm, and fertilize other female-phase shrimps even when incubating their own embryos,' says University of Louisiana biologist Raymond T. Bauer, who calls them simultaneous hermaphrodites."[34]

And by way of contrast to the drab and dreary regime of heteronormativity, Bruce Bagemihl opens his book *Exuberant Biology* with the following idyllic reflection on the nature of desire:

> In the dimly-lit undergrowth of a Central American rain forest, jewel-like male hummingbirds flit through the vegetation, pausing briefly to mate now with a male, now with a female. A whale glides through the dark and icy waters of the Arctic, then surges toward the surface in a playful frenzy of churning waters and splashing, her fins and tail caressing another female. Drifting off to sleep, two male monkeys lie in each other's arms, cradled by one of the ancient jungles of Asia. A herd of deer picks its way cautiously through a semidesert scrub of Texas, each animal simultaneously male but not-quite-male, with half-developed, velvety antlers and diminutive, fine-boned proportions. In a protected New Zealand inlet, a pair of female gulls—mated for life—tend their chicks together. Tiny midges swarm above a bleak tundra of northern Europe, a whirlwind of mating activity as males couple with each other in midair. Circling and prancing around her partner, a female antelope courts another female in an ageless, elegant ritual staged on the African savanna.

Bagemihl continues, giving the gist of a 750-page, abundantly documented text:

> On every continent, animals of the same sex seek each other out and have probably been doing so for millions of years. They court each other, using intricate and beautiful mating dances that are the results of eons of evolution. Males caress and kiss each other, showing tenderness and affection toward one another rather than just hostility and aggression. Females form long-lasting pair-bonds—or maybe just meet briefly for sex, rolling in passionate embraces or mounting one another. Animals of the same sex build nests and homes together, and many homosexual pairs raise young without members of the opposite sex. Other animals regularly have partners of both sexes, and some even live in communal groups where sexual activity is common among all members, male and female. Many creatures are "transgendered," crossing or combining characteristics of both males and females in their appearance or behavior. Amid this incredible variety of different patterns, one thing is certain: the animal kingdom is most definitely not just heterosexual.[35]

Bagemihl sketches an anthropomorphic picture, no doubt—one that takes an all too easy view of what might count as "sex" and how to compare human acts with animal behaviors. But his descriptions are useful in combating the oppressive heteronormative bias that understands sex as genetic selfishness, renders male and female as "opposites" that "attract," and equates heterosexuality with nature and homosexuality with death. Tracing the flow of different social happenings through images of nature, Bagemihl sees diversity as the condition of origins and preaches exuberance as a natural law truly fit for the conditions of late modern society. Acts of fetishism? Surely. Charms and talismans of a coming science that would at least be progressive once again.

Venus and Mars at the Fin de Siècle

Evolutionary Psychology and the Modern Art of Spin

For what is Nature? Nature is no great mother who has borne
us. She is our creation. It is in our brain that she quickens to life.
Things are because we see them, and what we see, and how we
see it, depends on the Arts that have influenced us.

OSCAR WILDE, "The Decay of Lying"

9   Biological Beauty
and the Straight Arrow of Desire

Sociobiology elaborates a set of tales about men, women, and the "nature" of desire. In these tales and through their organizing heteronormative conceit, natural selection slips from being what it is in the best Darwinian sense—contingency, the end result of a random sorting, a series of accidental adaptations—to become what it is in the worst Social Darwinian tradition: an active principle, a driving force, a divine design, a metaphysical telos, the very embodiment of culture in nature. These stories have proved especially appealing in certain quarters of science in recent years. But sociobiology—and perhaps especially its offspring, evolutionary psychology—is as much a phenomenon of popular culture as it is any kind of quasi-scientific enterprise. Bioreductivist claims and tactics circulate there, especially in the mass media.

In his 1950s classic, *Mythologies*, Roland Barthes showed that nothing exercises more appeal in the middlebrow media than the premise that common sense—socially shared prejudice—is deeply rooted in an irresistible, unchanging nature.[1] Modern media, and the stories about desire that they underwrite, would seem to continually confirm the darkest possible version of Barthes's observation. Recent science stories have broadly disseminated scenarios eerily reminiscent of nineteenth-century ideas about human nature, whose discredited eugenics filters back into present-day discourse like the return of the Spencerian repressed. Not to put too fine a point on the matter, but in the reductivist vulgate defined by sociobiology and evolutionary psychology, sex *is* eugenics, directly and without mediation. Nowhere is this clearer than in much-promulgated geneticist ideas about the biology of beauty, the subject of this chapter.

Such ideologies are haunted by the past, but they would not survive unless they possessed a kind of adaptability to changing cultural conditions, something that allows their plotlines to be told and retold, despite—or better yet, in the face of—changing circumstances. This adaptability is apparent in recent appropriations of bioreductive ideas about masculinity, femininity, and the "nature" of marriage. The heteronormative claims of sociobiology and evolutionary psychology have been taken up in different ways, to different ends, by different interests, in response to ongoing changes in the social and economic standing of men and women, as we'll see at the end of this chapter and in the three chapters to follow.

CARTESIAN MONADS AND GENETIC WONDERBRAS

June 3, 1996: With gushing fanfare and illustrated by prurient photographs of a pair of young, white, scantily clad models posing in their underwear, a front-page *Newsweek* article, "The Biology of Beauty: What Science Has Discovered about Sex Appeal," summarizes some of the latest evolutionary just-so stories.[2] Emphatically, the author intones: "Studies have established that people everywhere—regardless of race, class or age—share a sense of what's attractive." Devendra Singh, a University of Texas psychologist, matter-of-factly explicates: "Judging beauty involves looking at another person ... and figuring out whether you want your children to carry that person's genes."[3]

After a brief reprise of "courtship rituals" in the animal kingdom, the *Newsweek* article puts forward two basic claims. First, various researchers purport to have discovered a cross-cultural and universal norm of human beauty, a preference for left-right symmetry in facial and physical characteristics. Since asymmetrical development *might* reveal an underlying illness or developmental trauma, and since symmetry *might* indicate health, then our conception of beauty—*theoretically* the basis for our sexual attraction to other human beings—is *said* to be governed by the sociobiological principle of "inclusive fitness," according to which we seek mates who are likely to serve as good reproducers, good mixers with our own genes.[4] Second, it is claimed that men find women more attractive when they fall within the range of an "ideal" waist-to-hip ratio of .7 (i.e., an "hourglass" shape, with the waist seven-tenths as large as hips). This ratio is speculated to be the ideal proportion for conceiving and bearing children, as a "reproductive fitness" model might predict, but to date the evidence is lacking that this ratio is in any real sense "optimal."[5]

Micaela di Leonardo, one of the dissenting scholars interviewed by a

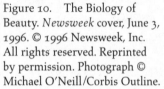

Figure 10.   The Biology of Beauty. *Newsweek* cover, June 3, 1996. © 1996 Newsweek, Inc. All rights reserved. Reprinted by permission. Photograph © Michael O'Neill/Corbis Outline.

Newsweek reporter for the preparation of this article, provides a behind-the-scenes look into the world of science journalism, particularly how mainstream reporters weigh evidence and expertise in the production of a news story such as this one. She recounts her difficulties getting across even basic points from introductory cultural anthropology and from gender/sexuality studies:

> I tried to explain to her [the interviewer] that human sexual attraction and mating patterns are extraordinarily various, and connected to human social and political institutions layered-over—and not reducible to—biology. I noted that this nouveau sociobiology (unlike E. O. Wilson's original, and embarrassingly silly, statements) makes no allowance for ubiquitous human homosexuality. I pointed out that it reintroduces the sexist (and anthropologically absurd) notion of a "bottom line" human nature in which men try to maximize their DNA reproduction through impregnating as many young, nubile women as possible, and women attempt to "capture" male parental support by enhancing their personal attractiveness. I argued that in most of human history we see instead very specific—and widely varying—fertility goals, that we have abundant evidence of widespread desires for few, not many, children in many sorts of societies, and that individuals do not make mating decisions as Cartesian monads but as social beings embedded in webs of kin, friends, and neighbors who

have enormous effects on sexual and marital choices. I pointed out that attempting to find some "essential" human attractiveness beneath skin and eye color, hair type, nose shape, and body type denied both culturally varying aesthetic systems and the long historical effects of Western imperialism. . . . I introduced her to the phrase "junk science."

The reporter was sympathetic, identified herself as a feminist, but kept returning to two points: her editor wanted a positive story on this "new science," and didn't contemporary American women's desperate attempts to improve their physical attractiveness through clothing, hair dye, makeup and surgery despite so many years of feminist activism "prove" that there was a point here? I tried to lay out the contemporary American political economy of gender, but she wasn't really listening. In my frustration, I finally exclaimed, "Look, Wonderbras are *not* genetic!"[6]

THE RECTITUDE OF THE STRAIGHT

The *Newsweek* article tells a story about certain all-too-human foibles: "Boys will be boys." "That's just the way men and women are." "It's in our genes." "It's natural." "It's normal." In this narrative, cultural and historical variations—in ideal weight, in proper vestments, in reproductive goals, in sexual practices, even in sometimes dramatic alterations of the body through scarring, piercing, stretching, reshaping, and tattooing—all become the epiphenomenal expressions of a deeper, more abiding sameness. Alternative explanations for shared aesthetic judgments—linked histories—are never considered. In this short history of the battle of the sexes, the Flintstones meet the Jetsons and find each other equally at home in middle-class tastes and suburban coziness.

It is not just that historical and cultural specificities are denied or that the contours of global history are extinguished in media reportage of sociobiologically inspired research. Contemporary variety, too, is suppressed. All those durable sentiments that pluralize or problematize the perception of beauty—proverbs like "Beauty is in the eye of the beholder," "Beauty is as beauty does," or even "Never make a beautiful woman your wife"— receive scant treatment here. Needless to say, the fetishes—attractions to this or that particular feature, whether symmetrical or not, to the exclusion of others—are never mentioned, although it is by no means certain that fetishism, broadly conceived, is a statistically rare form of sexual desire.[7] To the extent that they come up at all, social and individual variations serve only to foreground all the more clearly the ideal norm of symmetry and proportion. A universal ideal is supposed to lie at the heart of everyone's conception of beauty because it radiates in a straight line from ubiquitous, unchanging nature, expressing itself through essentially passive social media.

Such a norming of the norm is hardly a neutral description of objective reality. Rather, this linear narrative of "straight desire" might well serve as an object lesson in how one never simply "discovers" a norm. One actively crafts it through strategies of inclusion, exclusion, cooptation. Writing on the role of scientific norms in social "Ideas of Nature," Raymond Williams put matters this way: "A singular name for a multiplicity of things and living processes may be held, with an effort, to be neutral, but I am sure it is very often the case that it offers, from the beginning, a dominant kind of interpretation: idealist, metaphysical, or religious."[8]

Williams's observation would seem to be borne out in the material at hand. The so-called science of beauty is little more than the expression of a heterocentric metaphysics: natural man naturally desiring natural woman, and vice versa. Heterosexuality is "ex-nominated" (to use one of Roland Barthes's favorite terms) in the sense that its assumptions are buried in the terms of description and the logic of the analysis. Nowhere named, it is everywhere implied. So dissimulated in sex acts and reproduction, so suffused in the self-evidence of the senses, the rectitude of the straight body is thus made to appear natural, universal, and incontrovertible: a given, an unquestionable, a biological doxa. A fact of nature. With vicious circularity, the resulting research constantly serves as a demonstration of the very heteronormativity it also secretly posits.

SHAPING THE WESTERN BODY IDEAL

For all the Madison Avenue PR, the "new" findings of the "new" science trumpeted in a host of popular and scientific journals hardly seem "new" at all. Without recourse either to genetic biology or to polling techniques and guided only by a penchant for mathematical formalization, the fifth-century B.C.E. Greek sculptor Polykleitos not only stated that beauty is symmetry and proportion, he actually developed formulae for measuring beauty according to a general ideal, as exemplified in his prototypical statues, the (male) Doryphoros (Spear Bearer) and the (female) Wounded Amazon. According to Polykleitos's formulae, the well-apportioned arm, for example,

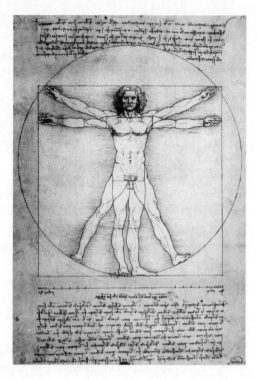

Figure 11. Leonardo
da Vinci, The Vitruvius-
Polykleitan canon, c. 1485–
1490. Venice, Accademia.
© Bettman/CORBIS.

is supposed to be the length of exactly three hands, a proper face should be divisible into three horizontal thirds, a well-shaped body should be the height of so many heads, and so on.

Polykleitos's text on human beauty, the *Canon*, like his original bronze statues—also known as the *Canon*—have been lost, but his aesthetic judgments have carried forward in countless reproductions and imitations, and his ideas have been passed along (with some variation) in quotations by Vitruvius and Galen, and in studies by da Vinci and Michelangelo. Polykleitos established the "body canon," a set of general rules that governed the production of art in the Greco-Roman world, a set of value judgments that shaped Western ideas about human beauty during classical antiquity and the Renaissance.[9]

But Polykleitos did *not* discover universal aesthetic norms good for all cultures at all times. As art historian and evolutionary psychology enthusiast George Hersey admits, preclassical art and sculpture display more varied forms of human bodies. The bodies depicted in the Middle Ages don't measure up to classical concepts, either—and twentieth-century modern art has tended to abandon Polykleitan strictures.[10]

In view of the periodic waxing and waning of classical ideas about beauty, it is good to remember the plural meanings of "canon." The term refers to a reed that might be used as a ruler to mark off an object. It also refers to a "cane" or braces that might be used to force a leather shield into shape. "Canon" thus conveys the idea of "measurement," but it also "stiffens that which would otherwise be without structure."[11] Perhaps more than anything, the term means "something straight." Not simply descriptive, it is prescriptive: the term invokes both "rule" and "regulation." The history of Western art shows that the perpetuation of canonical ideal forms relies on a regulatory institutional apparatus, on standardized instruction and disciplined production.

On this point, it is good to remember how—and why—that other source of enduring ideas from the fifth century B.C.E., Plato, conceived "form": as an object blanched of color and relieved of its material content, as an idea deprived of body, as a deeper, more enduring truth beyond the tainted evidence of sense and appearances. In Platonic philosophy as in modern Western culture, in airy idealist as well as vulgar materialist guise, "form" and "norm" serve as object lessons, as *instructive* ideals. As such, and because they predicate the general over the particular, the abstract over the concrete, they are not only invested with ideology from without, they are also constituted from within, as snapshots of, an ideological imperative.

THE DAPPLED DETAILS

Like one of those "fake universals" ridiculed by Alfred Kroeber—"shelter," "religion," "trade"—the idea of universal and timeless standards of beauty is something of an empty category. It cannot help clarify our understanding of any particular beautification practice in any real depth—indeed, it tends to distort them all.[12] Upper-class Victorian women ingested small amounts of arsenic to give their skin that much-admired pallid appearance; antecedent Mesoamericans strapped boards to their infants' foreheads to reshape their malleable craniums into something more conical in form; various body parts—lips, earlobes, necks, genitalia—are stretched to enormous proportions in scattered cultures, while piercing and scarring techniques alter the body's look and feel in many societies, our own included. Around the world, people offer similar rationales for these perplexingly varied practices. Most frequently, they are said to enhance physical and sexual attractiveness. They are done in the name of "beauty." But on the cultural and historical ground, what is *meant* by "beauty" and which "enhancements" are

Figure 12.   Wasp-waisted Melanesian warrior, Bismarck Archipelago.
Late-nineteenth-century photo by
Dr. C. G. Seligmann, from *Peoples
of the World* (1922). Courtesy
of Julian Robinson Archives.

affected vary widely. In practice, not only do people alter and adorn their
bodies according to aesthetic principles irreducible to some monotonous universal norm (and clearly incompatible with utilitarian models of "health"),
but many of these alterations have little apparent connection to "reproductive fitness," and they often outright interfere with reproduction—examples
of male genital surgery (subincision, or splitting the underside of the penis)
in Aboriginal Australia and female genital surgery (Pharaonic circumcision
or infibulation—removal of the clitoris and labia minora, with the labia
majora sewn closed, leaving a small opening at the vulva for urination and
release of menstrual blood) in East Africa being among the more dramatic
cases in point. Whatever ideals or goals these varied conceptions of beauty
express, they do not embody the singular and monotonous yearning of genes
to reproduce themselves.

When anthropologists have attempted to draft their descriptions of
other cultures' aesthetics in terms of supposedly universal conceptions of
beauty, they have most frequently staged unwittingly comic or parodical
performances of their own descriptive expertise. Bronislaw Malinowski,

Figure 13. Cover of a 1933 brochure by Howard Y. Bary. National Anthropological Archives, Smithsonian Institution (DPA 07094801).

for example, claims that the "European observer soon finds that his standard of personal charm does not essentially differ from that of the natives."[13] He then goes on to describe specific Trobriand standards of beauty that in fact seem far removed from any European conceptions of personal charm:

> The outline of the face is very important; it should be full and well rounded . . . (like the full moon). . . . The forehead must be small and smooth. . . . Facial painting is done in black, red, and white. . . . Biting off the eyelashes, the custom of *mitakuku* as it is called, plays an important part in love-making. . . . Eyes should be shining, but they should be small. On this point the natives are quite decided. Large eyes, *puynapuyna*, are ugly. There is no special treatment for the eyes, except, of course, shaving the eyebrows, which, together with the biting off of eyelashes, leaves them singularly naked to European taste. . . . The nose should be full and fleshy, but not too large. . . . A nose-stick used to be considered aesthetically indispensable, but it is now gradually going out of fashion. . . . Every ear must be pierced at the lobe and ornamented with ear-rings. The hole is made early in

Figure 14.   Sara woman with double labret, from Hugo Adolf Bernatzik's *Der Dunkle Erteil: Afrika* (1930). Julian Robinson Archives.

childhood by placing on the ear a turtle shell ring which has been cut and the ends sharpened, so that the points gradually work their way through the gristle. The resultant small hole is then gradually enlarged until a considerable opening surrounded by a pendulous ring is formed in the lobe. . . . Such a treatment of the ear is de rigueur; otherwise a man or woman would be said to have *tegibwalodila* (ears like a bush pig). Teeth, in order to be really attractive, have to be blackened (*kudubwa'u:* literally, black teeth . . .). . . . Body hair . . . is regarded as ugly and is kept shaven. . . . I am told that girls at the time of their first menstruation are tattooed round the vagina. This tattooing is called *ki'uki'u*, and is done, according to my informants, for aesthetic purposes.[14]

Malinowski claims that, after a time, such "artificial transformations" of face and body—"the shiny black teeth" revealed by "vermillion lips," "graceful scrolls painted in three colors over the face," and skin glistening with coconut oil—ceased to impress him as "mere grotesque masquerade."[15] Perhaps sensing the improbability of his own claimed cosmopolitanism, Malinowski gives the following impression of the way Europeans like himself were perceived by his Trobriand informants: "Europeans, the natives frankly say, are not good-looking. The straight hair 'coming round the heads of women like threads of *im*' (coarse pandanus fibre used for making strings); the nose, 'sharp as an axe blade'; the thin lips; the big eyes, 'like water puddles'; the white skin with spots on it like those of an albino—all these the natives say (and no doubt feel) are ugly."[16]

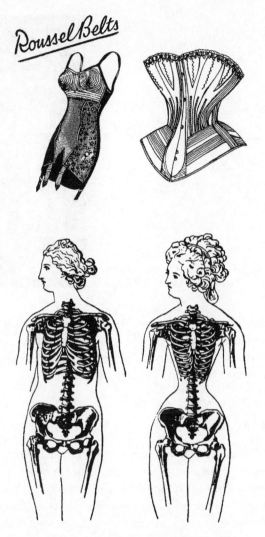

Figure 15. An 1860s spoon-fronted corset illustration, a 1935 *Roussel* corset ad, and an 1880s illustration of the effects of tight lacing. Courtesy of Julian Robinson Archives.

As Oscar Wilde has Vivian say to Cyril, "To look at a thing is very different from seeing a thing." Vivian adds: "One does not see anything until one sees its beauty."[17] The aesthetic banter of Wilde's character suggests how difficult it is to see well. Malinowski's own scrupulously detailed descriptions weigh against his thin claim that Europeans and Trobrianders

Figure 16.   Braden Quinn training.
Photo © by Michael Neveux.

share a common standard of beauty. At any rate, any conception that begins from "universal premises" would seem poorly equipped to understand
the specificity of actual aesthetics or the diverse transformations undertaken
on malleable human bodies in the name of "beauty."

THE ILLUSORY NORM

For all that, it would scarcely be surprising if it turned out that, on the average and across *most* cultures, what people regard as beautiful *tends* to average out along the lines of symmetry and proportion. But this lowest common denominator would hardly count as a very impressive finding. Although
the human form is extremely malleable—it can be stretched and reduced,
pushed and pulled, shaped and reshaped in many ways—it is not infinitely
malleable. We are creatures with a front and back, a top and bottom, a right
side and a left side. Eyes are for seeing and legs are for walking and it is best
to keep our heads up and our feet down. If our dimensions have any implications at all, it should scarcely surprise us that certain configurations of the
human form impose themselves, willy-nilly, on perceptions of beauty
among individuals and across cultures and histories. But this "imposition"
is on the order of a *consequence,* not a "cause" or "drive," and it is an artifact of the process of abstraction, averaging, and norming, not the expression of an essential form or ideal norm.

The biology of beauty simply reifies its own unchecked assumptions and procedural naiveté as "facts" of nature. Its claims are tantamount to saying that from Nunavut to Irian Jaya and from Oaxaca to Bangkok, masks the world over bear a resemblance to each other. Without a doubt, they do— inasmuch as masks must accommodate a wearer's eyes, nose, and mouth in roughly the same order and proportion and inasmuch as masks are realistic or fantastic extensions of the human face. But to generalize about the aesthetics of masks (or faces) is to choose to look for similarities rather than differences. It is also to choose abstraction over concreteness. Inasmuch as masks draw on different principles of imagination, embody different cosmologies, mark different social practices, serve different cultural functions, take up different orientations toward life, and signify different things, they resemble each other not in the slightest. Funerary masks are often solemn— they are meant to depict dead relatives. But other masks are comic, and some belong to the realm of child's play. Some masks are associated with specialized theatrical performance. Others allow wearers to comment mockingly on social relations. Certain types of masks are intended to resemble people or beasts with considerable accuracy. Other masks test the corporeal limits and physical boundaries between people, beasts, and things real or imagined.[18] As Franz Boas suggested a century ago, one learns nothing about any real masks anywhere in particular by posing the question of masks as a global one. The truth of the mask is concrete: It expresses a meaning within the logics, materials, and purposes available at some particular place and time. To consider masks in the abstract, apart from their relationship to some particular culture and moment, is to sever them from the context that allows them to mean anything at all.[19]

So, too, with concepts of human beauty, which also frequently involve inscriptions on the face, extensions of the body, imaginative elongations of the hair, colorful paint across the topography of the flesh, and masklike adornments. If the divers and sundry practices of beautification and adornment, in all their spectacle, gore, and glory, were laid out in one global constellation— foot binding alongside plastic surgery; beads, bangles, and feathers alongside bodybuilding—all the juttings and indentations would likely cancel each other out, and the resulting composite would likely come to resemble the "generic human forms" of "classical beauty," all symmetry and proportion. The resulting picture, however, still would not amount to an adequate sketch of universal, human "beauty," because all the idiosyncrasies, all the particulars, all the specificities would be lost, vaporized in a cloud of generality.[20] The same process of averaging that both extinguishes cultural specificities and denies history also abstracts and disembodies everything that is really meaningful about beauty, its perception, and its fashioning, in any actually existing place.

By the same token (and contrary to the generic compendia of supposedly universal traits brandished by sociobiologists in recent texts), one cannot discern a single or unequivocal human nature lurking beneath the dappled details of cultural diversity. All that we can say without distortion is that "human nature" is nowhere *simply given,* but that it is everywhere *endlessly elaborated* (as the chapters in the following section, "Varieties of Human Nature," themselves elaborate). If we were quadrupeds rather than bipeds, if we had thick hair covering most of our bodies, if we had the morphological characteristics of amoebas, undoubtedly our perceptions of beauty would be very different from what they are, with all their present variety and splendor. (We would also likely have very different cultural histories without upright walking, a need to keep warm, and more or less stable shapes.)[21] In this narrow, even frivolous, sense—which plainly admits no "test case"—it might even be said that certain biological givens "constrain" the parameters of beauty (and much else besides)—insofar as it is "in our genes" that we are creatures with a front and back, top and bottom, right side and left. All that has really been said is that all other things being equal (and this is to say everything), human beings, in that they are symmetrical and proportional, find the human form beautiful.

But what is most precisely *not* "in our genes" is the creative, intelligent perception that finds form, coherence, value, and beauty—not to say *attraction*—in human bodies, and in so much else besides. And what is not simply "given" in the basic and unreflective processes of our bodies is the active, improvisational desire that inheres in our every encounter with things, such that it makes a sense of nature and thus gives coherence to the world . . .

VISUALIZING SEX

The unabashed eugenicist tilt of biological research on beauty was distilled with spectacular vulgarity on the ABC news program *Day One,* which aired in April 1995 and included an interview with David Buss, author of *The Evolution of Desire.*[22] A portion of that program is described here by Martha McCaughey:

> As Buss elaborates in the interview, our evolutionary forebrothers
> who did not prefer women with high cheekbones, big eyes, lustrous

hair, and full lips did not reproduce. Buss explains that those men who happened to like someone who was older, sicker, or infertile "are not our ancetors. We are all the descendants of those men who pre- ferred young healthy women and so as offspring, as descendants of those men, we carry with us their desires." On that same show, *Penthouse* magazine publisher Bob Guccioni says that men are simply biologically wired to enjoy looking at sexy women: "This may be very politically incorrect but that's the way it is. . . . It's all part of our ancestral conditioning."[23]

Much has been written and said about the supposedly visual nature of men's sexuality and the supposedly "nonvisual" nature of women's. According to the old Darwinian formula, eager, aggressive men see beauty, whereas coy, choosy women calculate value.

Stephen Pinker reproduces the much-repeated proposition: "The male of the human species is aroused by the sight of a nude woman. . . . In for- aging cultures, young men make charcoal drawings of breasts and vulvas on rock overhangs, carve them on tree trunks, and scratch them in the sand." By extrapolation, pornography—a ten-billion-dollar industry in the United States alone, grossing "almost as much as spectator sports and the movies combined"—is much the same now as it always was, "a succession of anony- mous nude females eager for casual, impersonal sex." But, Pinker contin- ues, from an evolutionary perspective, "it would make no sense for a woman to be easily aroused by the sight of a nude male." Instead, she seeks "the best husband available, the best genes, or other returns on her sexual favors. If she could be aroused by the sight of a naked man, men could in- duce her to have sex by exposing themselves and her bargaining position would be compromised."[24]

I sometimes wonder whether evolutionary psychologists can see at all. Pinker overlooks the obvious in his brisk invocation of primitive art, his quick leap to modern pornography, his outlandish claims about what men and women universally feel. The substantial presence of homoerotic themes in modern and classical pornography might indicate some caution toward the unqualified claim that "the male of the human species is aroused by the sight of a nude woman." And men with erections, or even just erect phal- luses with no men attached, are by no means absent in the primitive art he selectively cites. Are these phallic figures to be understood as evidence of primitive homosexuality—or of the inherently bisexual nature of men? Per- haps one, perhaps the other. But perhaps neither. We do not even know whether it was men or women or both who left the cave drawings, wall paint-

Figure 17.   Upper Paleolithic
ivory carving, c. 25,000 B.C.E.,
discovered near Dolní Vestonice,
Czech Republic. Sometimes described
as an "abstract female form"
or as a "rod with breasts." From
Alexander Marshack, *The Roots of
Civilization* (New York: McGraw-Hill,
1972). © Marshack 1972.

ings, and fetishes Pinker selectively invokes.[25] We cannot even be sure that they were experienced as "sexual" representations at all, a point developed by archaeologist Margaret Conkey in her essay "Original Narratives."[26] LeRoy McDermott, an art professor, has suggested that the perspective on the female physique captured in the famous Stone Age "Venus" figures is consistent with the perspective of a pregnant woman looking down at her own body.[27] Folk herbalist Susan Weed, along a very different line, has suggested that such figures depict postmenstrual women: their large breasts are "pendulous, not protruding."[28] For that matter, it is not even always clear whether a figure is "phallic" or "feminine." A number of primitive fetishes embody precisely this kind of ambiguity. Viewed one way, a famous paleolithic figure looks like a female bust, with large, pendulous breasts. Viewed from a different perspective, it suggests a phallus with ample testicles. What one sees in these representations often depends on what one wants to see. (There is no reason to imagine that this is not precisely how the figures were designed.)[29]

We do know that pornography and people's taste for assorted forms of erotic images are not outside the history of men's and women's changing desires.[30] A famous Kinsey survey conducted in the 1970s suggested that most women were not stimulated by nude photos and drawings—a finding that has worked its way into both the sociobiological vulgate and into cultural feminism as a "natural fact." But much-less-noted restudies, conducted in the 1980s and 1990s, found that substantial numbers of women *are* in fact stimulated by male nudity. What happened in the intervening period was no mutation in women's genetic makeup, but historical changes induced by the tug of the sexual revolution and the tow of feminism, as Susan Bordo shows in *The Male Body*. In a relatively short period of time, Chippendale's strippers, *Playgirl* magazine, and the growing eroticization of the male body in advertising culture have socialized women into new ways of looking and new ways of seeing.[31] In fact, not just erotica, but bodies in every conceivable sense had changed.

Samuel R. Delany's discussion of how men have interacted with images in pornographic movie theaters on Times Square captures something of the historicity—and the complexity—of pornography, of what it might mean to see beauty or desire images:

> For the first year or two the theaters operated, the entire working-class audience would break out laughing at everything save male-superior fucking. (I mean, that's *what* sex is, isn't it?) At the fellatio, at the cunnilingus even more, and at the final kiss ["often on the actress's cum-streaked lips"], among the groans and chuckles you'd always hear a couple of *"Yuccchs"* and *"Uhggggs."* By the seventies' end, though, only a few chuckles sounded out now—at the cunnilingus passages. And in the first year or two of the eighties, even those had stopped. (No, *that's* what sex is: a four-part act, oral and genital, where everybody gets a chance to be on top. Anything *else* was what was weird.) Indeed, I think, under pressure of those films, many guys simply found themselves changing what turned them on. And if one part or another didn't happen to be your thing, you still saw it enough times to realize that maybe *you* were the strange one, and it behooved you to sit it out politely and put up with it, unless you wanted people to think *you* were strange.[32]

It was, of course, the absence of any depiction of male homosexuality "on the screen proper" that authorized it "to go on rampantly among the observing audience, now in this theater, now in that one. "The majority of that was guys like me," Delany writes, "who enjoyed sucking cock for our own pleasure, fellating other guys who were getting off on the straight screen action."[33]

I would frame the role of the "screen" somewhat differently: It is what allows men who enjoy getting blow jobs from other men nonetheless to define themselves as "straight." In any case, the relationship between image and meaning, idea and desire, is a complex one, involved, as it is, in the twists of social currents, the turns of historical happenings, and the coils of individual agency.

TURN ME ON, I'M A RADIO: "DESIGN" AND "MISUSE"

Attempting to inoculate the antecedent sociobiology, its derivative evolutionary psychology, and assorted forms of biological determinism against the criticism that they neglect social, cultural, and historical variations, Rutgers anthropologist Helen Fisher performs the sort of rhetorical dodge that has become de rigueur in sociobiological circles of late: "We are not packets of DNA or social creatures acting in complex networks of friends and relatives. We are both."[34] Science reporter Geoffrey Cowley echoes similar themes in his treatment of homosexual exceptionalism in the main text of the *Newsweek* article "The Biology of Beauty": "Homosexuality is hard to explain as a biological adaptation. So is stamp collecting. But no one claims that human beings are mindless automatons, blindly striving to replicate our genes. We pursue countless passions that have no direct bearing on survival. If we're sometimes attracted to people who can't help us reproduce, that doesn't mean human preferences lack any coherent design. A radio used as a doorstop is still a radio."[35]

A more muddled, misguided, and misleading set of statements would be difficult to imagine. In fact, sociobiology's inaugural premise of a "genetic biogram" is just another way of saying that at core we are *precisely* "fleshy packets of DNA." A common (and often sympathetic) illustration of sociobiological reasoning uses the image of a puppet—an automaton—to show how genes ultimately "pull the strings" that guide complex human behaviors. Despite such disavowals, the whole argument expressed by the *Newsweek* piece and the "science" that it (all too accurately) distills is that the perception of beauty and the operation of sexual desire are under genetic control, guided by unambiguously eugenic principles—a preference for "biological quality" in potential mates. Worse yet is the analogy with the radio, an instrument expressly engineered by human agents for a particular purpose. Such an argument by analogy not only embodies in the body a telos of evolution—a godlike design prior to and beyond our conscious designs, a grand scheme from which evolution follows—it also casts as "misuses" or "perversions" all uses of the body other than those for which it was "intended."

It is in the notion of "design" that one sees most clearly how mysticism crisscrosses scientism in the social history of nature: Like form to content, nature, the spirit-god, animates a particulate universe. It is exactly the same in sociobiology as in Ralph Waldo Emerson's meditations on metaphysical nature: "There seems to be a necessity in spirit to manifest itself in material forms. . . . [D]ay and night, river and storm, beast and bird, acid and alkali, preexist in necessary Ideas in the mind of God, and are what they are by virtue of preceding affections in the world of spirit. A Fact is but the end or last issue of spirit." In evolutionary psychology, as well, "every natural process is a version of a moral sentence. The moral law lies at the center of nature and radiates to the circumference."[36] Of course, Emerson says it with a good deal more style and grace than our latter-day nature mystics.

DAVID AGAINST D'AMATO

It is often when clinching the argument that sociobiology exposes its most dramatic contradictions. From the same *Newsweek* article, then, another such clincher: "Local fashions seem to rest on a bedrock of shared preferences. You don't have to be Italian to find Michelangelo's David better-looking than, say, Alfonse D'Amato."[37]

But of course, you also don't have to be a woman—or even a gay man—to reach the same judgment. Once it is admitted that men, even *straight* men, can see David's attraction, the argument that the perception of beauty is an assessment of genetic reproductive strategies becomes preposterous prima facie. Somewhere between D'Amato and David—image of a Semitic youth given flesh and substance by Michelangelo's desires—all the links in the argument break: sexual desire is not the same thing as the perception of beauty, and neither is reducible to reproductive yearning. No doubt beauty is "coherent," but just how this figures in *our* designs, rather than in the way nature is supposed to "design" our "human preferences," is the starting point for any question worth asking. No doubt beauty exerts many a pull over us. But the perception of beauty, its place in the human body, its relationship to desire and other forms of basic intentionality, not to say the connections among all these elements in that gravitational field, the flesh, are more varied and more marvelous than the advocates of a single, simple, natural, and genetically programmed conception of beauty have claimed—as they themselves sometimes have found, to their chagrin.

In August 1998, a study was published in *Nature* whose findings were summarized to the lay public as follows:

> In a new study of facial attractiveness, researchers from Scotland and Japan have found that, much to their astonishment, people of both sexes prefer feminine-looking men over rugged, manly-miened men.
>
> When shown a series of computerized photographs of young men whose images had been manipulated to make them look either more masculine or more feminine than the norm, viewers designated the artificially-feminized faces as somewhat more attractive than the average faces, and more appealing by far than the masculinized versions.[38]

This was indeed a curious development, since all the preceding literature on the subject had underscored an unabashedly heteronormative and re-productivist logic of sexual dimorphism: the manly man's thickened brow and protruding chin—supposedly signs of ample testosterone, the hormone linked in folk science to both male fertility and "male dominance"—were said to mark the sort of catch a genetically savvy woman might seek. And female attractiveness, contrariwise, was linked to a smaller chin and higher cheekbones: supposedly shaped by estrogen, such features were said to ex-hibit the very signs of fertility to the opposite sex, no less than the inflamed genitals of a bitch in heat. In this study, however, not only were men rated as being "more attractive" when their features were "feminized," but women with artificially "feminized" faces were identified as potentially "good mothers" less often than were "average" female faces.

By naming masculine beauty as "feminine" and by reducing the degree to which femininity might be seen as part of a stereotyped "good mother-hood," researchers and reporters have reached the point where words and meanings very nearly part company—perhaps for the better. A nonsensi-cal statement—"Good-looking men look more like women" (who presum-ably are "opposite" to men)—might come closer to insight than a sensible one written in the utilitarian logic of heteronormative mysticism: "A hand-some man and a beautiful woman are opposites held in natural attraction."

Perhaps, too, such logical convolutions written across the great mirror of nature evoke something of the ambiguity of gender roles and sexual expectations at this moment of epistemic flux.

Logical convolutions aren't entirely foreign to the advocates of heteronormative beauty, however. More recently, scientists have claimed that women go for the "masculinized" faces when they're ovulating (hence serious about maximizing their genetic chances) and for the "sensitive" types when they're not.[39] Susan Bordo quips: "Talk about infinitely malleable data. . . ."[40]

10   Homo Faber, Family Man

Part of what happens in the reading or retelling of any story—even a simple, well-known story—is a certain degree of opportunism. The reader of a story adjusts the text to her own wants and whims. The reteller changes some small detail to tell a different kind of tale. What counts as good order, good form, a good story—even good science—is open to interpretation and to history. What is held to be transparent about desire, or self-evident about bodies, or natural about nature all depends on the position one takes. Signifiers slide, references slip, and narrative worlds collide.

Evolutionary psychology embodies just this narrative opportunism—but so, too, do other sorts of narratives about gender differences, and over the course of the 1990s, a certain amount of slippage occurred within the prevailing tales of heteronormativity. For instance, the caption on the cover of *Time*'s March 8, 1999, issue put the skids under the prevailing narrative about "coy, choosy women" when it matter-of-factly announced that "the latest research into the secrets of biology and evolution reveals that women are tougher, stronger, and lustier than anyone ever thought." And the same magazine's famous June 29, 1998, issue captured something of the zeitgeist of an off-balance era; its photo spread depicted *Ally McBeal*'s Calista Flockhart as tipsy successor to a long line of feminist foremothers, her stance precariously perched on the edge of one shoe. Slippage, whether it's on a banana peel or on the hazardous ice of social change, can be the stuff of humor, and it is in popular comedies of the day that we can see most clearly how new social movements rescripted old story lines, how old ideologies wrestled with changing conditions, and how new narratives were erected on old assumptions about the nature of nature.

TOOLS

Flashback, 1990s: *Home Improvement*, a popular television sitcom, brings a kinder, gentler sociobiology to the masses each week on ABC. An echo of 1950s TV programming in an upgraded format, the show, which now enjoys a ghostly afterlife in reruns, provided light, topical treatments of family entertainment's perpetual themes, all those stereotypical foibles that afflict men, women, and their relationships with each other. A naive but likeable character, a buffoonish Everyman, protagonist Tim Taylor hosts a cable home repair show, *Tool Time*, where his on-the-air pratfalls serve as a show within a show: He is irresistibly drawn to bigger, more powerful machines, and he is inevitably entangled in minor marital difficulties emerging (in large part) from his cartoonish male emotional insensitivities. If (to invert Sherry Ortner's classic conceit) man is to nature (rugged, uncouth, and apt to belch) as woman is to culture (gentle, refined, and educated), then his wife Jill plays the role allotted to women by Victorian mores, family-values centrists, and modern cultural feminists alike: She is the domesticating civilizer, the family's keeper of higher moral and cultural values.[1]

When mild marital conflicts arise between Tim and Jill, matters are put into some semblance of analytical perspective by the avuncular neighbor, Wilson, a quirky but wise figure who dabbles in men's-movement motifs and is well versed in the exotica of a pop anthropology whose sensibilities lie at the confluence of contending masculinities: man the hunter, the naked ape, and the sensitive male. "The perennial difficulties between men and women," Wilson expounds in resonant voice, "lie in their essentially benevolent and complementary biological differences: Men were engineered by nature to make and use tools, to hunt, and to defend; this explains their enthusiasm for elaborate contraptions that make loud noises, their roving eye, and assorted masculine traits." "Contrariwise," he might continue on any given show, "women are oriented toward emotion, home, and hearth; they were made to—well, to complement those incomplete animals, men." Wilson invariably concludes his homilies with the familiar heteronormative moral on sexual complementarity: "One is neither better nor worse than the other, and each is necessary but not sufficient. Happy is the family that can live with, moderate, and harmonize the way men are and the way women are." Wilson thus gives prime-time voice not only to warmed-over sociobiological scenarios on male hunting but also to sentiments on gender expressed in both scholarly and middlebrow cultural-feminist publications from Carol Gilligan's *In a Different Voice* and Sarah Ruddick's *Maternal*

Figure 18. Wilson (Earl Hindman, left) and Tim Taylor (Tim Allen, right), ABC's *Home Improvement*. Touchstone Television © 1993. Courtesy of Photofest, New York, N.Y.

Thinking to Lillian Glass's *He Says, She Says.*[2] As I will show momentarily, this feat is rather more easily accomplished than one might expect. On the sociobiological side, Wilson simply downplays male aggression. On the cultural-feminist side, he plays up women's supposed affinity for nurturing, nesting, and networking.

An alter ego whose face is always partially concealed, Wilson serves as mouthpiece for comedian Tim Allen, who plays Tim Taylor and is the comedian on whose act the show is based. The same conception of differences between men and women is also the pivot of Allen's off-show comedy shtick. A typical routine might take off from supposed differences in the ways men and women use TV remote control devices—the former channel-surfing while the latter objects—to play out just-so stories about human evolution and the gendered implications of tool use. Thus, in what can only be called stand-up amateur paleontology, eons of evolutionary influences are refracted in the manipulation of TV remotes: men scan, prowl, are on the move, while women sit still and remain centered. There is no subtle way to put it: *Home Improvement* is evolutionary psychology for the masses, sociobiology tweaked by focus group responses and groomed according to market share.

THE CAVEMAN WITHIN

Home Improvement took up the preoccupations of a social phenomenon that spanned the 1980s and early 1990s, the men's movement, whose seminal text, *Iron John*, finds comic echo in the figure of the Tool Man.[3] The essential dilemma: how to claim or reclaim one's "wild" but benevolent masculinity in a world reshaped by feminism. *Home Improvement* first aired its version of authentic manhood in the fall of 1991. Meanwhile, Rob Becker's *Defending the Caveman* made its debut in San Francisco in 1991 and arrived on Broadway in 1995.

The play's promotional material boasts that Becker's "humorous exploration of the battle between the sexes" was composed during "an informal study of anthropology, prehistory, psychology, sociology, [and] mythology." According to *Variety*, this polemical defense of masculinity outlines "the idea that there are real reasons for our differences, possibly even genetic reasons whose origins are buried in the millennia of human evolution." Like Allen's comedy, and like the pseudoscholarly ruminations of men's-movement texts in general, Becker's *Caveman* spins parables about sex, biology, gendered tools, and men's and women's psychologies, elaborating and reinforcing certain widely circulated notions about the "real nature" of men and women.

Reviewers and theatergoers alike claim—a little too eagerly, if you ask me—to see themselves in Becker's on-stage explorations of the heterosexual mystique. This "celebration of the male and female of the species," says the *Chicago Sun Times*, amusingly catalogues "the instantly recognizable traits that differentiate the sexes." And in *Theater Week*, Simi Horowitz reports: "Men have told me that the show explains them to their spouses in ways they've never quite been able to articulate before."

Indeed, a number of reviewers hold this experience of *reconnaissance* to be therapeutic: "It is comforting to be assured that we are part of an ongoing family—the cave people from inner space," writes Clive Barnes in *The New York Post*. "Becker sends you out of the theater with a smile on your

Figure 19. Scene from Rob Becker's *Defending the Caveman*.
Photo by Joan Marcus. Courtesy of Photofest, New York, N.Y.

face. You feel less alone" *(Variety)*. "I think some healing takes place when
a couple sits in a darkened theater, laughing with hundreds of other couples, realizing they're not alone" *(Playbill)*. "Caveman should be seen by
anyone who wants to understand the opposite sex," gushes John Gray, author of the popular advice book *Men Are from Mars, Women Are from
Venus*. Similar sentiments are urged by Lillian Glass (author of *He Says,
She Says: Closing the Communication Gap between the Sexes*), Warren Farrell *(Why Men Are the Way They Are)*, and Anna Beth Benningfield (president of the American Association of Marriage and Family Therapists).[4]

A "genuine word-of-mouth hit," according to the *New York Post, Caveman's* uplifting moral is that "the struggles between a man and a woman
can be understood by thinking of them as coming from two different cultures. . . . [A] whole generation is catching on to the idea that we can work
for equality between the sexes AND we can bring harmony to our relationships by understanding our differences" *(Playbill)*. Out of contrasting
gender styles, then, "springs the potential for bridge building rather than
perpetual warfare" *(Chicago Sun Times)*.[5]

THE NEW, IMPROVED MASCULINISM

Home Improvement was lodged in the Nielsen ratings top ten for much of
the 1990s and topped the list for a shorter period therein. *Defending the*

Caveman went on to become the longest-running solo play in Broadway history. (So intense was "cavemania" at its peak that in 1996 Mayor Rudolph Guiliani renamed New York City's West 44th Street "Caveman Way" in Becker's honor.) These and other popular-cultural phenomena from the period (including a spate of sci-fi horror and alien invader films concocted around geneticist themes) embody the conservative influence of sociobiology and evolutionary psychology in American culture. Such forms of entertainment "interpellate" their audience, in the strictest sense of Louis Althusser's term. They "call out" to the viewer, as he supposedly really is, aggressively buttonholing him and insisting that he "recognize" himself in this particular image of the timeless, abstract, natural man.[6]

But such comic dramatizations also afford a revelatory window on changes in American culture. In their brisk traffic between folklore and science, they provide a space for the exploration of collective anxieties about changing sex roles. The perpetual rewriting of the script for the family scene in these comedies allows for (and facilitates) changing notions of just what is said to be essential, timeless, and wild about man and men. The ubiquitous humor of sudden role reversals, the plays on flawed masculinity, the shtick about the homoerotic implications of male bonding, and the comic situations that gauge the gap between the ideal self and the real self: by such plot turns and comic devices, these forms of entertainment demonstrate the emergence of new ways of seeing ourselves, of finding ourselves in nature, of telling stories about our practices. In so doing, they remind viewers—in spite of themselves—that nothing stands still, not even the eternal verities, not even the most conservative images of timeless nature.

On the face of it, *Home Improvement* seems to return to those cultural motifs I intuitively loathed as a child in the 1960s, themes drawn around a manhood at once "natural" and "struggled-for"—the locus classicus and moral center of a steady stream of Westerns, war movies, and father-son stories. But in fact, the program seldom indulges the older forms of masculine propaganda, where "a man's gotta do what a man's gotta do," and when it ventures close to such themes, its treatments depart—sometimes dramatically—from the linear fables and narrow stereotypes of a bygone era.

Despite the "masculinist" (as opposed to "feminist") label sometimes claimed for Allen's comedy, his TV show really affords a series of object lessons on how to domesticate the savage male. It is Tim who is always learning from his mistakes; it is Tim who has to moderate, adjust, and give ground. Jill is, at bottom, an essentially modern woman of liberal sensibility with career goals of her own. The boys are taught to solve their problems by talking them out, not, according to the long Hollywood tradition, by fighting them out. And although the program usually eschews controversial topics such as homosexuality, it also largely avoids cheap shots and one-sided mother-in-law jokes. (Taylor *tells* such jokes, but they rebound on him and the audience understands them as uncouth.) Tim the New Man is always learning to live with his feelings and vulnerabilities and to make room for the feelings of others. He even learns to value, support, and unconditionally love the youngest of his three sons, who is temperamentally disposed toward practices stereotypically understood as "feminine," not "masculine," and who the pop-archetypal father fears might be gay.

Eons of evolutionary adaptation make Tim a man of a certain definable sort—but they also allow him to seek a common ground with Jill. He is a wild man, surely—the character's signature barks and growls signal his wildness—but over the course of eight seasons, he becomes increasingly receptive to his feminine side. (In the program's final season, after playing house and having tea while baby-sitting his niece, Tim gets in touch with his sensitive side and wishes for a daughter—because, he realizes, his enthusiasm for male things has caused him to miss out on a valued sort of experience.) Tim's true nature thus equips him admirably for life at the fin de siècle, an era in which the desperate bid for a fixed, clear-cut identity slips up on the fact that just what is supposed to be fixed or clear-cut about that identity changes from one season to the next.

All of this might well qualify as some kind of political advance, even if counted as an intellectual retreat. At last, the man and the woman are talking! Moreover, the compass of respectable manhood is wider, more tolerant, and more open here than in earlier forms of family entertainment—or earlier paradigms of sociobiology and evolutionary psychology. And "nature" has likewise dilated: it is less constrictive, more generous. It admits more benign variation. It is less like a retributive divine law and more like a loving, suburban home. In properly functional families, men and women are distinct, but equal, and at least some of their better traits are interchangeable.

Who could have foreseen that this *Make Room for Daddy* of the 1990s, pitched as counter to the man-eating, working-class bitchiness of *Roseanne*, would eventually become a stalking horse for men's-movement feminism in popular culture? And who would have imagined that evolutionary scenarios of *any* ideological bend could play out on network television (as entertainment, no less) without triggering organized protests in a nation where creationists still vie with evolutionists for equal treatment in public school systems?

"MADE FOR A MAN"—"BUT I LIKE IT TOO"

Along with such shifts in the basic terrain of American popular culture, much has also changed in the framing of sociobiological arguments since the 1970s. Of course, the most outlandishly misogynist scenarios still occasionally break out in sociobiological circles around discussions of rape and other forms of male sexual aggression.[7] Racist arguments in academic and institutional culture, no less than in popular culture, perpetually return to the biological determinism they require in order to exist, and these claims frequently take refuge in the more or less benign image of post-1970s sociobiology—witness the often respectful and sometimes actively promotional media attention given Charles Murray and Richard Herrnstein's *The Bell Curve*, a book whose core premise is that blacks are genetically inferior to whites.[8]

But on the balance, savage, male sociobiology has undergone its own process of domestication. For example, most sociobiologists today (in polite society, at any rate) would be loathe to espouse Edward O. Wilson's bloodcurdling and ill-conceived "explanation" for genocide as a "natural" strategy of "territorial usurpation" and "genetic self-maximization." In the most popular works of sociobiology and evolutionary psychology, a certain bend in the direction of key arguments has occurred, and Wilson himself (the real Wilson, not the TV character) has turned to stressing more cooperative, less competitive themes in evolutionary theory.[9] In some quarters, feminist critiques, even feminist variants of sociobiology, have rubbed the sharp edges off paradigms that first stood as blunt arguments for the inevitability of sexual inequality and are now increasingly hedged as explanations for more benign patterns of sexual "difference."[10] Indeed, forms of sociobiological reasoning—often filtered through evolutionary psychology—have become the ultimate prop for cultural feminism, whose universal woman is keeper of the hearth and guardian of the heart.

In *Daring to Be Bad*, Alice Echols describes the historical shift from radical feminism, "a political movement dedicated to eliminating the sex-class system," to cultural feminism, "a countercultural movement aimed at reversing the cultural valuation of male and devaluation of female."[11] As Micaela di Leonardo and I put it: " 'Valuing women,' certainly a component of any feminist program, was transformed in the changing political climate [of the 1970s and 1980s] into a celebration of characteristics assumed to be inherent in women's universal nature—nurturance, altruism, cooperativeness, pacifism, and benevolent or absent sexuality."[12] In promoting such claims about women's essential nature, cultural feminism represents a return to the preoccupations of Victorian feminism, which similarly celebrated the pacific, civilizing mission of woman as against the violent, competitive nature of man, the sainted mother as against the stern father. Such themes provide ready points of transfer between feminist identity politics and modern bioreductivism (not to mention between pop feminism, junk science, the recovery movement, the "spirituality bazaar," and other elements of an irrational culture of credulity savagely skewered by Wendy Kaminer in books like *I'm Dysfunctional, You're Dysfunctional* and *Sleeping with Extra-Terrestrials*).[13]

The reduction of woman's universal value to her biological nature was already evident in the classics of cultural feminism. For example, mystical associations between blood, birth, and intuition were given a quasi-biological standing in Adrienne Rich's reflections on supposedly universal experiences of motherhood.[14] More systematic appeals to biology, first elaborated in sociobiological papers by Alice Rossi, have long since settled into a prevailing variant of evolutionary psychology—and are perhaps most clearly spelled out in Helen Fisher's recent book on women's "natural talents," *The First Sex*—a book that commingles feminism of a certain sort with evolutionary psychology of a certain sort. Woman's "coy nature" becomes the basis for a feminine sexual power. Woman's greater investment in the egg becomes the basis for her peaceful nature, her tendency to solve problems by talking them out, her superior social skills and networking abilities. Women's hormones likely contribute to the difficulties they experience rising to the top of status hierarchies—but, ironically, make them exceptionally wise rulers once they achieve positions of power. . . . So goes the familiar litany, first conceived in an oppositional subculture and now circulated with dreadful repetitiveness—and occasional creativity—in the machineries of mass representation.[15] Like iron filaments around a magnet, biological reductionist arguments constellate around cultural feminism's essentialist politics—for what better way to defend the notion of woman's value than to see the supposed female virtues everywhere, in all women, biologically scripted by woman's deep, unchanging nature?

WHAT EVERY WOMAN KNOWS

The universal woman of cultural feminism literally danced across the TV screen one evening on an episode of Fox network's cartoon sitcom *King of the Hill*. When Hank Hill, impatient with his niece's incessant weeping, undertakes to abbreviate Lou-Anne's grieving time after her breakup with a steady boyfriend, he is warned by his wife Peggy not to interfere with the natural grieving process. Peggy's admonitions, paraphrased here, simultaneously interpellate male *and* female viewers, urging them to recognize their own authentic selves, their natural strengths and weaknesses, in the interplay of complementary stereotypes: "I'm advising you, Hank, don't interfere. You don't know what you're doing. Men don't understand: You can't rush these things. Women have had thousands of years of experience with these matters."

Tribal drumming swells in the background, accompanied by hints of snake-charmer music. As Peggy speaks, images flicker into view like the stirrings of racial memories from the depths of the somatic unconscious. Cavewomen gatherers, keepers of the hearth, and exotic women of the Orient dance across the screen in succession—glimpses of woman's universal nature, decked out in New Age garb.

SPINNING THE DISCOURSE ON NATURE

Even the most reified and reifying of representations can lend themselves to different "spins," can be invoked in support of widely varied practices. This is an irreducible condition of what Bruce Mau calls today's "image economy."[16] The intersection of two essentialisms—cultural feminism and sociobiology—marks a certain double irony or contradiction therein. On the one hand, radical ideas (such as feminism), conceived in opposition to the status quo, are readily appropriated by the machineries of mass representation; they quickly become part of the dominant culture insofar as essentialist notions of gender, sexuality, or identity can be extracted from them. On the other hand, reductivist ideas (such as sociobiology), conceived full tilt against the current of historical happenings, nonetheless participate in the revaluation of cultural values; they register the heave and shove of social struggles and are thereby open to the effects of history. Today, more than ever, representations of nature are, in Foucault's sense, "polyvalent, mobile discourses."[17] Radical ideas are constantly appropriated by reactionary causes, and even the most conservative discursive formations move, like tectonic plates, beneath the visible social landscape, subject to underlying forces and headed toward uncharted constellations.

Should we, then, be won over by a reformed, soft-pedaled, even feminized sociobiology, which once posited so much brutality and nastiness in human nature and has now come to discern supposedly good things there?

And should we fight fire with fire, strategically employing a "good" (pro-woman, pro-gay) essentialism in place of a bad one?[18] Or, might "the good"—conceived as being already there, in eternal men and natural women—prove more dangerous than "the bad"? (I hope it is clear who the charmed circle of the "we" I invoke in this query embraces: gay men, lesbians, bisexuals, single mothers, transgenders, butch straight women, sissy straight men—all those legions whose affect or affections are excluded, denied, marginalized, or trivialized under the regime of heteronormativity: a teaming, motley mass who might someday yet constitute a real as opposed to a hypothetical queer coalition.)

Whatever its silliness, the secular-liberal Tool Man motif—even Robert Bly's confused archetypes of manhood, drawn from a not so serious study of world cultures—seems decidedly preferable to the Christian-conservative Promise Keepers phenomenon, which in 1997 drew several hundred thousand weepy marchers to Washington, D.C., to "stand in the gap."

Promise Keepers events were always staged in an all-male environment, with much cathartic weeping, hugging, and touching.[19] A front for the religious Right, Promise Keepers encouraged men to reassert their role as "heads of the family." Before attendance began to wane at Promise Keepers' glitzy events, the organization received a considerable amount of favorable media coverage for its commitment to male involvement in family matters and for its nonracist (even antiracist) organizing strategy. A much-burnished halo of progressivity prevented some liberals and even a few feminists from recalling the unapologetically hierarchical terms of the "male involvement" advocated by conservative Christians. Oleaginous equivocations about what it means to be the "head" of the family notwithstanding, Pauline Christianity has stood for the better part of two thousand years on the principle that the man is very much the boss. Promise Keepers' media-savvy multiculturalism also apparently stimulated a certain amnesia about the religious Right's long-term political strategy: the movement aims to build good will among socially conservative black and Latino churches in order to expand the electoral ranks of the homophobic Right opposed to feminism and civil rights.

Don't get me wrong. I have no doubt that at "Wednesday evening prayer meetings . . . all across America . . . people seriously and sincerely struggle with their feelings on issues like prejudice, tolerance and greed."[20] A white, rural Southerner born into an evangelical fundamentalist family, I've seen conservative congregations desegregate. I know individual believers who've struggled with—and modified—their beliefs about race relations, women's

Figure 20.   Kris Wilde from Boise, Idaho (right), and Tom Jensen from Bellevue, Washington, celebrate at a gathering of the Promise Keepers at the Kingdome in Seattle, May 23, 1997. Photo by Loren Callahan. AP/World Wide Photos.

rights, and homosexuality. I've also borne witness to the antics of cynical preachers and the machinations of right-wing organizations, who strategically cede ground on some issues so as to consolidate their position on others. That's part of the modern art of "spin." For a time, Promise Keepers had so mastered this slippery art that anyone who pointed out the organization's misogynist political agenda could be accused of being unfriendly to the needs of ordinary women and children. ("Ordinary" women and children, I take it, have need only for stable families, involved husbands, and active fathers—*not* for gender equality.) And those who criticized the organization's unabashedly homophobic pitch (No gays allowed! It's not God's way!) were accused of being insensitive to (straight) men's desires (for connection), or unwilling to accept Christian atonement for past racism.

Given the alternatives, the liberal Hollywood version of sociobiology, with its kindly, winsome dad, might seem preferable to the evangelical fundamentalist version of divine fatherhood. And by the same token, even cultural feminism might seem preferable to no feminism at all. But perhaps this is to put things the wrong way, to fail to see the salient interconnections between historically connected phenomena and in consequence to misunderstand the battles over nature that rage at our historic moment.

In the past, the conception of femininity embodied in Victorian feminism's ideal of womanhood (woman as carrier of moral virtues) was always
very close in sensibility to religious icons of femininity (the woman whose
role it was to domesticate and better the man) and not so far from the Victorian notions expressed by Darwin in his framing of the theory of sexual
selection (the woman whose "choosiness" betters the race). While hegemonic notions of male superiority stimulated each of these countervailing
conceits, none of these representations of female virtue really challenged
the logic of the system in which they occurred. Woman-centered religious
icons "empowered" women insofar as they supported God-sanctioned
forms of patriarchal rule. Darwin's choosy women embodied not a trans-
formational antithesis, but the logical mediator of and prize for male competition. In the social conventions that gave ballast to Victorian feminism,
the goodness of the woman enabled the badness of the man, like the social
dynamics of a latter-day couple locked in a codependent relationship.

By the same logic and from its very inception, cultural feminism opens
a wide venue for the renewed expression of heteronormative notions in public culture. Its core idea—the notion that women and men are, in their deepest natures, radically different from each other—supports the familiar ideology: men and women as complementary opposites, each necessary but
insufficient. In the resulting force field of gender, sexuality, and power, the
liberal Hollywood version of sociobiology reflects the feel-good and progressive side of what remains a decidedly heteronormative constellation of
meanings and values.

Today, naturalistic conceptions of male and female bodies characterize religious absolutism, pseudoscientific innatism, cultural feminism, and the new
masculinism because all four spring from the same soil, all four grapple with
the same predicament, and all four reflect a certain kind of longing for permanency and connection in a season of uncertainty. All four do a certain
identity work for men and women in crisis. And all four allow men and
women to pick and choose opportunistically which elements of the heteronormative configuration to keep and which to reject.

11 T-Power

Perhaps nowhere is the domestication of savage, male sociobiology more clearly expressed than in a spate of recent reflections on "the crisis of manhood"—a crisis, authors claim, that stems from men's supposedly reduced place in family structures in sexual modernity.[1] The subject of these partisan analyses is not your father's wild man (the playboy on the make, the cowboy who longs for independence), it is the sensitive man who yearns in his innermost nature to be a mensch, the guardian of family values and the keeper of civic virtue. But perhaps nowhere is a socially reactionary ideology more forcefully expressed than through this airbrushed image of the benevolent but frustrated paterfamilias, a latter-day Adam longing to reclaim his proper eugenicist role in a restorationist heteronormativity.

LAMENT FOR THE BELEAGUERED MALE

Writing in *The New Republic*, Andrew Sullivan laments the predicament of the endangered male, whose decline, he asserts, is an unintended consequence of feminism. "Greater opportunity for women is probably the most significant gain for human freedom in the last century. But with this gain has come a somewhat unexpected problem: How do we restore a sense of masculinity that is vaguely civilized? Take their exclusive vocations away, remove their institutions, de-gender their clubs and schools and workplaces, and you leave men with more than a little cultural bewilderment."[2]

Echoing similar claims by sociobiologist Lionel Tiger, Sullivan argues that the crisis of masculinity stems not simply from the rapid pace of changes in economic production and in gender conventions (a pace that most everyone agrees has left large numbers of men and women uncertain about their role expectations), but from a postfeminist denial of "any deep biological or psychological differences between the sexes"—that is, from something "unnatural" about the emerging system of gender and sexuality.[3] In the *New York Times Magazine*, Sullivan elaborates: "I don't think it's an accident that in the last decade there has been a growing focus on a muscular male physique in our popular culture, a boom in crass men's magazines, an explosion in violent computer games or a professional wrestler who has become governor. These are indications of a cultural displacement, of a world in which the power of testosterone is ignored or attacked." Sullivan wants to capture something of the sweep of recent changes in dramatic temporal terms: "growing," "boom," "explosion." But like most authors who invoke "nature" to gauge the state of gender relations today, Sullivan builds a rickety case for his exposition.

Since no long-standing baseline exists, it is unclear whether one should say that there has been an "explosion" of violent computer games or just an explosion of *computer* games. Nor is it clear just what the historical baseline might be for "crass men's magazines," much less what defines their "crassness." It is possible that Sullivan does not see what is campy about Jesse Ventura's coarse and funny wrestling persona, but surely it is relevant to note that Ventura staked his political bid not on revanchist masculinism but on a pro-choice, pro–gay rights platform—hardly the conventional stuff of male braggadocio.[4] At any rate, it is no more astonishing that a wrestler could become governor of Minnesota than that a B-film actor once became president of the United States, or that while in office he engaged in macho posturing by quoting an Eastwood rogue cop ("Go ahead. Make my day"), or that his understudy left tough-guy bluster ("Read my lips") as his unhappy rhetorical legacy. Whatever evidence he might admit from the weird and wacky world of American politics, Sullivan knows full well that the epicenter of the "growing focus on a muscular male physique" is the gay community, where gym culture all but eclipsed bar culture in the 1990s, and that the measurable increase in images of bare, muscular torsos in mainstream society reflects a growing eroticization of the male body in advertising culture, much of it homoerotic in nature. In short, Sullivan's breathless litany draws together a variety of incongruous and not altogether credible phenomena into a singular "trend": it gives the effect but not the substance of sociology.

Sullivan's appeal to natural science is not much better. His citation of some female-to-male transsexuals' experiences with synthetic testosterone (it makes them horny) and the invocation of his own experience with testosterone treatments for HIV-treatment-related anemia (they make him cranky, short-tempered, and energetic) could easily stand as a comic parody of essentialist arguments—there is nothing very "natural" about the "nature" whose experiences he recounts. But the author is as earnest when he fetishizes his biweekly shot of testosterone as "a syringe full of manhood" as he is when he frets that "low-testosterone risk aversion" in the corporate world "may lead to an inability to seize business opportunities." (Businesswomen, watch out: What you keep bumping against is not so much a glass ceiling as a testosterone thing.) Sullivan's celebration of the magical properties of testosterone, including its tonic effects on a variety of vague symptoms, is so effusive that his *New York Times Magazine* essay prompted one reader to quip: "If they'd known about testosterone in the nineteenth century, it would have been a cure for neurasthenia."[5] It goes without saying that Sullivan's disquisition on what gives essential manliness its juice includes the obligatory gesture toward man the hunter, that cartoon caveman who needed an extra dose of verve to go out and bring down big game. . . . From such weak premises, Sullivan argues that manhood has a natural and unchanging essence and urges a specious conclusion: "Our main task in the gender wars in the new century may not be how to bring women fully into society, but how to keep men from seceding from it, how to reroute testosterone for constructive ends rather than ignore it for political point-making."[6]

Sullivan's sermonette on male troubles gives a picture-perfect snapshot of the sociobiological vulgate: the attribution of social practices and cultural meanings to genetic causes by way of neurohormonal links. An openly gay man who avoids extremist positions, Sullivan brings considerable respectability to the reductionist and conservative position he popularizes. Still, his exposition telegraphs a social ultimatum on behalf of angry straight men everywhere: Construct a society that steers testosterone toward "constructive" ends, he warns, or run the risk that manly exuberance will be steered, willy-nilly, toward destructive ends.

Other opinion leaders draw blunter points and urge a more general revanchism. "If you agree with the biologists that manliness is something permanent and unchangeable, in a gender neutral society you'll run into problems," asserts Harvey C. Mansfield, a Harvard political philosopher interviewed in a *New York Times* piece on the crisis of masculinity. "One [of men's biological attributes] is the desire to protect those who are weaker than they are: women and, especially, children," he continues. "Since women are in the work force, they no longer depend on men for support the way they used to, which can make men feel they are not useful and lead them into irresponsibility, violence and criminality." One scarcely needs to underscore the snarling threat embodied in this bizarre syntax, offered to the reading public in the name of "biology." I will simply note that what actually undergirds Mansfield's argument is a rather unrigorous appeal to folk psychology's hydraulic model of human nature: dammed up here, male energy will seek another outlet over there. The picture thus painted—of a benevolent but improperly channeled human nature run amok—does the speaker's ideological work for him. (And an ugly work it is, too.)

In the same *New York Times* piece, Lionel Tiger gives a précis of his influential book *The Decline of Males* when he blames the birth control pill for male troubles today: "The decline of men has everything to do with them being alienated from the means of reproduction."[7] Clearly, the logical solution to the "problem" Tiger poses would be to curb women's reproductive freedom. The easiest way to "restore balance" to the sexual equation would be to make women less independent, more reliant on male protectors and breadwinners.

RAGING HORMONES?

Those who speak of sex in the name of science today thus arrive at a very different conundrum than that of Darwin, who thought he discerned a moment in the misty time of origins when men had expropriated the power of sexual selection. Yesterday, Darwin longed to grant white women some power in adjudicating civilized men's worthiness. Today, we are told, men have been all but routed in the war of the sexes—and this is said to hold especially true for white, middle-class men in northern, industrial societies. Dire consequences are said to follow from this most unnatural outcome: infertility, falling birth rates, a demographic crisis, outbreaks of barbarism ...[8]

Allusions to urban street crime and to events like the Columbine massacre, in which two alienated boys slaughtered classmates at a high school in Littleton, Colorado, figure prominently in Sullivan's, Mansfield's, and Tiger's frettings over the frustrations of the dispossessed male. But surely it is a salient fact that violent crime rates, including juvenile crime rates, have fallen precipitously in the years that correspond to the supposed days of male rage. (Perhaps, then, what the evidence actually shows is that a little male crisis—or, at any rate, something very close to full employment— is good for the health and safety of society.)

Tiger's contention that this is "the first time in history" women have ever controlled the means of reproduction is simply bad anthropology.[9] Women, with or without the knowledge and approval of men, have in fact practiced contraception, abortion, and infanticide in a wide range of cultures throughout human history. Indeed, Ju/'hoansi men somehow manage to imagine a productive role for themselves despite the fact that women pro- duce more calories than they do while also controlling the means of repro- duction. As anthropologists Marjorie Shostak and Richard Lee document, Ju/'hoansi women practice abortion. They also give birth outside the camp, attended only by other women—so that when Ju/'hoansi women practice infanticide, as they occasionally do, it is none of the men's business.[10] The idea that foraging relations of production structure asymmetrical social re- lations (female dependency on a male breadwinner) is no less a figment of the male sociobiologists' imagination. Richard Lee describes it as axiomatic in Ju/'hoansi society that no healthy adult really "depends" on anyone else for sustenance *or* protection: Everyone is mobile, and everyone has access to the tools and materials necessary for survival. (Perhaps Ju/'hoansi no- tions of individual autonomy provide as good a role model as any for mod- ern conditions, with their emphasis on nomadism, social equality, and per- sonal freedom.)[11]

No doubt hormones, including testosterone, both affect and are affected by how people feel and act. But basic research is all over the map when it comes to men, mood, and testosterone. Some studies find correlations be- tween testosterone and aggression, domination, or risk taking, and others do not. Some studies find correlations for certain age groups, but not for others.[12] The fact remains: "In humans, just as in monkeys and mice, most attempts to link testosterone levels to aggressive behavior have failed."[13]

In her thoughtful, expansive critique of evolutionary psychology, Na- talie Angier summarizes the current state of research on testosterone and male behavior:

> Some studies have found that among male prisoners, the more violent the offense committed, the higher the man's testosterone. Other stud- ies have failed to find such a correlation. Among young adolescents, boys rated as "tough leaders" by their peers have been shown to have high levels of testosterone; yet one boy's "tough" is another's "tough luck," for the same study showed that those boys who had spent their childhoods getting into fights and trouble had fairly *depressed* levels of testosterone. As a rule of thumb, a man's testosterone will rise right before a challenge, like a football game or a chess tournament, and if the man triumphs, his testosterone will stay high for a time, but that if he loses, it drops, and it has difficulty getting up again. . . .

But gearing up for a challenge does not always elicit a testosterone spike. Young men competing in a video game tournament show no detectable testosterone shift, neither beforehand nor in the wake of virtually exterminating their opponents. Before a male parachutist leaps from a plane, his testosterone *drops;* it sees what's coming, we can suppose, and swoons at the thought.

Angier then goes on to underscore problems with the interpretation of certain findings. For example, a man's testosterone level often falls when he commits to a relationship, as when his partner is about to give birth.

> Some scientists interpret these results to mean that men in a monogamous relationship don't need their testosterone. . . . It does them no good if they're going to stay put and be a loyal lover and a devoted father. They don't need as much testosterone as they did when they were on the hunt and might be forced to rattle their chain mail at a rival en route. There are alternate storylines, though. Testosterone can fall when a man is under stress; that's presumably part of the reason why it dips with a drop in social status, or with the loss of a chess game. Is commitment not stressful? . . . And impending and new fatherhood—are they stressful?

Studies that attempt to correlate testosterone levels with dominant or aggressive female behaviors are as variable, their meanings no less precarious. Women do not appear to "rev up" on testosterone the way *some* men do in preparation for *certain modes* of conflict, competition, or confrontation. But this hardly means that women are incapable of aggression or bravery. I quote again from Angier's thorough review:

> In one study, the higher a female prisoner's testosterone level was, the more likely she was to have committed a violent crime like murder than a nonphysical offense like embezzlement. In another study, the correlation did not hold. Investigators have found that female inmates with high testosterone display more domineering and intimidating behavior than low-T prisoners do, and, conversely, that the low-Ts are "sneaky," manipulative," and "conniving," acting like "snakes in the grass," according to the assessment of the prison staff. But let's bring some perspective to this study. . . . The high-T women in the sample population were also younger, on average, than the lower-T women. Youth has its perquisites. When you're young, you have quite a bit of muscle tissue. You still think death is exciting and provisional. As a rule, people in prison have a history of bad habits—too much smoking, too much drinking, too many drugs in too many corrosive combinations— and so the older you are, the weaker, sadder, and more shopworn you are likely to be. Better to connive in the grass than to confront in the flesh.[14]

In short, the ubiquitous association of testosterone with the manly virtues and vices—domination, assertiveness, and risk taking—is not actually supported by existing scientific data. What careful studies actually suggest is that men's testosterone levels respond to events in the world, including social cues. As Stanford neurobiologist and primatologist Robert Sapolsky shows in *The Trouble with Testosterone*, what can be documented about the relationship between male mood, aggression, and testosterone is actually the opposite of what usually gets asserted. Aggression—or rather, some forms of it—elevates testosterone secretion, not vice versa.[15] It remains unclear just how this works or what it means. Why testosterone should flow when men prepare for a chess match but not when they brace themselves for a video game is a matter of some wonderment. Contrary to the usual associations of androgens with violence, testosterone levels drop precipitously when recruits are about to be sent off to battle, no doubt in response to stress and depression.[16] And one of the strongest social correlations (but note: not *causes*) established by careful experimental studies is a surprising one: Men with low levels of testosterone tend to be irritable, tired, and nervous, whereas men with higher levels of testosterone tend to be more alert, more optimistic, and *friendlier* than men with lower levels.[17]

The friendly hormone? ("Hi-ya, Big Guy!") It's not likely you'll be reading about *that* in odes to the wonders of T-power or threnodies on the debased condition of modern manhood. But perhaps such an association might occasion a different set of claims about the "normal" disposition of properly evolved males and how they ought "naturally" to act in the modern world.

MALE TROUBLES

In all fairness to Lionel Tiger's vastly influential text, *The Decline of Males*, it must be noted that the author indulges a notion or two at odds with mainstream conservative family values. Under headings like "What Does Mother Nature Think about Single Mothers?" and "Youth Chooses Darwin?" for instance, Tiger suggests that unwed teen mothers are rebels against an unnatural, antinatal culture.[18] But in all seriousness, it must also be noted that one might well wish to avoid being the object of such praise, seeing how it casts all the usual suspects—working-class kids and minority youth—as puppets in a Punch and Judy version of nature.

Lionel Tiger's reactionary screed recalls the preoccupations of the antecedent American fin de siècle: an era in which men, or at least middle-class white men, were also perceived to be suffering from a surfeit of feminizing civilization. As Gail Bederman shows, this earlier crisis of manhood was connected to dramatic social changes then under way. First, with the rise of big industry, the way men worked was changing. The sharp decline of middle-class self-employment necessarily implied a crisis for the ideal of high-minded, self-restrained manhood it had once sustained. At the same time, an emergent women's movement challenged men's monopoly over education and their "natural" authority in the political sphere. Immigrant labor organizations and political machines further undermined the power of white, middle-class men in every major city. No doubt large numbers of previously secure men were rendered nervous, if not outright disoriented, by these social changes. It was about this time that a mysterious new illness captured the popular imagination: the aforementioned "neurasthenia," or nervous depletion. In the last decades of the nineteenth century, medical authorities warned that this vague ailment—a symptom of "too much brain work"—had reached epidemic proportions, draining the nervous energies and sapping the vitality of white, middle-class men everywhere.

The cure? White, middle-class men were enjoined by doctors, politicians, and reformers to recapture a sense of their own natural wildness. White, middle-class men would inoculate themselves against the diseases of a debilitating civilization by emulating, in small doses, the supposed practices of black-skinned and brown-skinned savages, who were seen as being closer to nature. A new regime of body practices thus grew up around competitive athletics, virile blood sports, totemic men's clubs, strenuous physical exercise, and outdoor exertions. It goes without saying that those who advocated a therapeutic reclamation of men's "natures" were actually engaged in a reconstruction of manhood, its ideals and values.[19] It also goes without saying that present-day understandings of nature and struggles over sex owe something to the antecedent era's epidemics and cures. Now, as then, the crisis of manhood is connected to wider social changes in the body politic—with a politics of sex, surely, but also of race and class. And now, as then, just how one sees the crisis, what values one attaches to it, and what one proposes to do about it, are matters of politics, not science.

THE NATURE OF DESIRE AND THE DESIRE FOR NATURE

And I am struck by how convoluted the politics of nature can be. Reading Andrew Sullivan's paeans to virile virtue, punctuated by lamentations for lost boys, I cannot help but think I know whence comes such warm fellow feeling. It expresses an identifiably *male* enthusiasm for strapping lads, an ardor for exultant youths in the full glory of their manhood. I am almost ready to enlist my services in the cause, for I, too, have always burned with desire for blokes with mischievous grins and aloof demeanors: for close-cropped brunettes in uniform who look at you through dark sunglasses— flat, reflective surfaces—a rifle slung lazily across the shoulder. It's all in his attitude, his natural demeanor, the easy way he inhabits a certain kind of hard, indifferent maleness. Impossibly, I want both to be like and to possess this man, the kind of man who's not afraid to put me in my place if I get out of line, the sort of cocky jock whose pose, when he leans against the door frame, suggests that he was just irrepressible enough to have spent much of his adolescence in trouble, but whose sensitive eyes let you know that he was never quite so bad as to have actually landed in a reformatory or in jail.

If I correctly understand the mood and tone of the present ruminations on the crisis of manhood, it makes little sense to cast such objects of homoerotic desire as the ideal subjects of a heteronormative evolution, much less as models for a eugenic masculine future. But there, perhaps—in the longings that give shape to what one sees without themselves being visible, in the circulation of desires just below the surface of claims about the "nature" of desire, in the very wildness of the wild man, lies the chiasmus or "blind spot" of heteronormativity today.[20]

Every male role model is at once the object of narcissistic emulation and of homoerotic desire. The one thing is virtually inextricable from the other.[21] Every story about one's own true self catches its teller up in questionable stories about others—and in stories told by others, some of them circulated long ago.[22] Every image of a timeless Adam and an eternal Eve distills frustrated longings for permanency while unwittingly registering the most subtle of historical changes. And every hard and fast claim to identity— cultural feminism, the new masculinism, gay essentialism—participates in a shifting constellation of signs and practices.

12 Nature's Marriage Laws

In the varied contests over whether and how to manage gender, sexuality, and family life, claims about the biologically-given "nature" of desire provide template and rationale for legislative agendas—but seldom as transparently as in Jane Brody's *New York Times* science report, "Genetic Ties May Be Factor in Violence in Stepfamilies." In an astonishingly vulgar redux of sociobiology's crudest arguments, Brody quotes Dr. Stephen Emlen, an evolutionary biologist at Cornell University, who asks "whether men are really so different from, say, male lions." Emlen claims that "when taking over a new family, the male will kill any offspring still present from the female's prior matings." "Genetically speaking," the author of the article uncritically elaborates, "stepparents have less of an investment in unrelated children and may even regard them as detrimental to their chances of passing along their own genes, through their own biological children."[1]

The *Times* report on killer stepfathers was published in the midst of one of the 1990s' periodic waves of pro-marriage (anti-divorce) agitation in state legislatures.[2] Just in case any readers might miss the drift of the sociobiologists' narrative of rescue, the author underscores the political and legislative implications of her reductionist piece: "The matter is especially pressing now when rates of divorce and remarriage are at an all-time high."[3]

In fact, it is not so clear that sociobiological stories accurately describe the behavior of lions, as Anne Dagg has shown in a review of the ethological literature.[4] Much less is it clear what the behavior of big cats on the savanna has to do with human social action—or with the institution of marriage. And never mind the fact (noted, but not fully acknowledged in the *Times* piece) that adopted children genetically unrelated to *both* parents fare quite well by all accounts, although according to genetic reductivist arguments, they should face maximum infanticidal risks. The problem with sexual modernity, according to the now sweet, now bitter tales of evolutionary psychology, is that it contradicts human nature, and the solution is to bring social law back into concord with natural law—or, alternatively, to protect us from our own worst nature, which is the law of the jungle.

Descriptions like Brody's might give rise to a sense of pity for those poor, brute creatures of the wild, heterosexuals, so lacking in the traits usually associated with higher order human capacities (adaptability, reason, empathy), so in need of moral guidance and benevolent social restraint. But humanitarian compassion is misplaced if it gives cause for misguided action. Better, instead, to take a hard, dispassionate, and skeptical look at all that excites apprehension and compassion in narratives that call themselves "scientific."

HARDWIRED TO PAIR BOND?

In "Women's Health," a special supplement for the *New York Times,* the irrepressible Natalie Angier ponders the age-old question: "Men: Are Women Better Off with Them, or without Them?"[5] Her article on marriage takes a more benevolent but no less heterocentric view of human nature.

"Marriage has a beneficial effect for both men and women, but the effect is much stronger for men," chatters Jeffrey Sobal, an associate professor of nutritional sciences at Cornell. He goes on, savoring sexual stereotypes and sounding for all the world like an enthusiastic amateur matchmaker: "Men . . . may kick and scream about getting married, but then it does them a world of good." Linda Waite, sociologist at the University of Chicago, contends that although marriage benefits both sexes somewhat differently—it "protects women against drinking" and "helps shield men against depression"—it benefits both "equally." Citing evidence from cross-cultural surveys in seventeen countries, Waite beams: "There's something about being married that makes people work better. . . . We're hard-wired to bond."

Waite's cheery aside gives away the naive biologism that undergirds a wide swath of quantitative research on marriage and family life, her own included. It is the usual package of culture-bound assumptions about heteronormative nature, sexual complementarity, pair bonding, and all the rest—the familiar, dreary formula that unites morality with utility under the number "two." But Waite's effusions on "the transforming power of the marriage vow," like Angier's breezy essay on domestic tranquility, skips gingerly over a number of key issues.[6]

Is it really the case that married couples necessarily "bond"? An abundance of ethnographic evidence and folkloric opinion, not to say country-and-western songs, testifies to the prevalence of loveless and emotionally distant marriages. And far from being the timeless "keystone" of "universal social institutions," the idea that marriages ought to be founded on "trust, love, loyalty, commitment, and meaning" is of historically recent vintage.[7]

Is this "bonding" really the psychological source of a well-being that generalizes to other aspects of health, income, longevity? It might actually be the case that health is the "independent variable," a good predictor of the probability that a person will fall in love, live long, and prosper. Or, it might be that the mundane but considerable *economic* benefits of marriage support most of the other aspects of well-being. And since marriage confers adulthood in most societies, the real seat of well-being here might actually be nothing more mysterious than the social approval and advantages bestowed by conformity—an approval conveyed in the form of higher status, higher wages, more job security, better health care, and so on. On this point, Waite's research fails to take into account the basic findings of sociology and social psychology on stigma and moral evaluation in everyday interactions. Peers, employers, supervisors, and others actively reward "regular Joes" perceived as being like themselves. They penalize those perceived as being different. It is not clear how these influences could ever be rigorously isolated in sociological research on the benefits of marriage. It is not clear that Waite wishes to do so.

The self-evidence of some of Waite's measures of good are also open to question. Most everyone would agree that longevity is a good thing—but if given the choice, would you opt to live two more years well, or four more years poorly? Would you define this "wellness" in terms of urban excitement and global adventure, or would you prefer settled suburban domestic calm? And would a 20 percent increase in your salary be sufficient inducement for you to find contentment in a marriage you could have sworn was unhappy? At any rate, is monogamous heterosexual marriage—the "one woman, one man" formula prescribed in the Defense of Marriage Act of 1996—necessary for psychological bonding and domestic bliss? Waite claims that live-in couples do not get the magical benefits bestowed by marriage, and perhaps this is so (at least for those domestic pairs whose partners haven't "emotionally committed" to the relationship—it is not clear how or whether researchers could identify subpopulations of live-in lovers). But if coupling is good, might not threesomes be better? And if "having someone to talk to," "having someone to have sex with," "having someone to come home to," and "having someone to nag you" all promote happiness, is it really more beneficial to subsume all of these functions into one relationship, or might they be successfully aggregated into relationships with different people? Why not also survey, then, for the relative benefits of polygamous unions, same-sex couples, companionate marriages, cloistering, communal living arrangements, spinster sisters living together, friendship circles, and so on? (By all accounts, cloistered nuns—who live with other women and are "married" to Jesus—enjoy extraordinary longevity.)

Waite heaps study upon study to build a case for the "benefits of marriage," but in the absence of good answers to methodological questions like these, it is not clear what she has actually demonstrated. Until statistical research on the supposed benefits of marriage vows can credibly test for the effects of different cultural contexts and varied institutional structures, all the data in the world will do little more than to measure the residual force of heteronormative prejudice and conventionality—even as growing numbers of people seek happiness and well-being in hybrid or innovative living arrangements.

●

NOT CREATED EQUAL

In "Gender Specifics: Why Women Aren't Men," the lead essay of the same women's health supplement, Dorion Sagan lays out the biological rationale for heteronormative gender roles—the crux of many arguments about the benefits of marriage, implicitly conveyed in Waite's arguments about "the specialization advantage" and "the virtues of nagging."[8] The author's first rhetorical move is to out-feminist the feminists: "In contrast to the feminist premise that women can do anything men can do, science is demonstrating that women can do some things better, that they have many biological and cognitive advantages over men." After thus presumably establishing good will, Sagan continues: "Then again, there are some things that women don't do as well."

Sagan hinges the argument for innately different cognitive specializations on the corpus callosum: "One of the less visible, but theoretically very important differences, is the larger size of the connector in women between the two hemispheres of the brain." The author then goes on to free-associate how both alleged and established gender differences *might* relate to this poorly understood bundle of nerves. "Although there is no hard evidence, the larger connector may also account for a woman's tendency to exhibit greater intuition (the separate brain halves are more integrated) and a man's generally stronger right-handed throwing skills (controlled by the left hemisphere without distractions)."[9]

"No hard evidence," indeed. Sagan's essay is typical of its genre: Without any evidence at all, it attributes poorly measured and not altogether demonstrable gender traits such as "women's intuition" to an organic seat. As with other alleged brain-structure differences, it is by no means clear whether sexually dimorphic variations in the corpus callosum might cause gender differences or be caused by them. Worse yet, it is not even clear that the corpus callosum *is* a sexually dimorphic structure. As Anne Fausto-Sterling painstakingly shows, biological research is all over the map on whether there are measurable differences between the tissues sampled from the corpus callosums of men and women: Some studies say yes, others no. Some suggest a difference in shape; others report no such difference. Some studies suggest that differences, if they exist, may show up only at certain ages. Fausto-Sterling concludes her survey with an understatement: "The corpus callosum is a pretty uncooperative medium for locating [gender] differences." It does, however, provide fertile ground for modern biological folklore.[10]

WHY LESBIANS DON'T GO TO THE DOCTOR

The unabashed, even giddy heterocentrism of these medical reports and of the sources they invoke might help contextualize a phenomenon covered near the end of the women's health supplement: lesbian avoidance of health care services.

> It was not until Shelly Weiss was twenty-six and came out as a lesbian that she realized something else about her life: she had been avoiding going to the doctor. Later, when she had a child, with a gay man as the father, some of the reasons became clear. "Do you really think this is fair to a child?" Ms. Weiss remembers a resident asking as her newborn son was placed in her arms.

A growing body of evidence suggests that lesbians face not only "distinct problems" but even "increased health risks" when they seek medical services. "There are a number of women who come in here at age forty-two and have never had a Pap smear," observes Jayne Jordan, a physician assistant at Michael Callen-Audre Lorde Community Health Center in New York City.[11]

THE LESBIAN OYSTERCATCHERS' MÉNAGE À TROIS

Ideas about what's "natural" and "unnatural" count. They affect the actions of social actors, thereby leaving an imprint on the design of the social world.

Naturalized and naturalizing ideologies are perhaps especially significant at times of change and upheaval, during periods of institutional and epistemic flux, when basic social forms are pliable and social contests are still in play: their ready-made designs bend historical agency toward certain conventional forms. I don't doubt that married people are better off than singles, live-in lovers, and gay couples—I just think the doctor's unsolicited advice, banked on the good "evidence" readily available to any reader of the *New York Times* *Science Times*, gives good indication of just how the "marital advantage" works.

Nature, as Oscar Wilde wisecracked, is always behind the times.[12] Still, I do not want to wax essentialist about essentialism, to give a truncated view of the kinds of practices ideas of nature might support. "Nature" is, among other things, a thought experiment, subject to considerable variation. Brody's brief report on the "nature" of desire is a Darwinian nightmare, to be feared and loathed. Waite's picture of "nature" is a cozy domestic space, circa 1955: a narrative of return—"Hi, hon! I'm home!" Sagan's "nature" supports a contemporary variant of cultural feminism: the idea that women and men are naturally disposed toward different temperaments and talents. Other narratives on nature draw other lessons, give templates or blueprints for other kinds of practices. I thus set beside Jane Brody's harrowing speculations and Waite's agitation for marriage law reform the whimsical music of Natalie Angier (inspired, no doubt, by the strains of "Lesbian Seagull," from the movie *Beavis and Butthead Do America*):[13]

> Scientists have found that female oystercatchers, mid-sized wading birds found along the shores of many European countries, engage in a type of cooperative behavior rarely seen in nature, let alone in a bird known for its scabrous temper and a general hostility toward others.
>
> On occasion, two females that have spent weeks or months fighting viciously over a choice piece of territory, and the male settled therein, seem to decide, as though in a bolt of Amazonian revelation, that there might be value in sorority. They stop fighting and start preening each other. They softly pip-pip-pip, the oystercatcher's "Song of Songs."

The hero of naturalist narratives since the nineteenth century has been the enterprising individual: the plucky little man who strives against all odds to better his lot. Natalie Angier has rewritten the familiar utilitarian story. In her accounting, the prototypical individual is a calculating female—improvising institutions, making do, and living sexuality as the pursuit of one happiness among others:

> [The two females] agree to share the territory and the male. They take turns copulating him, and then they seal the compact by copulating

with one another. They nest together. And woe to any predator, or designing oystercatcher, who might venture into the triad's territory. The three birds will defend their home and nest with synchronized wrath.

The discovery of such an unusual and flamboyant form of female cooperation demonstrates how behaviorally supple even a sexually conventional species can be. . . . The results have oblique implications for human mating strategies.[14]

Implications, no doubt, even for coy woman and aggressive man living in a time of change.

Varieties of Human Nature

The View from Anthropology and History

A theory ought to be judged as much by the ignorance it demands as by the knowledge it purports to afford.

MARSHALL SAHLINS, *The Use and Abuse of Biology*

13   Marooned on *Survivor* Island

Well-worn clichés about the natures of men and women leap to the fore of John Tierney's *New York Times* article on the final episode of CBS's summer 2000 hit, *Survivor,* a game show set on a desert island. "Men excel at simple, quick, competitive tasks," Tierney asserts, "which is why they have done so well on shows like 'Jeopardy!' and 'Who Wants to Be a Millionaire.'" Tierney attributes the source of this quickness to that mysterious substance usually linked in the reductivist vulgate to strength, aggression, and virility: "Their testosterone helps them focus."

The other sex, however, has its own set of special skills. "Women are better at seeing the whole complex picture," the author claims, invoking the ubiquitous heteroessentialism. "They notice social nuances and cultivate personal relationships, creating what anthropologists call lateral networks instead of status hierarchies. They're experts at cooperating, but they can also be quite ruthless." Having laid out the familiar set of claims about universal sex temperaments, Tierney then taps not an anthropologist, but a psychologist, Anne Campbell, to develop a litany of female stereotypes under the rubric of "indirect aggression," that down side of womanly nurture. Women gossip and spread rumors about their adversaries. They circulate stigma and practice ostracism. They are natural networkers, but they also "manipulate public opinion."

"These stealthy skills seem perfect for 'Survivor,'" whose contestants must cooperate to survive while voting one cast member off the island each week. And in fact, Tierney continues, "the women on the island have done lots of networking and plotting. But the mastermind has been a man, Richard Hatch, a gay corporate consultant who persuaded a homophobic retired Navy Seal to join him and two women in an alliance that gradually voted all the others off the island." (If I get the drift of Tierney's gender-essentializing patter, the homosexual in this case is an intermediary type who combines the worst elements of both sexes: male competitiveness with female underhandedness.) James McBride Dabbs, a social psychologist interviewed by Tierney on the significance of the supposedly odd alliance, matter-of-factly files Hatch's shrewdness under the heading of testosterone: "It's not the kind of coalition that women typically form," the psychologist proclaims after viewing an episode of the show. "It's more the ruthless utilitarian coalition men form to do a job, which might be catching an animal or taking over a company or winning a war or developing a new product." Having merrily skipped from the ancestral environment to the primal battlefield to the modern boardroom, Dabbs concludes with the familiar heteronormative send-up: The sexes belong together because, eager-aggressive to coy-choosy like yin to yang, they complement and balance each other. "'Women are more interested in complex, lasting relationships based on affection than getting a one-time task done,' Dr. Dabbs said. 'That's essential to the broader task of human living, even if it doesn't win a million dollars one night on television.'"[1]

Tierney's science fancy, along with a flurry of other such newspaper columns posted in the final week of the series, makes it clear how such events play in the national sexual and political Rorschach. Had Rudy (the retired Navy Seal) won the million-dollar prize, we'd all be reading about the sociobiological advantage conferred by manly military virtues (honor, discipline) in the struggle for life. If Kelly (the river guide who remembered the middle names of all the other contestants' children) had triumphed, evolutionary psychologists would be filling the pages of the *Times* with a fresh round of claims about the evolution of women's networking and social skills. A victory by Susan (the truck driver so butch her own husband calls her a "man trapped in a woman's body") would at least have provided the opportunity for some earnest crowing about the power of strong women or the benefits of androgyny. But alas, it was Richard, the fat middle-aged gay man—the scheming queen America loves to hate—who played the game the way the rules were actually written and who thus came out on top of this silly, neo-Darwinian farce. (In a moment of counter-Malthusian lucidity, Richard once looked the camera in the lens and said of another contestant: "She has got to go because she is bright and is strong, and she is a threat.")

Gay men, of course, really aren't supposed to win at anything other than the occasional ice-skating event—least of all a cut-throat sport vaguely construed around the principle of "survival of the fittest." And so in the end, the same personnel who regularly attribute biological or evolutionary significance to the most mundane of cultural happenings were at a momentary loss for words. The final postscript to the series thus reads: "It's Just a Game, Really."[2]

NATURAL SCIENCE AND SOCIAL THEORY

As with middle-brow reportage on pop culture, so, too, with much else besides. A study published in the *Archives of General Psychiatry* asserts that some forms of depression might have their roots in evolutionary history. "If I had to put my position in a nutshell," the author told the *New York Times*, "I'd say that mood exists to regulate investment strategies, so that we spend more time on things that work, and less time on things that don't."[3] The attribution of an evolutionary advantage to depression tells us nothing of great consequence about either evolution or depression—in fact, such ad hoc theorizing runs directly counter to what is actually known about depression[4]—but it does demonstrate the ubiquitous appeal of evolutionary storytelling among scientists and in the serious public sphere today.

Had John Tierney consulted an anthropologist or a social historian, rather than a psychologist, he would have heard stories of a substantially different nature—and stories in which "nature" itself is substantially different. Had he consulted those who study human behavior from the anthropological point of view, in particular, he would instead have heard stories about the varieties of human nature. I want to tell some of those stories here. In them, the human being—the *anthropos,* from which anthropology takes its name—is not reducible to a single set of evolutionarily determined natural characteristics, but is always already both the creature and creator of a wide and ever-changing variety of social practices and cultural beliefs that inflect and diversify whatever we all have in common as natural organisms, especially with regard to such matters as gender roles, sexual relations, and bodily identities. This section thus begins with an overview of basic arguments from the social sciences—with a précis of arguments about how and why to view "human nature" differently. Then, drawing on well-established studies and arguments, I want to show in some detail that heteronormative fables about universal human norms run counter to basic ethnographic and historical facts.

In drawing on the distinctions, meanings, and suppositions that are socially available at any given time, scientific ideas ever reflect the preoccupations of their age. Tales of "once upon a time" invariably turn out to depict happenings from the time of the now. The evolutionary psychiatrist's handy invocation of "regulation," "investment strategies," "time," and "work"—at a moment when the entire middle class seems to be on Prozac—is an extreme but by no means exceptional example of how cultural context shapes scientific theory. Even so, not every claim staked in the name of science has the right to be called "scientific." The scientific method is *supposed* to distinguish, however imperfectly, between knowledge and belief, between percept and precept, between fact and fancy. That is, as the very condition of its being scientific, science is *supposed* to test hypotheses experimentally and to revise assumptions in the light of evidence. Stories about human nature—about the evolution of the sweet tooth, the uses of depression, and how men and women got to be the way they (supposedly) are—do no such thing. They appeal not to evidence but to the perpetual world of common sense, relying on good order, good form, and "what goes without saying" to sustain their arguments. Such tales employ the rhetoric and the trappings of science, but the structure of the propositions they maintain more closely resembles that of ideology or superstition.

WHAT'S WRONG WITH A GENE'S-EYE VIEW

Methods, models, and procedures ought to be commensurate with objects in any scientific (or for that matter logical) accounting. Stories told ought to be consonant with the subject matter treated. Stephen Jay Gould's deceptively simple distinction between biological "potential" and "determination" is no mere quibble. It pinpoints the logical error underlying sociobiology, evolutionary psychology, and other theories that attempt to reduce complex, meaningful human activities to basic biological or genetic causes.[5] This is a reduction no real scientist would accept if it were carried out at another level of explanation—if, for example, the laws of atomic physics were invoked to explain the biology of gestation or metabolism.

The reader is invited to consider the logical and procedural disjuncture between the study of particle physics and that of organic biology. Everyone, even the most ardent of social constructionists, would concede that the laws of physics (as they are understood at any given moment in the history of science) represent valid attempts to describe and explain the behavior of physical matter in the universe. Moreover, everyone staging rational or scientific claims would also assent to the proposition that life forms are composed of physical matter. Notwithstanding this agreement on first principles, no reasonable person would try to explain animal gestation or human metabolism in terms of the way atomic particles behave, because phenomena at one level of organization exhibit characteristics not associated with phenomena at another level.[6] Living organisms behave very differently than lifeless objects. The laws of physics alone—an "atom's-eye view" of what goes on in living things—cannot adequately explain how *organic* systems work. A different set of rules and procedures not reducible to physics thus makes up the legitimate science of biology—a science that in no sense invalidates the procedures of physics, but that cannot be derived from them, either.[7]

The whole of the social sciences, as distinct from the life sciences, is built atop an analogous distinction between what Emile Durkheim called the "organic" (biological) and the "superorganic" (social) facts of existence. A "social fact," for Durkheim, is any way of perceiving, thinking, or acting that originates outside the individual and is endowed with some power to shape how the individual acts. That is, social facts exist at an intersubjective level that is superindividual but also more than organic: they are not so much instantiated *in* each individual body as somehow held in common *between* individuals. Social facts also come *before* the individual person, in the sense that every individual is born into a social context already structured by moral rules, collective beliefs, mutual obligations, and cultural meanings in general. Religious beliefs, for example, inasmuch as they rely on a collective body or congregation of believers, are plainly social facts, despite the altered somatic states (trance, ecstasy) they sometimes induce in individuals. Education and "the manner in which children are brought up"—feeding schedule, toilet training, discipline, the rules of hygiene—constitute an important assemblage of social facts because they have to do with the manner in which individual beings are transformed into social persons. And language, being the social fact par excellence, is acquired neither spontaneously (by the individual) nor "naturally" (as a simple part of the growth process) but *socially.* That is, the rules and meanings of language are learned and internalized through a process of social interaction.[8]

This "superorganicism" of language was perhaps best expressed by the linguist Ferdinand de Saussure, who used the terms "arbitrary" and "unmotivated" to describe the linguistic sign. What Saussure meant is basically this: Nothing in the realm of necessity—nothing in the world of things themselves—gives the signifier "tree" its signified, the mental concept of a tree, as marked off from what we think of as bushes, vines, and other sorts of plants. A different signifier, the acoustic image *"árbol,"* will do just as well. Or the semantic domain, the collection of signifieds, might be otherwise organized, amalgamating concepts that in English are distinguished as "trees" and "bushes" under a common term. Significance, then, is not "fixed" in the term—"tree," "bush," "vegetation"—but is held in contingent, oppositive relations *between* the terms—between "tree" and all the words for what are not thought of as trees. Put another way: The meaning of the sign derives not from the "thing" it designates, or even from the sign alone, but from comparisons and contrasts internal to the semiotic system, which is itself "external" to the organism proper. It is thus "convention"—an implicit agreement among speakers about how to oppose and relate terms—not "necessity," that sustains the structure of language.[9] Or rather, one might say that convention is the prime necessity for language to mean anything at all.

The same structured convention and struggles over it—not a genetic code or natural law—hold everything social in its place. Culture—the uniquely human system of public meanings—is elaborated out of signs, and every complex human behavior (tool use, ritual, social labor, and collective belief, surely, but also gender roles and sexual culture) presupposes cultural meanings, hence linguistic communication. So structured, the organization of any society mirrors or extends that of language in general: Its regularities are "arbitrary," "unmotivated" by the inherent nature of things themselves, and "relational." Its structure consists not of fixed parts, but of bundled relations between (changing) parts. As Durkheim famously argued, social facts owe their existence to other social facts, circulated as parts of a social whole. The "functions" they serve are related to social exigencies, not organic necessities. And according to an old anthropological saw, culture is sui generis: The meanings internal to culture derive from culture, not from nature.

No less a figure than Alan Sokal—ardent defender of positivism and objectivity in the "science wars"—affirms the basic, logical differentiation between the objects and methods of the natural sciences and those of social theory:

> Special (and difficult) methodological issues arise in the social sciences from the fact that the objects of inquiry are human beings (including their subjective states of mind); that these objects have intentions (including in some cases the concealment of evidence or the placement of deliberately self-serving evidence); that the evidence is expressed (usually) in human language whose meanings may be ambiguous; that the meaning of conceptual categories (e.g., childhood, masculinity, femininity, family, economics, etc.) changes over time; that the goal of historical inquiry is not just facts but interpretation, etc. . . . To say that "physical reality is a social and linguistic construct" is just plain silly, but to say that "social reality is a social and linguistic construct" is virtually a tautology.[10]

HUMAN NATURE IN MOTION

Marx grasped the essential point, somewhat in advance of (and, I think, more profoundly than) other social theorists, when he penned the constructionist paradigm in a nutshell: "The human essence is no abstraction inherent in each single individual. In its reality it is the ensemble of social relations."[11] What Marx meant by this is that "the nature of individuals . . . depends on the material conditions determining their production."[12] But Marx did not simply replace the essentialist concepts of his day with a straightforward or static form of social determinism. As opposed to thinkers like Saussure and Durkheim, Marx would not even define "society" or "culture" as that which imposes rules upon the individual in a wholly external manner: "Above all," he warns, "we must avoid postulating 'Society' again as an abstraction vis-à-vis the individual." "The individual," he adds paradoxically, "*is the social being.*"[13] Dialectically tacking between what is "inside" and what is "outside," Marx attempted to capture the dynamism and irreducibility of human existence in certain double-edged propositions. "Nature," he writes, in and of itself "is nothing for man," yet it provides the "foundation" for human existence in the sense that it exists for human beings as a "bond" with other human beings (that is, as a *social* and *socialized* nature). Or in yet another maieutic turn of phrase, human beings are said to take on culture *as* nature: "The nature which develops in human history . . . is man's *real* nature." So powerfully are human dispositions linked to social practices that for Marx, even basic anatomy is caught up in the making of history: "The *forming* of the five senses," he writes, "is a labor of the entire history of the world down to the present."[14] In another passage, he evokes the full dialectical flow of acts and facts, sense and labor: "This activity, this unceasing sensuous labor and creation, this production [is] . . . the basis of the whole sensuous world as it now exists."[15]

In such pyrotechnic prose, Marx poses claims about the relationship between culture and "human nature" that are both stronger and more dynamic than those forwarded by the classical traditions of social theory. First, Marx asserts that human beings are social creatures in their innermost nature. (Merleau-Ponty summarizes the point this way: The human being "is not in society the way an object is in a box; rather, he assumes it by what is innermost in him.")[16] Second, Marx refuses the temptation to imagine this social nature generically, as a set of stable givens, timeless norms, or performative scripts. (It was on precisely these grounds that he critiqued the radical humanist Ludwig Feuerbach, who had invoked generic "Man" instead of "real historical men [and women]" in his arguments—"Man" really being German all along.[17]) Third, Marx understands that social relations are "arbitrary"—or better yet "contingent"—and "conventional," but he moreover sees their *volatility*. Whereas more conservative theories understand conflict as a secondary or accidental aspect of the social order, Marx expressly argues that tension and struggle are implicit in certain kinds of social relations. Finally, Marx is relentlessly opposed to philosophies that begin from the idea outward, but he is also waging war against mechanistic materialism (the eighteenth- and nineteenth-century ancestor of scientistic reductions today), which begins with "the thing" and treats human beings as things among other things, subject to the same natural laws as other material things. The "matter" treated in Marx's materialism is never considered separately from the consciousness that, through human action, commingles with it and transforms it. That is, "the thing" is always "sensuous human activity, practice."[18]

"The essence of [hu]man[ity]," Marx concludes, "is freedom." Marx thus outlined a *praxical* approach to culture and human nature—what Donald Donham has called a theory of "humankind in the active voice."[19] According to this conception, what is most "essential" about human beings is neither "within" them (in what people *are*) nor "outside" them (in what social structures they inhabit), but in what they *do:* it is, as Terry Eagleton puts it, something "in our material nature which allows us to exceed and remake it."[20] That is why Marx's philosophical investigations always begin with the problem of the "human object," for the relationship between hand, eye, and object reveals that the human glance no less than social labor is creative, dynamic, transformative.[21] That is also why so much in Marx's oeuvre revolves around history and historicity. It is only in the course of historical happenings that one can catch a glimpse of the dialectic in motion—to see how people, in producing products, also make their social world, hence themselves. And that is also why Marx would brook no traffic with origins stories: "for the socialist," he claims, every day is the moment of origins, because "the *entire so-called history of the world* is nothing but the creation of [hu]man[ity] through human labor."[22] From a Marxist perspective, then, it follows that concepts and methods from the natural sciences serve as obstacles to social knowledge if they sustain the illusion that an arbitrary convention is somehow more "natural," more necessary, than its alternatives, if they give the impression that the status quo is based on immutable natural laws, if they freeze the flow of historical happenings into a fixed, fast-frozen picture.

14   Selective Affinities

*Commonalities and Differences
in the Family of Man*

The idea of commonality is crucial in modern reductivist narratives: It outlines a fixed, fast-frozen picture of what it means to be human.[1] Sociobiologists and evolutionary psychologists invariably claim that certain human
behaviors are consistent across cultures and throughout history. Men, it is
said, "get the job done." Women, on the other hand, supposedly think and
act within webs of social relationships. Gender roles, sexual mores, and family relations in all societies are said to be so consistent with these cartoon
cutouts of human nature that they *must* have a biological, even genetic, basis. Such claims allow modern reductionists to call upon the authority
(though not the accountings) of anthropology, even while leaping over the
Durkheimian interdiction that inaugurated modern social science by sepa-
rating "social facts" from "organic [biological] facts."[2] But as the previous
chapters have begun to suggest, behaviors deemed "similar" may not really be "the same." It all depends on how one looks at the question.[3] It all
depends on what one counts as a resemblance. It all depends on how much
information one takes in or refuses.

DIVISIONS OF LABOR AND SPACE

Arguably, a sexual division of labor is "universal"—or at least very broadly
distributed—in the sense that most cultures seem to associate some tasks
and tools with men and others with women. But this does not mean that all
such divisions of labor are "the same." In some societies, hard labor and
heavy lifting is men's lot. In others—Western arguments about the implications of men's "upper-body strength" notwithstanding—it is decidedly
women's work. In one culture, men raise vegetables, while women raise pigs.

In another, it is women who perform most agricultural tasks, while men tend the livestock.[4] Handicraft production is men's labor in some cultures, women's work in other cultures—and can be done by either men or women in yet other cultures. "Knitting, weaving, and cooking sometimes fall into the male provinence, while such things as pearl diving, canoe handling, and housebuilding turn out to be women's work in some settings."[5] No common thread unifies these various divisions (and combinations) of labor.

Similarly, the case could be made that the floor plans of culture tend to be mapped according to a gendered understanding of space—that men and women, masculine and feminine, are usually associated with different places. Not everyone making this case has harbored conservative ideas or reactionary designs. For example, a variant of the second-wave feminist theory that flourished in the 1960s and 1970s took for granted the notion that women's worlds are tied to home and hearth, while men's activity embraces a wider world of social relations.[6] In the first draft of an anthropologically oriented feminism indebted to Simone de Beauvoir, this social arrangement was said to follow universally from basic reproductive roles: Women's lives were thought to be curtailed by pregnancy, lactation, and child care, while men's lives were said to be free from such constraints. To a generation reared in middle-class U.S. suburbs during the 1950s or 1960s, it might well have seemed that women's activity was logically and universally associated with a devalued domestic sphere, while men were ontologically free to act in an esteemed public sphere. But these associations already seem far less salient for the generations that have taken the social stage since second-wave feminism. At any rate, what one finds on the ground in various cultures fails to support ready generalizations about the universality of male and female spheres, much less the public/private dichotomy onto which these ideas were mapped. The senior Iroquois women described in early explorer and colonial accounts ran the household, but they also elected elders, decided matters of war and peace, and controlled agricultural production and food distribution.[7] In any number of indigenous highland cultures in Latin America, it is women who control far-flung commerce in the marketplace, while men monopolize local ritual authority. Anthropologists have tracked the perambulations of women traders on several continents.[8]

In evolutionary psychology, the selective invocation of gendered divisions of space and labor provides the material for a straightforward evolutionary parable. It is asserted that the biological legacies of hunting and gathering give women a greater affinity with domestic affairs (hearth-close gathering of fruits, nuts, and vegetables) and men a more direct association with public affairs (big-game hunting in socially organized groups across long distances).[9] But contrary to speculative scenarios about what hunters and gatherers might have once done, modern Ju/'hoansi women forage over substantial territories, either with infants in tow or with small children left in the care of other adults back at the campsite. The Ju/'hoansi women studied by anthropologists in the twentieth century, moreover, space their births so as to avoid being hobbled by too many small children at any one time. Far from being specialists in nurture, Ju/'hoansi women make a hefty contribution to basic subsistence. Their labor produces far more calories (but less protein) than men's. Ju/'hoansi women also actively participate in most aspects of "public" life, including political decision making, and there is no real evidence that their gathering activities involve less shared knowledge, collective reflection, or social organization than men's hunting does.

The Ju/'hoansi described in the ethnographic record do practice a basic "division of labor," but that division is not as pronounced as is sometimes claimed. Women "gather" vegetables, roots, fruit, and nuts, but they also catch small game (e.g., rodents). Men engage in many of the same subsistence activities as women, in addition to which they also hunt large game. And just what happens to the man's big-game catch—the supposed payoff in bio-reductive stories about the evolution of heterosexual flirtation, courtship, pair bonding, and the nuclear family?[10] The protein-rich meat does *not* go to an individual man's wife and children, as the familiar fables suggest. It is circulated through an elaborate distribution network to be shared by the entire band. Since the products of men's labors are effectively "socialized," it scarcely matters whether a woman's husband is a good hunter or an inept one. A woman and her children are not dependent on a male "breadwinner."[11]

Even this division is by no means universal for foraging societies. Among the Australian Tiwi, both men and women engage in hunting and gathering.[12] And not only do Agta women of northeastern Luzon in the Philippines hunt game and fish, they do so until late into pregnancy, and they resume hunting and fishing shortly after giving birth. Like other foraging women, they practice contraception and abortion in order to space their births, and they receive assistance from networks of other adults to help with child care.[13]

As such examples show, the notion that the biological legacies of male hunting and female gathering imply strictly gendered social spheres and restricted female access to public space simply does not square with ethnographic observations, nor does it take into account basic human ingenuity. Much less is it clear that early human groups consistently practiced the division of labor that has been attributed to them—male hunting, female gathering.

Ethnographies of foraging and other small-scale societies also show that the idea of a "domestic sphere" marked off from a "public sphere" is itself a most problematic notion. It might seem self-evident that people distinguish between their own intimate dwelling and the rest of the world, but it is one thing to locate residency at the center of public life—as it is in cultures where daily life happens around temporary campsites or in joint-compound courtyards—and quite another to configure "home" as a private and personal respite from the ravages of a troubled world. Likewise, it might seem tempting to imagine that all cultures share a universal sense of how gender, family, kinship, and place go together, but it is a very different thing to situate production, ritual, and polity *within* domestic kinship units, as most lineage systems do, than it is to locate production outside the home and consumption inside the home, as modern Western societies do. In short, the string of associations sometimes asserted as having universal resonance— male/female : public/private : social/domestic : competition/nurture : high prestige/low prestige—turns out instead to be a culturally specific and historically contingent arrangement. Such a gestalt ultimately depends upon modern relations of production and consumption that distinguish between "home" and "work." Not just the associations, but even the supposedly fixed terms they connect prove highly variable.[14]

Between the topology of the body and the spaces of culture lie a virtually infinite number of possible realizations. The facts of biology, in fact, predict nothing of consequence about the designation of male and female spheres in human societies—but the meanings circulated in and around gendered activities have much to do with what is inventively construed of bodies, their limitations, and their implications.

"THE FAMILY"

It is often asserted that the heterosexual nuclear family is the basis for every human society. But this cozy domestic universalism simply is not supported by the evidence from cultural anthropology and social history. From the ethnographic and historical record, it cannot be established that such a "family"—consisting of a male husband, a female wife, and their dependent children, all living in one settled place—is the normal, natural, or universal ground for human social forms.

Consider a basic element in the nexus of familiar associations: the question of residency, the idea of a "home." People undoubtedly have to live somewhere—but that "somewhere" is not always a single, settled place populated by a stable cast of characters resembling modal North American families. Perhaps the Mundurucú would find it exotic that in America, husbands and wives are expected to occupy the master bedroom of a single-unit dwelling together, for in this Amazon rainforest culture, husbands and wives sleep apart in sex-segregated longhouses, a common form of residence in lineage-based societies.[15] Indeed, it is by no means a given that children live with their parents. Some societies practice forms of residency in which children live in places apart from one or both parents. The Ndembu of Central Africa might thus find something unseemly or irresponsible about the desperate American insistence that sons ought to live in the same place as their fathers, since, under the rules of matrilineal inheritance, a boy is heir not to his father but to his mother's brother.[16]

Even in modern Western societies, there's more than one way to constitute households, and basic questions of residency and kinship are further complicated by careful accountings of what people actually do, where they actually live, and how they make families. People who count themselves as "family" do not always eat, sleep, and socialize in the same place, as Carol Stack discovered in her classic ethnography of black families in an American inner city.[17] Examining the credibility of claims about the distribution of "stem," "nuclear," and "extended" families, Jack Goody touches upon problems inherent in the casual use of such terms as "household" and "family" when he describes the following scene from a not so exotic society:

> As I sit here in south-west France, an aged farmer walks up the hill
> to have his midday meal at the house of his already retired son and
> wife—who has officially taken over the farm so that her husband can
> draw a pension. The dwelling groups are separate; the consumption
> groups are largely combined; kin relations are close. Whether they are
> living apart . . . or together is of limited significance.[18]

Like residency, "marriage" would seem to be a crucial term in any definition of "family." People often make global pronouncements about "marriage," as though it were a self-evident institution. But in fact, it is no easier to specify any universal characteristics for marriage than it is to discern a universal form of residency. Marriage may involve a formal union marked by a public announcement or a ritual (a "wedding ceremony"). Or it may have the informal character of a union gradually acquired or consolidated over a period of time. What North Americans and Europeans call "common-law marriage" is actually the prevailing form of union in many parts of Latin America.[19] Depending on its cultural context, marital unions can involve a host of different persons in a number of possible combinations: a man and a woman, a man and several women, a woman and several men. . . . It might or it might not carry the expectation that one or all parties involved refrain from sexual relations outside of marriage. It might or it might not carry presuppositions of intimacy, permanency, and/or companionship.

Evolutionary tales about sexual selection, the biology of beauty, and marital pair bonding turn on familiar (modern, Western) motifs about how heterosexual courtship and amorous love are supposed to play out in the life cycle: Eager boy meets coy girl. Boy proves himself worthy. Boy and girl fall in love, get married, have children. They all live happily ever after. . . . No doubt such a plotline seems very natural (not to say ideal) to many men and women in modern American society. But the problem with this timeless tale is that in a great many societies up until the present era, marriages were arranged not at the whim of lovers, but in the interests of extended networks of kin.

John Boswell thus gives a short précis of two very different plotlines of marriage in the Western tradition:

> In pre-modern Europe marriage usually began as a property arrangement, was in its middle mostly about raising children, and ended about love. Few couples in fact married "for love," but many grew to love each other in time as they jointly managed their household, reared their offspring, and shared life's experiences. . . . By contrast, in most of the modern West, marriage *begins* about love, in its middle is still mostly about raising children, and ends—often—about property [i.e., in a divorce settlement], by which point love is either absent or a distant memory.

Boswell adds: "It is nearly impossible to formulate in a precise and generally acceptable way what is meant by 'marriage,' either by modern speakers or in ancient texts."[20]

The array of anthropological terms for descent, affiliation, and marriage testifies to the wide variety of kinship forms acknowledged to exist outside the West. In "patrilineal" societies, people trace descent through male lines, in "matrilineal" societies through female lines; "avunculocal" residency refers to housing with one's mother's brother; in "polyandrous" systems, a woman takes more than one husband, whereas in "polygynous" societies, a man takes more than one wife; "group marriage" (a rare but interesting form discussed later in this chapter) refers to marital unions conjoining multiple husbands to multiple wives. . . . [21] These diverse arrangements can scarcely be boiled down to generic motifs about heterosexual courtship, amorous love, and monogamous pair bonding. Above all, the maps of kin relations, residency patterns, and marital forms traced in anthropological studies show that what Western speakers mean when they indicate "family"—

a specific combination of gender, residence, affiliation, affect, and genealogy—turns out to be a unique constellation in the varied firmament of kinship, not a universal form.

(Lawrence Stone puts matters into succinct historical perspective when he connects "the rise of the family" in early modern Europe to the "decline of kinship"—a shift that inaugurates many of our current associations.[22] Ellen Ross and Rayna Rapp go on to show how modern conceptions of sex, family, and personal life in cultures of the North Atlantic are contingent on a series of separations associated with the rise of industrial capitalism: the separation of family from kinship, "of family life from work, of consumption from production, of leisure from labor, of personal life from political life."[23])

In fact, even the standard anthropological typologies understate the degree of cultural variation in marital and kinship practices. The ethnographic record recounts that Australian aborigines elaborate their complex networks of kin by a variety of means, some involving homoerotic play; that the Ju/'hoansi invent webs of kin relations out of nothing more substantial than the accident of shared names; that so-called fictive kin relations take on paramount importance in much of the modern world; that in some cultures, a man's erotically charged relationship with his brother-in-law eclipses his relationship with his wife; that female same-sex marriages (and the occasional "ghost marriage" of a woman to a dead man) are institutionalized among the Nuer of Sudan and in a wide belt of African societies; that in the past, a large number of Native American societies allowed anatomical males to marry other anatomical males, provided one of them underwent ritual transformations to become a gender-mixed or intermediary "man-woman"; and that bonds of same-sex friendship, publicly announced and ritually marked, amount to something very much like "marriage" in a great number of cultures, as do some forms of same-sex group affiliation.

In short, people forge kin connections with each other according to all manner of inventive rules and devious principles.[24] Recognition of this many-splendored variety would go a long way toward correcting narrow and emaciated depictions of marriage, family, and kinship, both in biological just-so stories about the centrality of heteronormative nuclear families and in political discussions of gay marriages and civil unions.

A scrupulous author prefaces his cross-cultural survey by noting that whether one sees the family as "universal" will depend on how one defines "family." He concludes that no credible definition can really be inferred from all the various ways people form unions, count kinship, and take up dwellings.[25] And as Jane Collier, Michelle Z. Rosaldo, and Sylvia Yanagisako note in their survey "Is There a Family?" many languages do not even have a word for "family" as distinct from household, extended kin networks, or lineages.[26] An odd thing it would be to leave so supposedly "crucial" an institution linguistically unmarked.

It is sometimes asserted that, notwithstanding the great variety of kinship forms and linguistic designations, "family" is still universally present, albeit in disguised or unnamed form. Universalists are ultimately compelled to issue a very qualified claim: At least one male father and at least one female mother—however these persons are spatially distributed and whatever their actual rights and responsibilities—might in some sense be said to exist everywhere progeny exist. But to conclude from this that "family" is a social universal is to confuse the biological facts of sexual reproduction with claims about social institutions—a confusion of the first degree. The question is not whether human beings sexually reproduce according to biological principles (a trivial truth), but whether those biological principles predict anything of great significance about the shape of social institutions.

A more credible rendering of basic lessons from the comparative study of social practices goes something like this: People everywhere tally relations of affinity, consanguinity, and residency. That is to say, they engage in practices that anthropologists gloss as "kinship" in some form or another. They form unions—they get "married"—in some way or another, they count blood relations in some fashion, and they take up dwelling with certain other people according to some set of rules or procedures. But what qualifies as a marriage, what counts as a blood relation, and what rights, duties, and living arrangements follow from these accountings—in short, the actual manners in which human beings make their lives—do not lend themselves to much in the way of meaningful generalization. And what meaningful generalizations might be credibly proffered are of a social, not biological, order.

On these as on other questions of institutional import, the social facts come before the biological facts. Or perhaps it should be said that human beings, because they are social creatures and by dint of their sociality, take up questions related to biology in ways that cannot be reduced to biology.

IS THE MAN MANLY? IS THE WOMAN LADYLIKE?

Gender temperament and psychological needs are notoriously difficult to gauge, even within a single society. I am not always convinced that straight

American men are really motivated by the feelings and needs that are attributed to them in both popular and scientific culture. But given the variety of institutional and living arrangements the world over, given the wide variety of cultural beliefs about men and women, it seems hilariously ethnocentric to imagine that everybody in all cultures is guided by the same emotional compass.

Is there, for example, a culturally universal system of masculinity: the Darwinian man who acts, is competitive, and pursues women, the man whose thoughts and deeds go straight to the object? Invoking social rather than biological facts, David Gilmore has come close to saying so: he argues that an ideal of manly action, configured in terms of efficacy or agency, is very widely distributed across cultures and throughout history. (To his credit, Gilmore notes that there are prominent exceptions to this pattern, but he also argues that it represents a widespread, even "normative," tendency.) Gilmore links his assertions to the social facts of fatherhood, to the notion that men provide for and protect their families.[27] But Susan Faludi has detected a notable shift in just how this "efficacy" might be imagined and experienced, even in American culture, even over a relatively short term: from post–World War II ideals of teamwork and loyalty to a more contemporary emphasis on competitive individualism and an "ornamental" masculinity that vaguely resembles the manhood Margaret Mead describes as prevailing among the Tchambuli. (Tchambuli women are understood as practical, level-headed creatures. Their men, when not head hunting, engage in aesthetic pursuits, trade catty remarks, and sit around looking pretty—quite contrary to Darwin's snapshot of male ferociousness and female vanity.)[28] Of course, even a moment's reflection ought to suggest that manhood is constructed and lived differently by the bourgeois gentleman than by the factory worker, by the pure Brahmin than by the polluted outcaste, by the Ladino townsman than by the Indian peasant. . . . One wonders just how salient the idea of a universal conception of manhood could really be, once it is acknowledged that manhood is subject to varied realizations—indeed, that efficacious agency has been altogether *withheld* from an unprivileged majority of men in a majority of stratified societies for much of human history.

And is there such a thing as a culturally universal system of femininity—the universal woman whose thoughts and deeds are ever immersed in social webs? Those claiming so have often produced the facts of motherhood as their trump card. It is woman's role as mother that would seem to fix her in a caring, nurturant role. But this is simply to transpose the question: Is there such a thing as a biologically given "maternal instinct"—or even just a universally valid form of relationship between mother and child, some core connection indifferent to the effects of history or the variability of culture? Many have thought so.[29] It is not an altogether implausible idea. But the case cannot be made without construing everything around a suspiciously ad hoc form of argument: Whether feminist or patriarchal, proponents of the notion of a maternal instinct strain to count off the most varied motherly acts as instances of a universal natural tendency. Thus, it is said that even when mothers kill their infants, they express a maternal instinct, a "special bond." A quick survey of world cultures might suggest instead that the supposedly singular notion of motherhood is actually subject to the most varied imaginings—indeed, that what counts as "good mothering" (or even, for that matter, as simple "mothering") is not altogether interchangeable from one culture to the next, or from one historical moment to the next.

What one makes of motherhood will depend in part on what one makes of childhood. Much of the basic repertoire of mid-twentieth-century anthropology explores the different notions cultures have had about childhood and the different practices these ideas have inspired.[30] In some cultures, mothers swaddle their infants, motivated by the belief that constriction is good for the body. In other cultures, mothers allow their infants' arms and legs free movement on the premise that exercise is good for the development of a healthy child. Toddlers are kept close to adults in some cultures. In others, they are allowed to venture farther away before being retrieved. As Gregory Bateson and Margaret Mead succinctly demonstrated in 1952, bathing rituals vary a great deal from society to society, reflecting very different suppositions about health and supporting very different ideas about proper forms of maternal nurture.[31] Nancy Scheper-Hughes has remarked upon the folk beliefs of Irish mothers, who understand a "good," "well" child as one who solicits little attention.[32] Others might find this belief, and the maternal distance it implies, to constitute prima facie evidence of bad mothering.

Even within a single culture, maternal practices and beliefs about the proper treatment of children can vary dramatically. I know mothers who hold their babies every time they cry on the premise that too much weeping might harm the child's health and others who refuse to do so on the premise that such indulgence would "spoil" the child. I also know mothers who spank in order to instill a sense of discipline and others who do not for the exact same reason. Each justifies her practices—even opposite practices—in terms of "good motherhood."[33] Spanking itself, the most mundane form of good, maternal discipline a generation ago, is now increasingly viewed as bad mothering, even child abuse. Changing attitudes toward corporal punishment provide one measure of the *historicity* of motherhood, of how rapidly basic notions of good mothering have changed in this century. Another measure: It seemed a flash of insight when Benjamin Spock suggested to progressive parents two generations ago that mothers ought to feed their infants when they were hungry, rather than at predetermined intervals. Modern mothers, it would appear, required some instruction on what is "natural" for them to do.

THE UNIVERSAL MOTHER

Conservatives often have invoked the supposed facts of motherhood and the universal human norms they are said to imply when contesting social changes since the 1960s. Francis Fukuyama cites Lionel Tiger and Robin Fox to this effect when he claims that the mother-child relationship is less variable historically and cross-culturally than the father-child relationship, and when he asserts that, whatever their variable kinship arrangements, societies normally mobilize social resources to protect the mother-infant bond.[34] Such a proposition rightly scales back the usual sociobiological claim that the nuclear family is universal. But it dramatically understates variations in mother-child relationships while woefully overstating the degree to which many patriarchal societies have actually protected or nurtured the mother-infant relationship.

The lactation of the mother is commonly cited as an irreducible natural fact, a fact that is said to carry certain institutional implications: the association of women with domestic work and child rearing. It is probably true that in most human societies up until the age of infant formula, small children who were actually *wanted* were generally kept close to lactating women. But that is not to say that the lactating women were necessarily the children's own mothers. Anthropologists report that breasts are offered indiscriminately to crying children in a number of small-scale societies. Sarah Blaffer Hrdy reflects on such widespread practices in elaborating her concept of collective maternity or "allomothering": "From Efé net-hunters in the Ituri forest to the fisherfolk of the Andaman Islands, allomaternal suckling was a mutually beneficial courtesy extended by coresident women— affines, neighbors, and blood kin."[35] It is not just in "exotic" societies where such practices weigh against any idea of a singular, universal "maternal instinct." By common estimates, in 1780, at the peak of a wet-nursing culture in Europe, only 5 percent of the 21,000 births registered in Paris were nursed by their own mothers.[36] And in the face of scientific-sounding folktales about the magic of mother-infant "bonding" and recurring anecdotes about how mothers automatically "let down" their milk upon hearing their infants' cries, Nancy Scheper-Hughes's extensive research in Northeastern Brazil shows that there is nothing "automatic" about mother-infant bonding. Even a practice as apparently "natural" as breast feeding is culturally learned, not biologically instinctive.[37]

With the basic facts of motherhood, as with other aspects of culture, one finds variation. In some societies, mothers are uniquely responsible for the care of their own infants and small children, in others, child care is shared by networks of women. In fact, in some societies, basic child care is shared by mothers and fathers or is a substantially communal undertaking consisting of women and men. For example, Barry Hewlett reports that among the egalitarian, indulgent Aka Pygmies, fathers provide about the same amount of child care as do mothers in a context where small children are held and nurtured by a large cast of adult men and women.[38]

In sum: Even the strongest link in the argument for a maternal instinct— the "special" connection between infantile dependency and maternal lactation—does not preclude a range of possible child-care arrangements: wet nursing, shared nursing, the use of supplements in between breast feedings, paternal nurturing. . . . And even the most direct imaginable link between the lactating mother and the dependent infant proves surprisingly variable.

John Boswell's history of child abandonment in Europe throws light on the dark side of this variability in infant-parent relations—and on the niggardliness of the social subsidy the supposedly protected pair sometimes receives. Boswell acknowledges the paucity of reliable figures on child abandonment rates during the Middle Ages. But the first century for which reliable records are available yields the following figures:

> In Toulouse in the eighteenth century the rate of *known* abandonment as a percentage of *recorded* births varied from a mean of 10 percent in the first half of the century to a mean of 17 percent in the second half, with the final decades consistently above 20 and sometimes passing 25 percent for the whole population of the city. That is, in the late eighteenth century in Toulouse, one child in every four was *known* to have been abandoned. In poor quarters the rate reached 39.9 percent; even in rich parishes the rate was generally around 15 percent. In Lyons between 1750 and 1789 the number of children abandoned was approximately one third the number of births. During the same period in Paris children *known* to have been abandoned account for between 20 and 30 percent of the registered births. In every city in France in the eighteenth cen-

tury where it is known, the rate of abandonment was 10 percent or better. In Florence it ranged from a low of 14 percent of all baptized babies at the opening of the eighteenth century to a high of 43 percent in the nineteenth. In Milan the opening of the eighteenth century witnessed a rate of 16 percent; by its closing it was 25.[39]

France and Italy, Boswell notes, were relatively prosperous countries. It is difficult to imagine that abandonment rates were lower in poorer countries. For Europe as a whole, "fragmentary evidence suggests very similar urban abandonment rates from 15 to 30 percent of all registered births."[40] The temptation might be to call such a situation "abnormal," but there is no reason to believe that these conditions were either more or less "normal" than those of any other time and place. My own research in Nicaragua, where local culture passionately venerates the cult of sainted motherhood and the sanctity of the mother-infant relationship, suggests that women and dependent children are *especially* disadvantaged under conditions of social duress and economic scarcity. (And when has Nicaragua *not* known such conditions?)[41] An abundance of studies suggest that women in patriarchal societies are at elevated risk for spousal violence when they are pregnant.[42] As such examples indicate, the image of a normative society heroically mobilizing resources to the aid of women and their dependent children is at best a rose-tinted view of how many societies have actually functioned.[43]

IS MAN TO WAR AS WOMAN IS TO PEACE?

In vernacular no less than scientific culture, the manliness of the male, the ladylikeness of the female, turn on notions of male aggression and female nurture. And so as decisive "proof" of temperamental sex differences, advocates of universal propositions about gender often claim that warfare is, necessarily, men's business.[44] It is true that when organized killing is conducted by states, men tend to be the ones who perform actual combat roles. But this is just another way of saying that, up until now, hierarchically organized, militarist states tend to be quite patriarchal. The one practice is predicated on the other, and neither necessarily follows from *biological* considerations.[45] By contrast, so-called primitive warfare—a highly ritualized practice not altogether comparable with the massively lethal forms of warfare waged by territorial states—sometimes has allotted roles in war to women.[46] It is surely a social fact of some weight that the armed social revolutions and prolonged insurrectionary wars of the twentieth century were waged with very substantial percentages of female combatants.[47] And postmodern warfare, beginning with Desert Storm, seems to have accommodated female combatants without disrupting military performance. The idea that men monopolize warfare does not withstand any more careful scrutiny than the notion that "warfare" itself is universal.[48]

Nor does the notion that there exists some natural antinomy between maternal nurture and institutional violence. Planetwide, mothers have sacrificed everything for the good of their children, but they have also played the role Bertolt Brecht scripted in *Mother Courage*, rearing generation after generation of children to be good soldiers in the interests of expansionist, warmongering states. The *Florentine Codex* records that the Aztecs celebrated the virtues of motherhood in explicitly militarist terms, dubbing the mother a brave soldier and treating parturition as an act of war. The mother's role was to "capture" a baby, and women who died in labor died heroic, "manly" deaths.[49]

Katha Pollitt has hilariously lampooned the idea—taken as an article of faith by Darwin, enshrined in sociobiology, and promulgated as gospel by some cultural feminists—that woman's role as mother gives her a natural or universal affinity with nurturant peacemaking. Such a claim, Pollitt suggests, can be made only by keeping a double set of intellectual books: All the things women do that conform to the desired maternal, peaceable picture are classed as "normal" or "feminine," while all the things women do that depart from the desired portrait are classed as deviations from the rule, exceptions to the norm, departures from the essence of true womanhood. Or worse yet, it is sometimes claimed that although women have acted in violent, nonnurturant ways, they were acting thus at the behest of men.

These are not serious claims, but deflections designed to insulate the icon from facts to the contrary. The universal mother, lauded in evolutionary stories and celebrated in woman-centered subcultures, was really the idealized middle-class suburban mother all along.[50]

THE FACTS OF REPRODUCTION

Contrary to standard universalist claims, it is not really possible to deduce specific facts about social institutions (including sex roles) from facts of biology. Indeed, a serious perusal of the ways in which people have lived suggests an even more radical principle: It is no easier to extrapolate meaningful statements about concepts of conception—the social facts of reproduction—from the biological facts of reproduction.

The biological "facts of fatherhood," for example, are sometimes said to be more attenuated than those of motherhood: "Mama's baby, papa's—maybe." But societies organize this ambiguity in dramatically different ways, if they worry about it at all. Patrilineal societies often underwrite paternity through the close regulation of female sexuality: the purdah (female seclusion), the veil, Pharaonic circumcision, marital residence with the groom's family . . . [51] The matrilineal Nayar (a warrior caste in Kerala, India), however, ingeniously insured legitimacy—as they defined it—by practicing an elaborate form of "group marriage": All the girls of a lineage effectively married all the boys of another lineage or set of lineages.[52] Under this arrangement (which runs counter to all sociobiological fables about female fidelity and male jealousy), women could take a series of sex partners, but no matter who slept with whom, the offspring were legitimate, and caste inheritance rights were preserved.

It might seem clear enough that, up until very recently in human history, the conception of a child has depended upon sexual intercourse between a man and a woman. But just what one *makes* of such a "fact" (or its implications) is an open question.

A yellowed page from the field of sexual studies is instructive here; it chronicles the failure of one widely disseminated theory that attempted to extrapolate social rules and cultural meanings from basic biological principles. Sexologists and anthropologists sometimes used to offer a simple, utilitarian explanation to account for the homophobic design of many sexual cultures. In their accounting, taboos on homosexual relations serve to foster heterosexual relations, supposedly spurring high rates of reproduction, presumably to offset high rates of infant mortality.[53] According to the logic of this approach, the ban on same-sex relations serves a straightforward biosocial function: It holds ecological duress, economic scarcity, biological need, and marginal rates of return in a precarious balance. Plausible as such an accounting might seem to those who assume the worldview of heterosexual utilitarianism, nothing in this Rube Goldberg device quite connects. Nothing in the facts of reproduction implies that either a man or a woman has sex *exclusively* with a member (or members) of the other sex—nor, for that matter, does biological exigency require that *everyone* reproduce. Under dire circumstances, a culture might just as easily achieve higher reproductive rates by enjoining everyone to copulate more frequently, or by holding bisexual orgies, or even by prohibiting married couples from having sex except at certain times during a woman's menstrual cycle.

The magical association of homosexuality with sterility, decay, and death is an enduring trope in Western cultures, and it is this symbolic association, uncritically applied, that animates conventional functional and materialist explanations for sexual culture. But not every society construes utility and social good in heterosexual terms. Some cultures construe utility and social good in homosocial, even homosexual, terms.

Opposite the conventional Western conceptions of nature, the Bedamini of Papua New Guinea associate same-sex intercourse with principles of fecundity, regeneration, and growth. "It is believed that homosexual activities promote growth throughout nature, particularly in gardens, while excessive heterosexual activities lead to decay in nature as well as in social groups. The balance of these forces is dependent on human action. . . . The Bedamini do not . . . experience any inconsistency in the cosmic equation of homosexuality with growth and heterosexuality with decay."[54] By means of similar logic, the Etoro believe that "heterosexual relations in a garden will cause the crops to wither and die, while homosexual relations will cause them to flourish and yield bountifully." This culture, which one might call "homonormative," therefore prohibits heterosexual intercourse for some 205 to 260 days a year, during periods when its depleting effects might interfere with gardening or other aspects of the productive economy. And even during the period of time when sex between a man and a woman is not prohibited, such intercourse "should only take place in the forest, never within a garden, in a garden dwelling, in the longhouse, or in the general vicinity of the longhouse. (Even in the forest one cannot be entirely at ease, for the Etoro maintain that death adders are offended by the noxious odor of intercourse and are particularly likely to strike a couple thus engaged.)"[55]

Contrary to the propositions once widely circulated in the name of sexual science, David F. Greenberg's careful review of the available cross-cultural evidence suggests no relationship between pro-natalism and prohibitions against homosexuality. In fact, some of the most pro-natal cultures on the planet accept, even celebrate, same-sex love, while a number of anti-natal cultures prohibit same-sex relations.[56] Taboos on homosexual behavior, where they exist, thus index ideological and political considerations—what people *make* of sex—not biological exigencies. In categorizing desires and disciplining bodies, such prohibitions achieve social, not biological, ends. At any rate, the broad institutionalization of same-sex practices in cultures around the world testifies against narrow, reductivist accountings of sexual culture—practices documented in the famous Ford and Beach study of Human Relations Area Files and later revisited in encyclopedic works like David

F. Greenberg's *The Construction of Homosexuality* and Gilbert Herdt's *Same Sex, Different Cultures.*[57]

For that matter, it is not even clear that every culture construes even the narrowest "facts of reproduction" just so. Malinowski's Trobrianders appear to make no direct, causal connection between sexual intercourse and child conception.[58] But through institutions like the couvade, other cultures imagine a direct paternal involvement not just in pregnancy but also in gestation and parturition. In these societies, the father is subject to some combination of food taboos, ritual precautions, and a period of prepartum or postpartum seclusion.[59] Bruno Bettelheim's interpretive classic, *Symbolic Wounds,* treats the couvade as a manifestation of womb envy—a form of male pregnancy: "The man wishes to find out how it feels to give birth, or he wishes to maintain to himself that he can."[60]

The notion of male pregnancy, like that of male menstruation, occurs in a number of cultures.[61] F. E. Williams recounts a Keraki lime-eating ceremony designed to prevent the young boys who are sodomized by older boys and young men from becoming pregnant. According to Keraki theories of sex and conception, the boys' bellies would eventually swell in pregnancy if this precaution were not taken.[62] In another part of the world, the Mehinaku recount a simple myth about male pregnancy:

> A man [Kiyayala] went fishing with his comrade. . . . Kiyayala had sex with his friend through the rectum. His friend got pregnant. His abdomen grew big. Then the child, a boy, was born. "*Aka, aka,* it hurts," cried Kiyayala's friend. . . . He picked up the little boy and held him to him. The boy grew up. A woman gave him her breast. He grew to be big and handsome.

Thomas Gregor cites this simple narrative as a cautionary tale—a reason the Mehinaku give why men should *not* have sex with each other. But all in all, the outcome seems less than disastrous. With the help of an allomother, the offspring "grew to be big and handsome."[63]

Myths of male pregnancy sometimes take on cosmological significance, especially, perhaps, in patriarchal societies. Creation myths in many cultures recount how a man gave birth to the world, either by way of same-sex intercourse or through male parthenogensis.[64] In a pioneering paper on the subject, Alan Dundes suggested that a wide array of creation myths (including the one found in the Book of Genesis) might be interpreted as tales of male anal birthing.[65] Myth and ritual on many continents celebrate the fecundity of a phallic anus or anal phallus: a magical male organ that gives birth, thunders, and whose flatulence is sometimes linked to impregnating winds.[66] Not all cultural notions of male conception involve either same-sex intercourse or patriarchal world-origins. An ample heterosexual lore in the United States treats male "sympathetic pregnancies" as a function not of sociobiological sexual competition but of male-female companionship and proximity: Because he is so intimately involved with his partner's changes and feelings, the father, like the mother, becomes "pregnant," gains weight, and goes into imitative labor.

It is not just male bodies and effluvia whose actions are subject to varied understandings. What people make of the female role in reproduction is likewise subject to some variation. Folk biology in many societies associates menses with gestation, but whether menstrual blood is associated with pollution and witchcraft or with fertility and positive magic varies according to culture, circumstance, and perspective.[67] And just what happens when conception happens is also subject to varied opinions. Here is a sampling of ideas from one small area of the world:

> For example, the . . . Kaliai of New Britain assert that the fetus is composed exclusively of the father's semen, with the mother's role confined to providing a place of protective growth during gestation. . . . Among the . . . Mae Enga of the New Guinea highlands . . . the fetus is believed to be formed largely by *maternal* blood. . . . [T]he Kwoma of the Sepik . . . hold that male semen and female blood are necessary to conceive a child, but that the anatomical derivatives of these substances are unfixed and highly variable. . . . Kwoma also believe that preeminent conceptual transmission is of prestigious patrilineal bush spirits from a father to his eldest son.[68]

Like the Kaliai of New Britain, Aristotle thought that semen contained "the seed"—the person in miniature—and that the woman's contribution to reproduction was essentially passive: He believed the womb served as a "field" in which the homunculus grew. With such examples in mind, Cynthia Eller aptly warns feminists against matriarchal myths and universalist claims derived from the supposedly self-evident, awesome power of the womb. Far from being self-evident, women's role in reproduction has been minimized, even denied, in a wide swath of cultures.[69]

Concepts of conception in many parts of the world vary considerably from contemporary Western biomedical models of heterosexual reproduction. Not all such beliefs redound against women. Natalie Angier notes the prevalence of the belief of "partible paternity" among foraging and horticultural societies in lowland South America. Ache foragers of eastern Paraguay believe that a child can have more than one father and that "the multiple ejaculates of different men make for better and sturdier children than the discharge

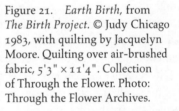

Figure 21. *Earth Birth*, from *The Birth Project.* © Judy Chicago 1983, with quilting by Jacquelyn Moore. Quilting over air-brushed fabric, 5'3" × 11'4". Collection of Through the Flower. Photo: Through the Flower Archives.

of one fellow alone can." Pregnant Ache women thus enjoy adventurous sex lives—and their children are supported not only by their mother's husband but by a cast of dutiful fathers. The Barí of Venezuela and Colombia believe and act in a like manner. Sociobiologists monotonously proclaim that male jealousy is "natural" and "universal," but according to anthropologist Stephen Beckerman, Barí men "don't seem to mind" their wives' promiscuity. "They don't seem to object. They're not demonstrably jealous." Such examples militate not only against sociobiological fables about universal sex temperaments but also against sociobiological lore about what sorts of family patterns or sexual habits might be "adaptive" or convey some survival advantage. Barí children with just one father have a 64 percent chance of surviving past the age of fifteen, whereas those with multiple fathers have an 80 percent survival rate.[70]

COY EGG AND EAGER SPERM?
MORE "FACTS" OF REPRODUCTION

"Look," a somewhat exasperated colleague recently insisted during an informal exchange with me around the topic of evolutionary psychology, "it stands to reason that sex has different implications for men and women, since it is women who get pregnant, whereas men don't. And it's a basic fact," he continued, playing the part of the hard-headed realist, "that women are more likely to die during childbirth than men. And so," he concluded, as though my ethereal queer relativism had got on his last good nerve, "isn't what the evolutionary psychologists argue at least *plausible*— the idea that natural selection has left men 'eager' to have sex and women 'choosy' about it?" My interlocutor then invoked his care for his young daughter, and his abhorrence of the idea that she might someday experience an unwed teen pregnancy, to urge me to reconsider my categorical dismissal of bioreductive sexual lore.

My interlocutor's argument not only turns the anthropological clock back a hundred years but also, I suspect, hails me as an uncomprehending creature outside the charmed circle of heteronormative reproductivity. I'm struck less by the vulgarity and melodrama of his claims than by the abstract and disembodied approach to sex conveyed in his reasoning, with its appeal to the self-evidence of basic facts, its elision of all the varied things people have thought about sex and reproduction. Far from providing the occasion for self-evident truths or stable understandings, the question of reproduction actually seems ripe for mythical thinking and magical investments. (In a place no more exotic than my native North Carolina, for example, I have met many people—some of whom work in hospitals, are familiar with medical models of reproduction, and understand how ultrasound works—who believe that a child's sex is not fixed until moments before birth, and who thus urge expecting mothers to eat certain foods and to avoid engaging in certain activities so as to influence the sex of the child.) But even if one takes it as a given that the prevailing biomedical understanding of reproduction is the correct one—and even if one also buys the premise that basic biology, like the force of gravity, somehow draws the stuff of anthropological exotica down to a common ground—my colleague's arguments still remain, ultimately, silly. Rather than saying that his arguments attribute *too much* force to biological corporeality, I should say instead that his line of reasoning actually empties human bodies of any lived experiences in order to constitute them entirely from within an abstract heteronormative logic. "Has your daughter no mouth, hands, or anus?" I responded, perhaps a bit too bluntly.

To put matters more delicately: I can imagine—and cross-cultural sex research documents—dozens of ways in which anyone's daughter might enjoy a sexually adventurous life without incurring any risk whatsoever of unwanted pregnancy. This hardly takes a lot of imagination. It's all a question of how one learns to use one's body, and how others use theirs. I have no doubt that there are situations under which my colleague's line of reasoning holds—but that is to say, his associations follow from cultural regimes of sexuality, which instruct young girls in a very narrow range of carnal enjoyments, not from the evolutionary biology or psychology of sex. As even a cursory survey of the ways human beings have construed their bodies shows, the "facts" about human sexuality and reproduction are not hard, cold, immutable objects, but artifacts of what we do and how we represent: consequences of our actions and framings. In any claims about the biology of sex, what one sees as "plausible" will thus depend on what one counts as, and how one imagines the practices of, "sex."

Darwin's hypothesis about sexual selection, then, at least seems plausible: The woman who plays coy and waits to have sex with a man *might* have some reproductive advantage over the woman who plays fast and loose. But the "plausibility" of such reasoning rests on certain strategic denials and elisions (and on the institutional apparatus that supports them). It seems equally plausible that the woman who engages in varied and nonreproductive sex acts with many men—in effect "screening" them for one or two long-term partnerships—might also enjoy the advantages of natural and sexual selection over her less adventurous rivals. By the same token, it seems plausible that the eager, promiscuous man might enjoy certain biological advantages over other men who fail to broadcast their seed widely. But it also seems equally plausible that the gentle, considerate man who nurtures a limited number of offspring with a single mate or two might enjoy a reproductive advantage over those men who abandon large numbers of ill-fed and poorly nurtured children to a string of hapless women.

Just as there might be more than one strategy for enhancing one's "reproductive fitness," a plausible Darwinism could logically support opposite conclusions about human nature. But even enriched by multiple scenarios, a good Darwinism is worse than a bad social science at explaining human acts, prodded as they are by cultural meanings in social and institutional contexts. On the ground of human existence, cultures make the most varied and sometimes contradictory meanings around human anatomy and body fluids, out of what happens when intercourse happens—indeed, they are not even in agreement over what counts *as* "intercourse" or how a child is conceived and delivered. The question, then, is not whether men, eager to insure that the offspring they support are really their own, are jealous "by nature," but rather, how paternity is defined: Is it singular or partible? Essential or transferable? And has the man anything to do with the pregnancy at all? Nor is it a question of whether a woman can obtain the support of a male breadwinner for herself and her offspring. The salient questions for human relationships are: How are practices like subsistence and child care socially organized? Are women party to or excluded from the means to subsist? Is subsistence an individual undertaking or is it a communal activity? Is food hoarded or shared? And just who is supposed to be involved in child care: the biological mother, a network of doting allomothers, both mom and pop, or the whole bustling neighborhood? In any serious reckoning of the varieties of ways people have adapted to the conditions of life on this planet, these are the questions that matter. They are of a social and political, not biological, order.

Biology is no doubt "present" in all such social phenomena—one might say that biological problems are here the very site of cultural elaborations. But it is on these sites that we see most clearly the limits of attempting to explain the one thing by recourse to the other. No set of biological "facts," "laws," "drives," "needs," or even "tendencies" can really account for such varied cultural notions and social practices as those recorded in the ethnographic and historical records. The oft-invoked "limitations" of basic biology turn out to have no positive bearing on the meaningful forms taken by human culture or on the structures of social institutions. Nor is the understanding of human practices much advanced by a Pooh-like simplicity that declares the centrist position: "I'll have them both" (biology *and* culture). For even in those areas of life that might be said to lie on the border between organic and social facts—the "transitional phenomena," as it were—the "transition" in question is of a very certain order: It is culture that organizes and gives meaning to biological experiences, not the other way around. Or perhaps it might better be said that in human affairs, meanings are made around material questions in such a way that cultural contingency precedes biological necessity, so much so that what is considered "biologically necessary" is itself often a function of cultural conditioning.

KINSHIPS REVISITED

I stress the complexity of human beliefs about body fluids, sex, and gestation because this complexity provides poor ground for the generic formulae and universal norms that are often invoked in the name of science. It is the social facts themselves that rebut reductive pseudoscience. Nowhere is this more evident than on very basic questions of kinship.

Everything in sociobiology, evolutionary psychology, and other forms of bioreductivism follows from the premise that human sociality obeys the principles set forth in William Hamilton's theory of inclusive fitness, also known as "kin selection."[71] Succinctly put, Hamilton's model attempts to reconcile Darwinian notions of individual selfishness with various kinds of "altruistic" behavior. It suggests that when animals favor individuals to whom they are closely related, they favor their own selves, inasmuch as close relations carry the same genes. Thus the apparent "altruism" of the worker bee, who slaves away for the anonymous hive, is actually self-interested behavior: It turns out that she can pass more of her own genetic material to successive generations by rearing her sister's offspring than by having her own offspring. One biologist expressed the formulaic thinking about human sociality in a simple aphorism: "I would gladly die for two brothers, four cousins, or eight second cousins," each set of kin carrying the requisite number of genes necessary to "replace" the self-sacrificing individual. Put in mathematical terms, r (the coefficient of relationship, or the measure of shared genes) between any ego and his sibling is $\frac{1}{2}$ (meaning that siblings carry one-half of each other's genes). Between cousins, $r = \frac{1}{4}$, between second cousins, $\frac{1}{8}$, and so on.

Hamilton's elegant model goes a long way toward solving the riddles posed by "altruistic" or "social" behaviors in species like bees and ants, where what appears to be "unfit" from the perspective of an individual organism (the self-sacrificing, nonreproducing worker bee) might actually be "fit" from an "inclusive" or aggregate point of view. In other words, Hamilton shows that natural selection can act not just upon individuals but also upon groups of closely related organisms (thus the term "kin selection"), giving rise to phenomena of some complexity in the animal kingdom. It is this complexity and Hamilton's calculations that equip modern reductionists to argue that social facts come under the sway of biological facts—to assert that what people do when they act in kin groups is ultimately governed by genetic self-interest. But as appealing as it might seem to translate human social relationships into the pristine formulae that apply to haplodiploid species, what holds for bee and ant "societies" carries no special implications for the organization of human societies. To invoke an enduring distinction from the social sciences: Bees "behave." Humans "act."[72] Between the one thing and the other lies a world of difference.

And in fact, the extrapolation from bee and ant "society" to systems of human social organization and kinship is counterempirical, and extravagantly so. As Marshall Sahlins shows in his brief survey of kinship practices in the Pacific, actually existing systems of kinship and the concepts of heredity they organize never really conform to the genetic principles sketched out in biological models of reproduction. In many instances, the forms of altruism and solidarity orchestrated by human kinship systems run directly counter to the supposedly inevitable logic elaborated in the theory of kin selection.

Patrilocal residence, for example, implies that a man cooperates with and thus favors the reproductive success of people with whom he may be only distantly related in genetic terms (including cousins several degrees removed, where $r = 1/8$, $1/32$, $1/64$, etc.). Meanwhile, his biological sister, along with her children (close genetic relatives: $r = 1/2$ and $1/4$, respectively), belong to another residence, another patriline, that is not the beneficiary of his cooperative action and that may actually be the target of competition or hostility. Indeed, since some cultures count coresidents as "close" and noncoresidents as "distant," irrespective of genetic relationships (r), the cultural reckoning of kinship sometimes bears not the slightest resemblance to genetic theories of kin selection. In other cases, claims of consanguineous descent from a "common ancestor" are patently fictitious—or go back so many generations as to be virtually meaningless. The coresident relatives of a representative district in the Solomons trace their common ancestor back nine generations, which translates into a very unimpressive degree of genetic relatedness among members of the cooperative kin unit. (Under the terms of the theory of kin selection, the coefficient of relationship is as little as $1/1024$.)[73] Sahlins also reviews such well-documented social phenomena as Tahitian adoption (25 percent of all children born are adopted), Nuer "ghost fathers" (the "father" is not a living person), and the Rangiroan tendency to adjust kin terms to fit social practices. "Genealogy," he pithily summarizes, "is deduced from kinship, rather than kinship from genealogy."[74]

There can be little doubt that kinship practices are "utilitarian" to the extent that social cooperation (almost any form of cooperation, however contrived) is mutually advantageous to participating kinfolk (however defined). But to cast this "utility" in biological, reproductive, or genetic terms is to distort everything significant about how human beings actually cooperate, how they really make meanings, and how their actions transform the conditions of existence. In human life, utility subserves culture, not vice versa.[75] The question is not how biology expresses, shapes, or affects culture, but how human activities take up biological problems in the light of cultural aims.

15 The Social Body

At this point, even friendly interlocutors or readers are likely to exclaim (as several in fact have): "But Roger, surely you don't mean to argue that human beings are a tabula rasa—that *everything* is a social construct, that there are *no* significant differences between the sexes, that genetic or neurohormonal differences between male and female (or gay and straight) account for *nothing* in the way of general tendencies or predispositions . . . ?"

Even in her critique of evolutionary psychology—a book heralded by Barbara Ehrenreich as the "chief manifesto of the new 'femaleist' thinking"—Natalie Angier expresses just this sort of impatience with perspectives from the usual critics of bioreductivism (feminists, progressives, and perhaps especially social scientists). She complains that

> they disappoint when all they can do is say nay. They pick out flaws,
> they grumble, they reject. Hormones don't count, appetites don't count,
> odors, sensations, and genitals don't count. The body is strictly vehicle,
> never driver. All is learned, all is social construct, all is the sequela of
> cultural conditioning. Critics also work from a premise, often unspoken,
> that human beings are *special*—maybe better, maybe worse, but ulti-
> mately different from the rest of evolution's handicraft.[1]

Undoubtedly, all social constructionists (and in fact most social scientists and humanists) *do* believe that human beings are "special" in the sense set forth with some eloquence and clarity in Clifford Geertz's famous essay, "The Impact of the Concept of Culture on the Concept of Man." That is, all constructionists believe that "there is no such thing as a human nature independent of culture."[2] Human beings use language (whose meanings are arbitrary, relational, and conventional); they elaborate cultural systems of meaning and complex institutions (whose variety and specificity I have been trying to evoke in the preceding chapters); they relate to their environment, not by tooth and claw, but through the use of tools and in various forms of social labor (whose historical development Marx called *"real"* human nature). . . . Such human reliance on culture, and the plasticity of basic human behaviors it implies, are not "secondary" features of human existence, "added onto" a basic biological substratum. Reliance on culture is a central fact of human biology—the *real* essence of our "evolutionary psychology."

The story of human evolution, in this telling, is the story of how hand and eye, tool and sign, became caught up together over the course of millennia in a "positive feedback system" that shaped virtually every aspect of the human biology: hand, eye, and the connections between them, surely, but also the brain, particularly the forebrain.[3]

> As our central nervous system—and most particularly its crowning curse and glory, the neocortex—grew up in part in interaction with culture, it is incapable of directing our behavior or organizing our experience without the guidance provided by systems of significant symbols. What happened to us in the Ice Age is that we were obliged to abandon the regularity and precision of detailed genetic control over our conduct for the flexibility and adaptability of a more generalized, though of course no less real, genetic control over it. To supply the additional information necessary to be able to act, we were forced, in turn, to rely more and more heavily on cultural sources—the accumulated fund of significant symbols. Such symbols are thus not mere expressions, instrumentalities, or correlates of our biological, psychological, and social existence; they are prerequisites of it.

Geertz underscores the implications of this arrangement for the nature of human existence: "Without men *[sic]*, no culture, certainly; but equally, and more significantly, without culture, no men."[4]

The result of a long evolutionary process, then, is an animal uniquely reliant on culture for the conduct of its basic and characteristic activities, a species uniquely "programmed to learn" (as the title of a classic text correctly applied the cybernetic analogy)[5]—that is to say, a being for whom what is most essential and necessary is external and arbitrary. Mediated by meanings and technologies, human activity is creative, open-ended, and thus not commensurate with other kinds of animal behavior. The existentialists, cribbing Marx, framed this free and indeterminate relationship between biology and culture in a gnomic expression: "Doing precedes Being." Or, as Simone de Beauvoir sententiously distilled the existentialist perspective on human nature: the human being is *"l'être dont l'être est de n'être pas,"* the being whose essence lies in having no essence.

Reflecting on the unique relationship between necessity and contingency in human affairs, Maurice Merleau-Ponty put matters this way:

> Everything in man is a necessity. For example, it is no mere coincidence that the rational being is also the one who holds himself upright or has a thumb which can be brought opposite to the fingers; the same manner of existing is evident in both aspects. On the other hand everything in man is contingency in the sense that this human manner of existence is not guaranteed to every human child through some essence acquired at birth, and in the sense that it must be continually reforged in him through the hazards encountered by the objective body. Man is a historical idea and not a natural species. In other words, there is in human existence no unconditional possession, and yet no fortuitous attribute.

No slouch on the materiality of the body, Merleau-Ponty concludes that human existence "is the transformation of contingency into necessity."[6] In similar terms, Geertz summarizes the "open-endedness" and plasticity of human nature this way: "We are, in sum, incomplete or unfinished animals who complete or finish ourselves through culture."[7] Rhetorical flourishes notwithstanding, human beings *are* special.

THE FACTS OF THE BODY

For all my insistence on taking this difference fully into account, I trust I have not presented a case for disembodied abstraction—for mind over matter—but rather for reckoning with material facts of the body in a manner that accommodates the empirical evidence on human cultural variation. From this evidence, it is plainly impossible to infer sociobiological fables about a singular, uniform, and homogenous human nature. But to make matters yet more complicated, the ethnographic and historical evidence, which demonstrates innumerable ways of solving the basic problems posed by existence, also problematizes even an appeal to the basic "facts" of the body as such.

Bodies are material, and they really exist—but precisely by dint of this materiality, its mode of active existence, much depends on what is *made* of them. Hormones, odors, and appetites *do* count—but their effects are always called forth within a cultural context, which is to say, they count in dynamic and nonreductive ways. The irreducible facts of reproduction are everywhere much the same (or at least they always have been, up until the era of test-tube babies), but in the varied transformations of social life, it matters less that they are biological than that they are creatively articulated within a framework of arbitrary meanings and contingent practices.

What I will try to show in this and the following chapter, by way of a series of examples, is that even what seems most "natural," "given," or "universal" about the body is enmeshed in social facts and human acts, ongoing transformations of contingency into necessity. Even those actions that seem most firmly rooted in biology or that give some appearance of universality display the unique character of human action.[8] The "facts" of the body are but the last issue of cultural *acts*.

THE MEANINGS OF THE SMILE

Consider an example close to the point—a good indicator of what is unique about the most elemental human action. The lyrics to a somewhat homo-erotic Crosby, Stills, Nash, and Young song, "Wooden Ships," asserts that the smile is something "everybody everywhere does in the same language." The sentiment expressed by Woodstock's hippie-poets is essentially correct. An accumulated body of evidence suggests that, with a few exceptions, people the world over are in broad agreement over the meanings of certain facial expressions—even complex or compound expressions.[9] Are we to conclude that the meaning of the smile is "given" with the acultural facts of nature? Not really.

A review of research on facial expressivity prompts Norbert Elias to conclude that the face functions as something of a "signaling board" in human interactions—a form of communication less variable and more stereotyped than language—but Elias goes on to argue that even in such basic behavior one finds not "inflexible instinct" but a "plastic core." Infants smile without ever having been taught to do so. Adults smile, grin, wince, sneer—or withhold facial expression altogether—according to cultural meanings and social conventions. Barring deliberate deception, a certain kind of smile might always indicate good feeling or goodwill. But whether one smiles at an inspired rendition of *Madame Butterfly* or at the conclusion of a successful raid on a neighboring village is strictly a cultural question, and as such it is subject to varied motivations and expressions.

(Traveling, I've often been struck by the stern and somber faces that greet me on the streets of certain foreign cities and by how quickly those faces melt into smiles over beer in a pub or cafe. I've also often heard the residents of these very cities express wonderment over the odd fact that Americans give their smiles "too easily," sometimes beaming at the world for no apparent reason—a thing no rational being would do.)

It is *precisely* this "malleability" of the smile—and of the emotions it expresses—that make it significant in human interactions. As quirky as the Mona Lisa's or as broad as the Cheshire cat's, a smile "can be deliberately used to convey to others a rich variety of shades of feeling. It can be a hesitant, a withdrawn, a broad, a triumphant, a supercilious and even hostile smile," as Elias says. "In all these cases a learned and deliberate steering of conduct merges with an unlearned form of steering one's face muscles."[10] To make matters yet more complicated, people almost never smile as in the posed and iconic little photographs used in survey research. In real life, they combine smiles with facial expression and physical gestures in subtle ways that are often difficult for nonnatives to comprehend.[11] That ever-present companion of a certain kind of American smile, the conspiratorial wink, is by no means universally understood, even in the Western Hemisphere.

If it cannot quite be said that culture assigns random meanings to the smile (e.g., using it to convey sadness), it also cannot be said that the meaning of the smile is immutably "fixed" or "hardwired" in the body.[12] What can be said is this: Culture appropriates, organizes, and gives meaning to the basic biology of the smile. It is this symbolic appropriation that gives the smile its social expressivity and allows it to mark subtle nuances of meaning. What is most "essential" about the smile is thus not "within" the biology of the smile, but in the way the smile is taken up in "external" social relations—that is to say, its *cultural* dimension.

"BODY LANGUAGE"

The enigma of the smile is only the beginning of the carnal mysteries involved in simple acts of physical expression. In basic physical conduct, social manners interact with combinatory rules, often making cross-cultural physical communication—"body language"—a more difficult proposition than linguistic communication. For example, gestures coded as "masculine" in one culture are not necessarily viewed as manly in another. Small gestures kept close to the body are deemed manly in the United States. Small gestures might be read as reticence in a wide swath of Latin American cultures, where large, extended gestures are components in the competent performance of manhood.

Before doing fieldwork in Nicaragua, I had imagined that American gestures for "come here" were indexical signals, universally understood because they were rooted in the natural language of the body. Not so. Nicaraguans indicate a desire to draw someone close by extending the hand and fingers straight out from the body; the fingers, and perhaps the hand, gesture repeatedly downward and back toward the gesturing person. To the American eye, this looks rather more like waving—saying "goodbye"—than like beckoning. Pointing—thought by some to be a universal gesture meaning "in that direction" or "over there"—is highly discouraged in Nicaragua, where it is seen as a vulgar overture. Instead of indicating direction with the finger, polite Nicaraguans point with their lower lips and chin—a gesture usually read in the United States as an invitation to kiss.

Kissing itself is sometimes said to be an "instinctual" part of human sexual expression, supposedly derived from infantile suckling. Infants do indeed root and suckle without ever having been taught to do so, and it is not implausible to think that the enjoyment of oral or genital kissing is somehow linked to basic mammalian patterns of response. But there is no "kissing instinct," and in fact a wide swath of cultures do not practice the kiss.[13] To paraphrase (from another context) a masterful interpreter of body practices: Like swimming, dancing, or breast feeding, kissing—and making love in general—"must be learned, and the knowledge of 'how' to do it comfortably and well (though with many cultural variations) can be lost."[14]

It is not a question of whether to choose between the absolute and the relative—the gravity of the body or the weightlessness of the idea—but of how to situate the absolute within the relative, and the relative vis-à-vis the absolute. Not that one ever quite encounters the one or the other, in good Cartesian style: evolution's blank, dumb handicraft (all biology) versus devious human ingenuity (all culture). In human affairs, the one thing is always entwined with the other. As Merleau-Ponty puts it, the human being "is not a psyche joined to an organism but a movement to and fro of existence": an entwining of "corporeal form" and "personal acts."[15]

But biology and culture are not entwined just any old way. Precisely what is unique about human action is its capacity for turning contingency into necessity, for transforming the relatively absolute into the absolutely relative, for taking up biological givens in the light of social aims. If this is understood, and understood well, it leaves little room for confused theories that naturalize historical contingencies—or for idealist claims that unhinge social practices altogether from the material world.

MORE ON HUMAN EMOTIONS

By way of another example: Some scholars have suggested that the basic human emotional pallet consists of a small, fixed number of irreducible feelings: interest, surprise, joy, anger, shame . . . [16] Evolutionary psychologists have made much of this idea. Supposedly, the evolutionary effects of natural selection on the endocrine system have rendered us "prewired" to have certain feelings, and (non sequitur) to express these emotional dispositions in certain pre-given institutional forms. I am skeptical of the notion that anything so mercurial as emotion could be convincingly catalogued and classified—indeed, I suspect that researchers are committing a basic category mistake when, prodding human subjects to make intelligible distinctions, they "find" agreement among their informants about such matters.

But even if it could be convincingly demonstrated that human bodies come equipped with only so many basic emotional responses, such a finding would not really amount to much in the way of positive knowledge about social life, for everything significant about *human* emotions relates to their institutional context, their symbolic valuation, and to the jostling of interests within modes of cultural organization. That is to say, people the world over might very well understand feelings of pride and shame, joy and sadness, and we might clearly recognize pieces of our own selves, our own feelings, in the emotional responses of other people, other cultures. But all of the pertinent questions remain: How are such dispositions symbolically organized, what social occasions give rise to such feelings, and how are these feelings mobilized for institutional purposes? What sorts of conflicts or contradictions arise from the states of feeling so cultivated? Is one proud of one's sexual exploits, or ashamed of them? What constitutes a legitimate occasion for anger, as opposed to a capricious one? What forms of action variously express, suppress, or sublimate anger? And to what ends are feelings of communal affinity mobilized—on behalf of class solidarity, local interests, international good citizenship, or patriotic jingoism? Is fellow feeling broadly or narrowly distributed, and according to what rules?

Even if one grants the premise that evolutionary eons of group life have made a basic emotional repertoire part of our "plastic core," then, nothing of great consequence has actually been ceded to innatism. It would still remain for specific cultural regimes either to develop or to atrophy the basic emotional abilities, to elaborate some capacities while deemphasizing others. It would also still remain for human beings, as social actors, to say "yes" or "no" to the feelings naturally solicited under varied institutional domains.

This is exactly how Ruth Benedict distilled the gist of cultural relativism in the 1930s. Notably, Benedict reasoned by extrapolation from basic linguistic principles. The human organism is biologically capable of producing hundreds of vocalic sounds. Infants everywhere spontaneously produce this range of phonetic utterances, which indeed is "hardwired" into the vocalic apparatus. Languages, however, use only a small number—three or four dozen—of these potential sounds, and the actual sounds employed in speech vary from language to language: English phonemes do not even resemble those in closely related languages such as German or French, much less those in the Khoi languages, with their glottal flaps and alveopalatal clicks. By way of analogy with basic linguistics, then, Benedict conjures the image of a "great arc of human potential." "In cultural life as it is in speech," she argues, "selection is the prime necessity."

> In culture, too, we must imagine a great arc on which are ranged the possible interests provided either by the human age-cycle or by the environment or by man's various activities. [Since so much of Benedict's text is actually about the cultural organization of emotion, it is safe to also infer emotional capacities and responses here.] A culture that capitalized even a considerable proportion of these would be as unintelligible as a language that used all the clicks, all the glottal stops, all the labials, dentals, sibilants, and gutturals from voiceless to voiced and from oral to nasal. Its identity as a culture depends upon the selection of some segments of this arc. Every human society everywhere has made such selection in its cultural institutions.[17]

If Benedict's analogy with linguistic selection has any heuristic value, the sociobiological leap from "emotional motive" to "institutional expression" is deeply and irrevocably flawed. Inner feelings, like raw sounds, take on significance only in a structured social context. As Benedict's analogy suggests, emotions are not the "sources" of culture, any more than random vocalic noises are the sources of linguistic meaning. They are, at best, the ground on which cultural motives or linguistic meanings take root.[18]

Reflecting on the place of motives and emotions in culture, Marshall Sahlins expressly cautions against seductive tautologies and invitingly circular appeals to common sense—the commonplace sociobiological argument that social institutions "express" primitive emotional impulses. "There is no necessary relation between the phenomenal form of a human social institution and the individual motivations [or emotional dispositions] that may be satisfied therein. . . . Any given psychological disposition is able to take on an indefinite set of institutional realizations. . . . Conversely, it is impossible to say in advance what needs may be realized by any given social activity."[19]

War, so often conjured as a singular expression of men's innate "aggression," might just as well "express" men's need to cooperate—or their willingness to follow orders. Marriage, so often depicted in terms of domestic cooperation, might actually serve as "aggression" by other means. Because

any institution can serve a multiplicity of emotional needs, and, contrariwise, because the most varied of institutions could in fact serve very similar emotional needs, one can scarcely say in the abstract just what emotions
are expressed or just what needs are actually met by what institutions. Contrary to the self-referential logic of just-so stories about human nature, then,
the need we feel for a given institution and the emotions we express
through it cannot, by that very token, be employed as explanations for that
institution's existence. It is our needs and feelings that follow, in no small
part, from our institutions, from the situations they predicate, from the subjective experiences and social conflicts they imply.

In short, the meaning of the emotion is not "in" the feeling any more
than the meaning of the sign is "in" the signifier. It lies somewhere in between individual motive and social institution. To think otherwise is to turn
away from all that is salient about the human object of inquiry.

COLORS AND CULTURES

To broach another pertinent example: Perhaps the classification and mobilization of human feelings works much like the naming and invocation of
colors. Although the human eye can discriminate thousands of color percepts, "natural" (vernacular, nonspecialist) languages distinguish only a very
limited number of "basic color terms" (black, white, red, green, yellow, blue,
brown, gray, purple, orange, pink). And although the number of colors
named does indeed vary from culture to culture, those color terms appear
in a very regular order of succession. That is, if only two colors are named
in a cultural lexicon, the color categories will invariably be black and white
(or some approximation thereof: "light" and "dark," "wet" and "dry"). If
a culture names three colors, the three will always be black, white, and red.
Cultures that name four colors will name the same terms as their first
three—black, white, red—followed by either green or yellow. And so on,
according to the sequence discovered by anthropologists Brent Berlin and
Paul Kay (table 1).[20]

Their cross-cultural study tested an enduring conceit in American anthropology: the Sapir-Whorf hypothesis, a proposition that conjoins cultural relativism to linguistic determinism. Given the significance of the
Sapir-Whorf hypothesis in the development of American anthropology—
indeed, given its bearing on discussions of cultural variation today—it is
worth pausing to consider the propositions it gathers under a single rubric.

Named for Edward Sapir and Benjamin Lee Whorf, the hypothesis was

TABLE 1. Order of Succession for Basic Color Terms

"Stage"	I	II	III	IV	V	VI	VII
Basic Terms	Black white	Red	Green or yellow	Yellow or green	Blue	Brown	Purple gray pink orange
Number of Color Terms in System	(2)	(3)	(4)	(5)	(6)	(7)	(8, 9, 10, or 11)

never expressly stated as such by either the anthropologist (Sapir) or his student (Whorf), but the term "Sapir-Whorf hypothesis" came to serve as shorthand for an idealist variant of constructionism much invoked in an archaic anthropological lore. According to the Sapir-Whorf hypothesis, since Eskimos have several words for "snow," they must (non sequitur) have a different, more nuanced experience of snow than English speakers. People from cultures with only two color terms were said to be capable of seeing only two colors. And, in part on the basis of Whorf's whimsical translations, the Navajo were said to have a completely different experience of time than Anglo-Americans.[21] The propositions associated with the Sapir-Whorf hypothesis get at important features of language and culture—semantic domains can in fact be drawn up any number of different ways, and cultural values permeate linguistic systems in ways sometimes subtle and sometimes not so subtle—but they suggest that cultural distinctions can be drawn up "just any old way" while positing a linguistic determinism that resembles biological determinism (turned upside down). That is, under Sapir-Whorf a linguistic (rather than genetic) code is said to determine perception and experience directly. Such formulations give a theory of perception somewhere in between what Merleau-Ponty called "associationism" (the idea that what one already thinks determines what one sees) and "intellectualism" (the idea that a conscious subject "posits" the surrounding world)—premises that have by no means waned from the academic scene and that today perpetuate themselves in a number of guises—in sloppy or abbreviated claims that "the real" is "made up," that identities are "performed," that truth is a "fiction," that language "posits" the world, and so on.[22]

Berlin and Kay's results tilt against such implausible formulations. Cultures, it would appear, are actually somewhat "constrained" in the manner in which they construct colors. And contrary to the predictions of Sapir-Whorf, Berlin and Kay also found general agreement from society to society on the empirical referents of the basic color terms—for instance, a cross-cultural consensus on what constitutes "focal red."[23] In view of such results, a number of anthropologists publicly opined that the Berlin and Kay study tolled the death knell for cultural relativism: The "constraint" on color terms, the source of their consistencies across cultures, some argued, was "nature," not culture. Berlin and Kay themselves treated the pattern they discovered as an "evolutionary" sequence—a series of "stages" beyond the reach of signification and practice, a firm biological substratum supporting secondary cultural variation.

In his classic essay "Colors and Cultures," however, Marshall Sahlins puts the Berlin and Kay study into a properly nuanced perspective, arguing that while its results undoubtedly put to rest the idealist nominalism and anything-goes relativism of Sapir-Whorf, its findings actually allow for a mature and convincing demonstration of basic relativist, constructionist principles. I abbreviate his basic argument here. Sahlins shows that the principle underlying the sequence of color terms is not a "natural" or "evolutionary" succession of terms proper, but the structuring of contrasts within systems of meaning. Obviously enough, he reasons, there are no color systems with only one color term, because one term alone would be meaningless—it could not be contrasted with anything else. According to the basic semiotic principles spelled out by Saussure: no distinction, no meaning, no system. By the same logic, where there are only two color terms, the two terms could not be orange and red, for there is not enough contrast between such a pairing to ground a system of meaningful, unambiguous distinctions. It is the opposition between black and white that gives the maximum distinction, the maximum difference physiologically possible: the presence of all hues versus the absence of all light. So wherever there are only two color categories, language is compelled, as it were, to select *these* two, and no other. Any other pairing would be too ambiguous.

But why, one might ask, is the third term red? Red does indeed have certain physiognomic aspects that single it out from other chromatic hues. A red object appears to stand out against other objects, and red is the last hue to remain visible as dusk fades into darkness. As Sahlins explains, red is third in the sequence because the succession of color terms follows a basic logic: from most general to most specific (maximum distinctions before subtle ones), from more to less salient (red before other hues), and from simple to complex (unique to mixed).

Sahlins's argument about the relationship between biological apperception and cultural construction is a subtle one. Physiognomic apperception is by no means *irrelevant* to the cultural sequencing of colors—for colors to exist as semiotic categories, they must logically be subject to perceptual discrimination—but it is by no means "determinant," either. Or rather, to speak of biology as "determinant" of color terms is to misunderstand what people do when they distinguish colors. In color terms, what is significant is not a natural succession of independent categories, but systematic relations *between* terms. As in the use of vocalic sounds to produce meaningful utterances, "perceptual distinctions" are "engaged as the support for conceptual constructions."

Sahlins summarizes the relationship between contingency and necessity in color terms this way. "Berlin's and Kay's results are consequent on the social use of color not merely to signify objective differences of nature but *in the first place* to communicate significant distinctions of culture. Colors are in practice semiotic codes." Thus, colors "are engaged as signs in vast schemes of social relations: meaningful structures by which persons and groups, objects and occasions, are differentiated and combined in cultural orders." And "because colors subserve this *cultural significance,* only certain color percepts are appropriately singled out as 'basic,' namely, those that by their distinctive features and relations can function as signifiers in informational systems. . . . *It is not, then, that color terms have their meanings imposed by the constraints of human and physical nature; it is that they take on such constraints insofar as they are meaningful."*[24]

To name a color is an act of *relating,* not "recognizing."[25] "Red" carries meaning because it implies an oppositive relationship—at minimal—with "black" and "white," which are themselves oppositive terms. What is most absolute and necessary about color terms, then, is also what is most contingent and relative about them: It is the requirement that for anything to be meaningful, it must involve a difference—a demand that adheres to every human action and defines the human mode of adhesion to the world.

But even this relativist approach to color terms, cast solely in terms of color, provides little more than the scaffolding for a properly constructionist accounting of how color is mobilized in culture. It solves none of the more interesting questions about color, as Sahlins suggests when he links the relationality of color terms to the social relations and cultural distinctions they structure. Red undoubtedly stands out, but what does the color red actually *mean?* How is it engaged in concrete systems of social relations? Is red the color of bravery, or of sexuality? Of adultery, or of blood? Of dread, or of anger? Is it the sign of good luck, or does it distinguish clan A from clan B? And is whiteness arbitrarily coded to signify all that is good, clean, and pure, while blackness is arbitrarily associated with all that is evil, unclean, and impure?

Such questions do not belong simply to the realm of signification. They involve the play of meanings in the open-ended process of historical happenings and social transformations. Such questions are necessarily linked to practices of kinship that give occasion for one moiety to distinguish itself from another by recourse to a color scheme; to colonial histories that freight "black," "white," and "red," and "yellow" with cultural meanings and institutional connotations by no means implicit in the colors themselves; and to modes of production and reproduction through which colors like "blue," "pink," and "lavender" are attached to other socially elaborated distinctions.

The relationality—the relativity—of color terms embraces not just the terms themselves, but also what and how the terms designate.[26] As with any system of signification, there are no positive terms, only differences.[27] And as with any system of meanings, colors are caught up in complex and unpredictable ways with social practices and historical events. Like the ability to give and to understand a smile or to communicate an emotion, the ability to see and name colors is, as Merleau-Ponty would say, "no unconditional possession, and yet no fortuitous attribute." "To learn to see colours is to acquire a certain style of seeing, a new use of one's own body."[28]

I dwell on the question of color at length because this page from anthropology's luminous past illustrates a major deficiency in the current discussions of culture and human nature—and because what characterizes so much of the literature in sociobiology and evolutionary psychology is both a lack of theoretical sophistication and a refusal to grapple with the complex findings of anthropology, history, semiotics, and twentieth-century cultural theory in general. Even if the variation of basic color codes corresponds to a perceptual logic—that is to say, even if cultural colors are relative to some absolute physiognomic feature of the eye—this does not mean that color terms are "biologically fixed," that their meanings are neurologically "hardwired," or that their cross-cultural sequencing is "naturally given." In the first instance, the human eye itself acts by opposing, relating, systematizing, and in the second instance, color terms themselves are semiotic codes. The passage from one thing to the other—from seeing to naming—connects uniquely human actions on either side: on the one side, the physiognomy of the eye, an active, interactive organ whose operations are meaningful, creative, and intelligent; on the other side, the realm of symbolic representation, premised on other representations as they relate to the dynamic operations of the active eye.[29]

It is useful to keep in view this two-sidedness of meaningful activity. On the one hand, the "agreement" between cultures on how to mark color terms is an agreement not on the order of things, but on the order of distinctions. On the other hand, the "disagreements" between cultures on how to mark colors (and how many colors to mark) are likewise disagreements of a relative sort—that is to say, they are differences over how to make a difference.

It is also important to remember what the Berlin and Kay study shows and what it does not show. First, the cross-cultural evidence demonstrates considerable cross-cultural variation in color systems. Some cultures have two basic color terms, others three, four, and so on. That's the very stuff of cultural relativism. Second, the discovery of a form and order to the sequencing of color terms does not mean that any particular constellation of basic color terms (out of the eight sequenced) is more "normal" or more "natural" than the others, nor does it imply that some schemas are more "advanced" or more "desirable" than others. Two colors might serve as well as eight, and evolutionary phraseology notwithstanding, Berlin and Kay researched contemporary cultures, not early human societies. Third, the regular sequence tracked by Berlin and Kay does not necessarily establish outer limits on color terminologies. It does not even mean that there could never be a culture in which the distinction between fuchsia and mauve might constitute a "basic" opposition. The everyday use of industrial techniques and high-tech imaging equipment already seems to be expanding the horizon of color in culture.[30]

Undoubtedly, Berlin and Kay do trace the arc of a minimal "limit" on basic color terms—a limit whose contours lie at the fluctuating boundary between perceptual biology and semiotic culture. But the limit on color terms is not so much in the biology of perceptual discrimination as in how cultures take up perceptual discriminations as meaningful differences. And none of this implies any necessary meanings or social engagements for colors: It does not mean that "red" has to signify "stop" or that "purple" will invariably stand for royalty or that "whiteness" will always symbolize goodness. To claim otherwise on any of these counts is to misunderstand at a fundamental level the questions posed by the problem of color, its perception, its cultural coding.

MISPLACED ANALOGIES

These are just the sorts of basic misunderstandings that trumpet themselves as "new knowledge" today. When sociobiologists seize upon certain recurring features of social life—color perception, conception, parturition—in order to write fables about biological destiny and human nature, they make the classical error of assuming that since what is different about societies is cultural, what is shared must be biological. Likewise, when evolutionary psychologists trade in the prestige of Chomskian linguistics, or when they liken the brain to a computer, they betray a fundamental misunderstanding of the implications of their own analogies.

Boiled down to its essence, Noam Chomsky's theory proposes that the human brain contains, "hardwired" in its circuitry, certain "deep grammatical structures," and that all existing grammars are in some sense variations on those essential forms.[31] In three books that popularize evolutionary psychology, Stephen Pinker invokes Chomsky's as yet undemonstrated and perhaps undemonstrable hypothesis about "universal grammatical structures" in order to stage claims about cultural universals and panhuman sex roles (the coy, choosy woman, yet again, and the aggressive, promiscuous man). Although he invites uninformed readers to mock anthropological descriptions of exotic cultures, Pinker does not actually counter the findings of relativist anthropology with much in the way of empirical evidence. Instead, he develops his argument by way of analogical extension, occasionally supported by anecdote or speculative evolutionary scenarios, culminating in a frontal attack on that antiquated relic from anthropology's past, the Sapir-Whorf hypothesis. If it can be shown that human beings across all cultures share some understanding of things like facial expression, emotion, and color, Pinker reasons, then biology takes precedence over culture. And if it can be shown that anything at all about language is "universal," Pinker suggests, then by extrapolation, it can also be argued that linguistic and hence cultural relativism is wrong. Chomsky trumps Saussure—check. Evolutionary psychology trumps cultural anthropology—checkmate.[32]

The careful reader who is familiar with the history of anthropology might well experience some disorienting feeling of déjà vu upon encountering Pinker's arguments. We have been here before, after all, but without the unwarranted conclusions or the tidy burgher homilies on human nature. Let me describe the outlines of this "here," a major continent on the map of twentieth-century cultural anthropology.

Lévi-Straussian structuralism turned on the notion that the human brain is essentially a computerlike organ operable only by a binary code. And Claude Lévi-Strauss argued that this basic binary logic is always the same everywhere, a universalist argument if ever there was one, framed in terms somewhat narrower than what anyone has ever attributed to Chomsky. And yet Lévi-Strauss—a student of Franz Boas, Ferdinand de Saussure, and Karl

Marx—understood his empirical investigations, collectively, as a demonstration of relativist, semiotic, and Marxian principles. That is, his work tracks the same binary oppositions (up/down, high/low, in/out, hot/cold) across cultures and through history, but it also shows that those basic building blocks of human existence can be put together in any number of patterns, that they can be mobilized to very different ends. In Lévi-Strauss's opus, then, one encounters basic similarities in the context of larger differences: The irreducible elements of culture are everywhere the same, but cultures are everywhere different. The structure of the sign is always the same—it is arbitrary, unmotivated, relational, and conventional—but it is precisely this universal form that supports the myriad varieties of cultural experience. Lévi-Strauss never confused the complexity of cultural constructs with the simplicity of their building blocks. That is, he never conflated "potentiality" with "determination," or the basic binary code with culture in the round.[33] What Lévi-Strauss knew (and what the modern reductivists have forgotten) is that even if one accepts the premise that the mind is not a "blank slate," it does not follow that human nature is an orderly compendium of definitive traits, hardwired tendencies, or fixed predispositions.

With Lévi-Strauss, as with Chomsky, there is always the notion of what Norbert Elias calls the "hinge"—that indeterminate passage between the principles of biology and those of culture. Language, in the structuralist tradition going back to Saussure and Roman Jakobson, occupies just this liminal site. Giorgio Agamben interpolates Lévi-Strauss with Chomsky on this very point: "What marks human language is not its belonging to either the exosomatic [superorganic] or the endosomatic [organic] sphere . . . but its situation, so to speak, on the cusp of the two."[34] This "cusp" is a precarious but necessary cradling of the one thing in the other. Somehow, human beings *need* language—a need that must be counted as part of their essential, biological condition. Somehow, infants and small children appear "ready," even anxious, to develop their linguistic facility. Small children acquire verbal skills rather more quickly than a simple behaviorist model would predict, and it is not implausible to think that certain basic elements of language might be "hardwired" in the structure of the brain. Yet somehow, this language—neither unconditional possession nor fortuitous attribute—is always acquired "from without." It is always *learned*. It is never simply "given" with biology. Arthur C. Danto marks the irreducibility of one thing to the other with uncharacteristic bluntness in his speculations on Chomskian linguistics and future science: "There may be—no, I am certain there *will* be discovered—a microbiology of linguistic competence, which accounts for the fact that any child can learn any natural language. But that is not the same as saying that in every child every language—including languages for possible worlds that never are to become actual—is stored in DNA sequences: that there are Russian, French, and Yiddish genes."[35]

Lévi-Strauss's work approaches this "cusp" between language and biology not as the site where culture comes to serve biological exigencies by other means, but as the means whereby biology makes a qualitative leap beyond itself—as the moment when what was "outside" becomes constitutive of what is "inside" and whereby contingency is dialectically transformed into necessity. In the structuralist accounting, then, human beings step forth from a "state of nature" not by claiming "culture" per se but by drawing diacritical distinctions between the raw and the cooked, the wild and the domestic, the animal and the human, and so on—distinctions that in sum (and behind the backs of those who draw them) constitute "nature" and "culture" as correlative, not absolute, terms. Or rather, this ongoing transformation *is* nature for human beings.[36]

The consummate universalist, Lévi-Strauss uses a similar reasoning in his general theory of kinship. The "atom" of kinship is not the biological or reproductive family—the mother who keeps house in the suburbs, the father who commutes to work, and their Gap-clad offspring—but a polyvalent structure involving a large cast of characters in relations of mutuality, reciprocity, rights, and obligations. It is close to Lévi-Strauss's argument to say that kinship is a social rather than a biological phenomenon, but closer still to say that it is the distinctly human political act that takes biology in hand and produces society. Kinship systems, in all their variety, are essentially sets of rules for the association and circulation of human beings—a circulation that sets in motion larger social alliances.[37] What is most universal about marriage and kinship, then, is also what is most *social* about them, and also what is subject to myriad variations.

I never thought I'd pine for the halcyon days of high structuralism, but if what one wants is a sense of order and regularity amid variation and turbulence, one could do worse than to contemplate the serene chains of binary oppositions composed by Lévi-Strauss, associations that spiral, like the Milky Way, from Tierra del Fuego to the Bering Strait, a galaxy of signifiers that begin somewhere in the molecular structure of the brain and span the indefinite human past and an undisclosed future. . . . There is no doubt something bracingly ahistorical, even reductive, about Lévi-Strauss's method. It undeniably provides a poor procedure for understanding historical happenings and social transformations. Everything in culture is set up according to a basic binary logic, and that logic is attributed—reduced—to the absolute structure of the brain, to its hardwired neural circuitry. But if Lévi-Strauss makes recourse to the idea of a universal human interior, his arguments suggest effects very different from those of other twentieth-century essentialisms. Significance does not break into human affairs entirely from without. It is constitutive of them from within, as well. The human being is capable of receiving culture because the human mind already organizes information in relational, oppositive terms. What is most "biological" about human mental capabilities is thus active, interactive, and already implies—requires—social learning.

In this sense, one could speak of biological "limitations" on culture, but also and with the same breath of biological "open-endedness." At the curved cusp of the one thing and the other, a semiotic biology sets limits on language—the limits of the oppositive, binary brain on the logical mind—but without anywhere predicting how a given ritual will proceed, why a dream will occur, by what plot turns a myth will unfold, or what any one person will actually say or do. The diacritical logic of the sign itself might be a feature of the "limits" imposed by the structure of the mind, but such limits have no *positive* implications for the content of human meanings and institutions.[38] Indeed, the "limit" in this case is precisely what sets human nature in motion, allowing for the play of signs that are arbitrary, relational, and conventional—and for the realization of human practices that perpetually transform nature into culture, turning culture into human nature and changing contingency into necessity.

What Lévi-Strauss was really getting at was the idea that human actions—from basic perceptual acts to language, myth, and religion—somehow secrete culture, even when the actors least intend to do so.[39] And what Chomsky suggests is that how human beings put together meaningful utterances may be subject to a limited number of possible combinatory rules. By comparison, what evolutionary psychologists argue is much like saying that the circuitry of the TV somehow determines the content of soap operas or music videos.[40]

Chomsky's ideas about the biology of language might eventually go the way of other grand theories about universal structures—or they might someday amass sufficient evidence to be regarded as true. In the latter event, they would scarcely disprove what empirical studies have already demonstrated about cultural variety. And in fact, they would only supplement concepts and approaches already well established in anthropology and semiotics.

Whatever proves to be the case with grammatical structures, there is an enormous gulf between saying that linguistic (or perhaps cognitive) "deep structures" embedded in the brain might account for the parameters of linguistic variation (or even for the structure of the basic building blocks of culture) and claiming that these deep structures *positively* account for cultural meanings, or that they carry the seeds of predetermined institutional structures. Neural structures are not the "sources" or "causes" of meaning. They are the grounds on which meanings take root. Meaning is not "in" the neural structures, but in the open-ended interaction between perceptive bodies and a universe of signs. And social practices are neither independent of nor reducible to the realm of meanings. The relationship between the one thing and the other is open-ended: dynamically active and dialectically interactive.

16   The Practices of Sex

Through examples like facial expression, body language, emotional communication, color terms, and language acquisition, I have tried to outline an alternative understanding of how biology enters into culture: Meaningful human activity takes up our given biological "limits" insofar as they are useful in structuring cultural distinctions. Biology therefore is not irrelevant to these praxilogical transformations, but it is not determinant of them, either. As a result of this uniquely appropriative relationship between culture and biology, much of what appears to be "natural" or "necessary" thus is actually the biological consequence of a *social* arrangement, the course of biological development under social circumstances. To put it another way, human beings learn how to use their bodies. They learn how to see and to understand their bodies in the context of this usage.

Among the new uses of the body continually reforged in the hazards of experience is sex—a case in point that gets at the nub of the heterosexual binary that grounds every evolutionary just-so story about human nature. As we've seen, examples from the repositories of anthropological exotica illustrate the many-splendored variety of human lifeways, and the best thinking of social theory demonstrates the irreducibility of social institutions to biological instincts. Both empirical data and theoretical reflection thereupon also suggest how culture-bound is the heteronormative logic that gives two and *only* two stable, permanent, "opposite" sexes as the starting point for origins stories about the human design and as the fixed point of reference for universalist claims about eternal norms.

HOW MANY SEXES ARE THERE?

I hasten to add: All systems of social classification do indeed allow for two sexes, male and female. But this is not to say that all social systems give public validation to *only* two sexes. Some societies expressly enunciate three sex categories, socially distinguishing hermaphrodites as a third, mixed, intermediary, or alternate sex. Other societies seem to mark additional categories, allowing for various types of mixed, crossed, or fluid identities in between male and female.[1] Traditional Pueblo-Zuñis, like many other Native American societies, thus allowed for ritual transformations that would turn anatomical men into social women. Other Amerindian cultures allowed for the reverse transformation of anatomical women into social men.[2] The Hijras, a caste group of entertainers in India whom Serena Nanda has memorably described as "neither man nor woman," include persons who might be differentiated in the West as hermaphrodites, transsexuals, and eunuchs.[3]

Even in the West, cultures have scarcely been uniform in their accounting of sex. Classical antiquity, before the arrival of Adam and Eve, gave prominent space to intersexual themes: Hermaphroditism was a recurring motif in Hellenic mythology and in erotic art. Postmodern (one might optimistically say "post-Christian") culture has replicated classical themes, and that bastion of heteronormative dualism, American culture, has recently witnessed a proliferation of new folk genders: male-to-female transsexuals, female-to-male transsexuals, unreconstructed anatomical intersexuals claiming social space as such, not to mention those recent stars of the straight kinky porn circuit, hermaphroditic she-males.

In discussions of intersexualities, transsexualities, and cross-sexualities, it is sometimes said that there are two "biological sexes" from which some societies derive supernumerary "cultural genders." As Judith Butler suggests, such a shorthand accommodates the facts of cultural diversity while preserving a heteronormative logic in nature as anatomical "sex."[4] But "physical differences need not necessarily be seen as bipolar," as M. Kay Martin and Barbara Voorhies pointed out long ago in their chapter, "Supernumerary Sexes."[5] As Gilbert Herdt has more recently argued, the bipolar theory of sex, which purports to objectively describe the precultural facts of biology, is ultimately derived from Darwin's theory of sexual selection, with its reproductivist teleology of the body: Where biological reproduction is understood as the sole aim of sex, male and female will invariably appear as the privileged, primary, and "normal," if not exclusive sexes.[6] And as I have shown in the second part of this book, such a depiction of the biology of sex relies on the elision, omission, and exclusion of considerable evidence to the contrary.

The temptation might be to call hermaphrodites a third "sex" and two-spirit people or Hijras a third "gender," with some transsexuals falling on one side of the divide and some on the other. The operative distinction then would be between "biologically given" and "culturally inscribed" forms of intersexuality. This distinction logically echoes the familiar Cartesian categories—gender as "mind" and sex as "body," with the former superimposed upon the latter—but a synthetic perspective (unevenly developed in the social sciences and humanities since Merleau-Ponty's *Phenomenology of Perception*) begins with the premise that the human body is always "mindful" and, correlatively, that the human mind is always embodied.[7] In human experience, meaning clings to every carnal contour, and volition inheres in the slightest movement of the body. Anatomical "sex," inasmuch as it makes a difference in human affairs, is already "engendered."[8] And insofar as they are linked to systematic associations involving men, women, and others, physical genitalia are from the start party to institutional practices, involved in the play of magical meanings and in the work of metaphorical transformations. The question, then, is not how culture elaborates inventive fictions (gender) around the stable facts of biology (sex), but how temperament, vestment, and anatomy are all caught up together in the varied imaginings and practices of sex.

Heteronormative traditionalists—who cling to the two-sex system the way Flat Earthers once clung to a two-dimensional view of the world—sometimes counter that "supernumerary" genders *and* sexes are culturally marginal or statistically abnormal occurrences. Their implication is that what is marginal is also unimportant, that male and female remain the normal, natural, hence desirable, states of embodiment, default modes of sex *and* gender.

The numbers are actually unclear. Colonial conquerors and missionaries give widely divergent accountings of the number of "berdaches" present in traditional Native North American societies. Many of these accounts suggest that persons of mixed identity were quite common.[9] And although modern medicine focuses on "correcting" "anatomical abnormalities," hospitals do not actually record the incidence of intersexed births.[10] At any rate, it is far from certain that human categories populated by small numbers of people are by that token "marginal" or "insignificant." Hoary cosmologies on all continents give important place to intersexual or transgendered people and to the notion that anatomical sex is somewhat fluid. Those old pederasts present at the origins of Western civilization itself—the revelers convened at Plato's *Symposium*—described the original humans as hermaphroditic creatures belonging to a three-sex system. They imagined present-day people not as superior descendants, but as poor derivatives from the originals. Important ritual and economic functions were often allotted to two-

Figure 22.   Joel-Peter Witkin, *Man with Dog,* Mexico, 1990.
Courtesy of Fraenkel Gallery, San Francisco and Galerie Baudoin
Lebon, Paris.

spirit people in Native North America, and ethnographic sources on various indigenous cultures suggest that every household viewed it as essential to have at least one man-woman.[11]

   The spectacular increase of intersexed figures in recent visual, filmic, and
literary representations does not seem to correspond to any dramatic increase in the number of persons literally claiming mixed or crossed identities, but it does suggest that transgendered or transsexual themes serve an
important cultural role during times of transformation, flux, and anxiety,
an idea developed at length by Marjorie Garber.[12] The social significance of
trannies lies not in their absolute numbers but in the cultural work they do—
on afternoon talk shows, for instance. Since sex terms are "oppositive"—*x*

is defined as not-y—supernumerary sexes affect the entire constellation of meanings available. Transsexuality mediates, complicates, and qualifies the absoluteness of the opposition between male and female. In the presence of a third term, x now enjoys a number of possible relations to others, including not-y, not-z, x/y, x/z, not-x/y. . . . To put the issues in a pragmatic perspective: The categories "Male" and "Female" would not quite sit the same on the page if the U.S. census allowed a third category, "Other"—no matter how few or how many people marked "Other." At the very least, everyone would pause over not two, but three, official options.

HOW GENITALS SIGNIFY

There *is* something basic (and, to use the term in a different sense) "irreducible" about the number "two" in these meaningful distinctions and institutional practices. As I've shown, all systems of social classification allow for *at least* two sexes, male and female. And however many sexes a culture recognizes, male and female are in fact always among them. But this is not to say that it is the facts of anatomy that somehow prioritize the number two (a notion Clifford Geertz scoffed at in *Local Knowledge*).[13] (In fact, in a careful review of the genetic, hormonal, and anatomical aspects of sex, Anne Fausto-Sterling has suggested that the facts of biology, dispassionately considered, might actually lead us to conclude that there are five "naturally occurring" sexes.)[14] Rather, it is to say that human beings, by dint of the way they reckon with material facts and by virtue of the way they make meanings, perpetually reinvent "male" and "female" as basic social categories. Marshall Sahlins's arguments about basic color terms suggest social and semiological rather than biological reasons for this feature of sexual systems. I parallel his arguments here.

Sexes, like color terms, "are in practice semiotic codes." That is to say, sexual categories "are engaged as signs in vast schemes of social relations: meaningful structures by which persons and groups, objects and occasions, are differentiated and combined in cultural orders." Because sexes, like colors, "subserve this *cultural significance*, only certain . . . [anatomical] percepts are appropriately singled out as 'basic,' namely, those that by their distinctive features and relations can function as signifiers in informational systems." For the same reason that black and white serve as oppositive terms in minimal color schemes, the "basic" sex terms will always be male and female—because two is the minimal number capable of carrying meaning and constituting a system and because the anatomical contrast between penis and vagina supplies the least ambiguous of the possible distinctions. (I hasten to add that even this distinction is not an absolutely unambiguous one: the clitoris might be imagined—and has often been imagined—as a "little phallus," the phallus might be seen as an exterior clitoris, and so on.) *"It is not, then, that [sex] terms have their meanings imposed by the constraints of human and physical nature; it is that they take on such constraints insofar as they are meaningful."*

Like color terms, sexual distinctions structure physiognomic contrasts within cultural systems of meaning. Corporeal percepts organized as cultural signs, the sexes designate social distinctions, not natural things. Put another way: what is most "universal" about sex is a manner of *relating*, not "recognizing," a question of form, not content. This is also precisely what is most indeterminate about sex and what leaves the simplest sexual distinction open to the effects of history. Nothing in the distinction between male and female dictates how many other sexes might be discriminated, what their roles and relations might be, or what meanings might be said to follow from these distinctions and relations.

ON THE "FUNGIBILITY" OF SEXUAL SIGNS

"The body is essentially an expressive space."[15] On the question of sex, as on other questions of carnate significance, the meaning of the body is not "in" the body, but outside of it: in how bodies are compared, contrasted, and otherwise related to each other. Or rather (to put the terms in motion): Everything depends on how bodies are engaged in systems of meaningful practice.

"Penis" and "vagina" are, without question, physical organs, truly "given" with the facts of biology. It is also undoubtedly the case that people in all cultures distinguish between penis and vagina, as between male and female. These are the brute facts, and they are not irrelevant to cultural innovations around sex and sexuality. But to imagine that anatomical genitalia (or inner, precultural experiences of them) define some irreducible "bottom line" of human sexual identity is to give shallow treatment to all the ways the penis signifies and to what the vagina means. At any rate, those who would locate sexual identity "in" the anatomy (or worse yet, in the neurohormonal substrate) seldom grapple with the slippery matter of what these "natural" symbols actually signify or how they enjoin strings of associative relations with other body parts, other bodies.

A penis is a penis—but it is also exchangeable for other things in the world. American readers of Clifford Geertz's famous essay on the Balinese cockfight might well understand the symbolic equivalence between a man, his penis, and his fighting cock. (How not? These associations also hold in vernacular English, as in many European languages.) The same readers might puzzle over the implications of the postgame feast, wherein the victor's owner cannibalistically devours the defeated bird.[16] (What, exactly, might *that* mean?) They might also pause over the variety of ways penises have been understood, for even the most familiar cultural associations between things and the penis fail to constitute a singular, coherent set of connotations. "Snake," "worm," and "banana" might seem clear enough, since they bear a passing resemblance to the thing in question (although I cannot help but add that these things are very different from each other: One is regal and dangerous, the other is lowly and unimpressive, while the third is somewhat ridiculous). "Tool" might be said to resonate to a certain sort of supposedly manly activity—but "bird?" The image of a creature apt to take off in flight has always puzzled me. And *bicho*—in Latin America, any small animal? I always picture a rodent scurrying across the ground. To make matters yet more complicated, the nose is sometimes analogically associated with the penis. (Both *are* protuberances from the body.) Sambia men thus periodically induce nosebleeding as a male equivalent of menstruation.[17] Such practices give the formula penis = nose = vagina, a ledger of transformations neither more nor less astonishing than the one that renders manhood self-identical: penis = penis. And a Keraki myth tells how man first obtained the phallic bull-roarer so central to the secret men's cult, so pivotal to male identity: He stole it from woman. It was in her vagina, like an interior penis.[18] Simply put: The meaning of the penis is not "in" the penis, but in how the penis is taken up, distinguished, exchanged, imagined, and instrumentalized in human affairs.

As the Keraki myth suggests, women's parts are no less subject to analogical equivalences, magical transformations, and marvelous discoveries. Plato thought the uterus to be an animal within the animal—an autonomous creature that would go a-wandering in search of satisfaction if deprived of sex or left barren.[19] The "removable vagina" of Hawaiian legend allows for long-distance sexual intercourse.[20] A Vedic tale recounts that the Demoness Dirghajihvi (Long Tongue) had not one but many "mice"—vaginas on every limb.[21] Some Guiana myths attribute the origin of fire to an old woman's vagina.[22] Western sources often read a natural propensity for nurturance into the female anatomy. In a gesture consonant with those sources, Luce Irigaray's vulva-centered feminism makes much of the nonphallic, nonpenetrative, undivisive, and nonviolent action of the vagina: "two [labial]

Figure 23.   Keith Haring, *Untitled,* 1982. Vinyl ink on vinyl tarp,
120 × 120 inches. © Estate of Keith Haring.

lips in continuous contact." "Within herself," Irigaray suggests, the woman
is "already two—but not divisible into one(s)—that caress each other."[23]
But the vagina dentata, an image depicted in some cultural traditions, gives
teeth to a rather different conception of female anatomy.[24] In one place imagined as a gentle, passive organ, in another a voracious, insatiable creature,
the vagina is no less open to meaningful elaboration and transformative practices. So, too, with all our organs of perception, corporeal members whose
contours are culturally imagined, biological parts whose natural use must
be learned.

BIOLOGY, CULTURE, AND SEX

And here, I rest on one of the better established subtleties of modern anthropology: the fact that a human phenomenon is "biological" in nature does
not preclude its having a *cultural* character. To claim that sexual culture is

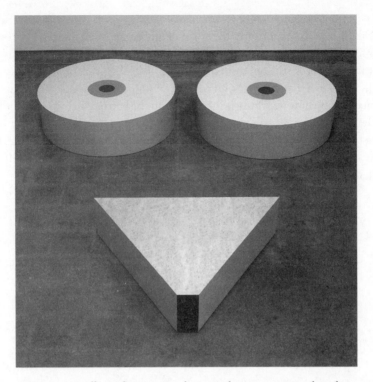

Figure 24. Millie Wilson, *Dressed as a Girl*, 1992–93. Mixed media: formica and wood. Approx. 120 inches × 82 inches. Courtesy of Millie Wilson.

arbitrary, relational, conventional—and sometimes volatile—is not to say that it is *random* with regard to human morphology. When it comes to sex, anatomy "counts"—but that is precisely to say that there is an accounting: an act, a practice, of distinction. Men, women, and others *are* "different," but they are also similar to each other—and therein lies the rub, the ambiguity, and the openness of sexual distinctions to the effects of history. What people construe as "different" and what they *make* of similarities is subject to the most varied formulations. How they reflect upon the physical contours and effluvia of their own bodies is less a question of "biological predisposition" than of social intentionality. "Human nature," such as it is, consists not of stable norms or precoded schemas, but of dynamic possibilities—even where the most "biological" of questions are at stake. And that is why even the best of biology is poorly equipped to comprehend what is neither "unconditional possession" nor "fortuitous attribute." Everything significant about human action depends on what one *makes* of anatomy, and on what implications are said to follow from a biology that gives both difference and similarity as a support for the play of meanings. "The social in effect confiscates the natural," as Terry Eagleton concisely puts it. "The human organism is reconstituted from the ground up by its insertion into the symbolic order of a specific historical culture."[25] Or rather, it is not quite a question of seeing how anything is inserted from the bottom up into some other order—for we might still imagine the "ground" as a stable one—but of understanding how ambiguous facts about organic nature are "taken in hand," transformatively realized in meaningful practices, and articulated in different social and historical contexts.[26]

Permutations on the "Nature" of Desire

The Gay Brain, the Gay Gene, and Other Tales of Identity

How are love, power, and science intertwined in the construction
of nature in the late twentieth century? What may count as nature
for late industrial people? . . . [W]ith what tools is nature constructed as an object of erotic and intellectual desire? . . . Who
may contest for what the body of nature will be?

DONNA HARAWAY, *Primate Visions*

17 This Queer Body

In the context of this prolific, contested, and oh-so-slippery "nature," those of us who once belonged to "unnature" engage in a certain kind of intellectual game, a deadly serious sport of claim and counterclaim regarding our place in the scheme of things. Here is something of the gist of that game, its give and take.

If my desires are not "choices," in the sense that a conscious and volitional "I" chooses from among a set of available options, then my desires must come to me from afar, or from a remote interior, with a will of their own. . . . The conceit of a "sexual orientation," as opposed to a "sexual preference," thus distills a partial truth about the experience of desire, something of its volatile relationship to being. *We do not choose our desires; if anything, our desires choose us. In this choosing, it is "I" who is the "chosen": I am specified and singled out by the desires that find me, move me, and in a sense also* make *me.* . . . And if the Good of late capitalism (like that of Plato) is still measured by its proximity to the ideal form of natural design, then it is the most logical of temptations to keep one's footing in this dangerous game by taking that one further step: *My desires, more than my conscious wants or thoughts, must then belong to nature—to the good form of my inner nature.*

A happy nature, home of radiant truth—*Oh, how I want to be, and be seen as, natural!* But a dark shadow stalks the light of this "nature," like the bad conscience that haunts a bad-faith wager. *For if "I" am never quite the master of my own wandering lusts, I am also never quite in a position simply to receive a desire that announces itself without my active participation— without something in me that desires to desire.*[1] This "nature," too, is in part what we make of it.

And so the desire to catch desire at its most authentic by the tail becomes a game of infinite regress. Nature and desire go off, hand in hand, leaving me in the lurch. In the end, I cannot become "natural" without annihilating the very thing I set out to embrace: a (my)self that strives and desires.

Thus the tortured dilemma of introspected sexualities—of desires subjected to introspective contemplation, not by their nature, but rather, under certain regimes of "nature." Claiming a self lies close to the brink of annihilating a self. Nowhere is this more true than among the high-tech, postmodern biomythologies of the late twentieth century—a social formation whose ideological dilemmas will likely remain in play for a long time to come.

In recent years, biology has been much invoked in public sphere discussions of the "causes" of homosexuality and in political arguments by advocates of gay rights. As we will see throughout the section that follows, a substantial number of people appear to believe that modern science has already demonstrated the existence of a "gay brain" or a "gay gene"—ultimate sinecures of an inner nature, of a fixed, immutable sexual orientation, and the ultimate scientific justification that whatever is, is right, because it is, simply, "natural." The spectral "presence" of these genetic anomalies and anatomical oddities, however, logically provokes angst as much as comfort among the proponents of gay biological essentialism. As the chapters in this section will show, there are good scientific reasons to be skeptical of such "discoveries," and good social reasons to be nervous about the effects of the perceived "discoveries" on the body politic.

OF SCIENCE AND SUPERSTITION

Taken as a gestalt, the February 28, 1997, issue of my local gay newspaper, the *Washington Blade*, indicates some of the intellectual and emotional forces in play in this game of identity. The reader is entreated to play along, to join the game in all its deadly seriousness, ducking between anthropologies, biologies, histories, and myths while weaving between anxious fears and wishful thinking.

The *Blade's* top news story for the week was "Atlanta Lesbian Bar Bombed." The bombing came on the heels of a similar attack on a suburban abortion clinic and preceded subsequent acts of terror. A fundamentalist Christian terrorist group, the "Army of God," claimed responsibility. No doubt, the organization's propaganda somewhere invoked Saint Paul's much-recited verses on "nature": "For this cause God gave them up unto vile affections: for even their women did change the natural use into that which is against nature: And likewise also the men, leaving the natural use of the woman, burned in their lust one toward another; men with men working that which is unseemly, and receiving in themselves that recompense of their error which was meet."[2]

Given second billing and along a single farthest-left column runs the headline: "The Real Amazons? Archaeologists Uncover Graves of Female Warriors." The story summarizes the exhumation in Ukraine of fourteen women's graves believed to date from between 400 and 200 B.C.E., in which what appear to have been women warriors were buried along with weapons of war and hunting tools.[3] The vouchsafing of familiar identities and practices through exotic anthropological conceits—a long-standing ploy in American culture[4]—is campily played up by the author, with comic allusions to Sappho's Isle, "lesbian Amazons," and *Xena, the Warrior Princess.* The story's comedy, more than its content, provides a useful model for how notions about classical antiquity might resonate in queer late modernity—and, now that I think about it, for how conceptions of human nature might inform current conceptions of culture.

Just below this story, at the bottom of the front page, runs the headline: "DNA Discoverer Backs Aborting 'Gay' Fetuses." (Much to the paper's credit, "gay" appears in quotation marks in the title.) The *Blade*'s report summarizes a February 16 story in London's *Daily Telegraph* in which James Watson argued that parents have a moral responsibility to bear children who are "as healthy as possible." Watson reflected, in particular, on the use of genetic tests to determine whether a fetus is carrying "Gay genes." Thus, on the same week that Scottish researchers announced they had successfully cloned a sheep,[5]

> the co-discoverer of DNA told a London newspaper that he supports the idea of women terminating their pregnancies if they find out the child might grow up Gay. . . .
>
> "Every genetic disease is different," [Watson] said. . . . "If you could find the gene which determines sexuality and a woman decides she doesn't want a homosexual child, well, let her" abort.
>
> Watson also said that, while scientists do not know for sure how "normal sexuality" is determined, he believes they eventually will.
>
> The next day, the *London Times* ran an interview in which Quentin Crisp, Gay author of *The Naked Civil Servant*, said he, too, supports the idea of . . . aborting "Gay babies."
>
> "If it can be avoided, I think it should be," said Crisp of homosexuality. He described his own life as a Gay man as very unhappy and "absurd."[6]

Less than a month later, on March 23, 1997, HBO aired *The Twilight of the Golds*, a made-for-TV movie based on a play of the same name by

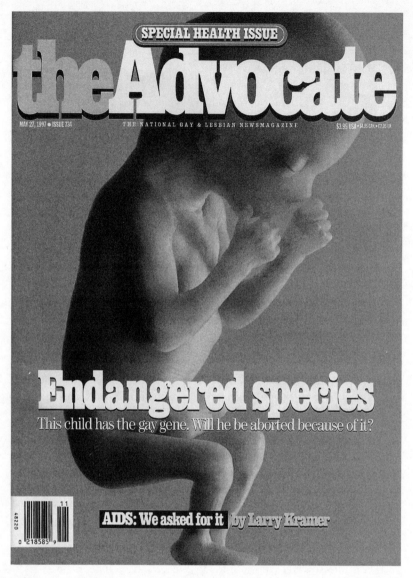

Figure 25. *The Advocate* cover, May 27, 1997. © 1997 by Liberation Publications, Inc. All rights reserved.

Jonathan Tolin. When a genetic test reveals that the son Suzanne is expecting will be gay—as is her brother, David—she considers having an abortion. Neither her husband, Rob Stein, nor her parents, the Golds, try to talk her out of it.[7]

Two months later, the *Advocate*'s cover story went one better than *Newsweek*'s famous "Is This Child Gay?" baby-face cover, reflecting the deep entrenchment of biological determinism in the identity politics vulgate. The *Advocate*'s cover depicts a pink fetus illuminated against a bright red background. The text reads: "Endangered species. This child has the gay gene. Will he be aborted because of it?"[8]

THE HOMOSEXUAL SPECIES

The idea that homosexuals (almost always meaning male homosexuals) are innately, biologically distinct from "normal" people is as old as the idea of homosexuality itself—which is to say, it is only a little more than a hundred years old.[9] This is not to say that same-sex intercourse was either absent or unrecognized and unremarked in earlier dispensations. Saint Paul is surely inveighing against it (or some variation of it) in the above passage, part of his larger polemic against Gentile apostasy. However, in archaic framings— Paul's included—same-sex intercourse was immersed in a much larger class of sex acts deemed "contrary to nature" and linked to religious heresy. These "unnatural" (nonreproductive) acts could also occur between men and women, within as well as outside marriage.[10] Same-sex intercourse was just one variant of many "unnatural uses" of the body. Taking up a quip by Michel Foucault, Jonathan Goldberg has thus illustrated how "sodomy" once constituted—and legally still constitutes—an "utterly confused" and imprecise category: Here it connotes one set of practices, there another; here it defines male-male intercourse, there it describes certain acts between men and women.[11]

Modern taxonomies, which classify sexual acts as either homosexual or heterosexual, along with the attendant notion of a distinctly "homosexual person" or "homosexual body," emerged in the nineteenth century at the confluence of medicine and psychiatry. Foucault marks the difference between archaic and modern conceptions of sexual practices this way: "The sodomite had been a temporary aberration"; he was a sinner—as all men were sinners, as all men were subject to temptation.[12] Foucault goes on to evoke the coming into existence of homosexuals at the dawn of sexual modernity in a striking passage: "An entire subrace was born, different—despite certain kinship ties—from the libertines of the past. . . . The nineteenth-century homosexual became a personage, a past, a case history, and a childhood, in addition to being a type of life, a life form, and a morphology, with an indiscreet anatomy and possibly a mysterious physiology." Thus, "the homosexual was now a species."[13]

Distinctly modern understandings of sexuality—with homosexuality figured as an innate organic condition, a psychiatric abnormality, a deviation from the heterosexual norm, and/or a distinctive way of being—are caught up in histories of empire, changing relations of production, and new ideologies of "nature." As Siobhan Somerville and Jennifer Terry have shown in their wide-ranging explorations of medical texts and psychiatric practices, the construction of a "homosexual body" expressly drew on theories of sexual "degeneracy" (literally, "to become unlike one's race"), on fantastic images of race/gender hybridity ("psychic hermaphroditism," "decadence," "subhuman" traits), and on notions of species "hygiene" (a blurring of same-sex desire with venereal disease, nervous conditions, and morphological irregularities).[14] By way of such associations, the making of the homosexual body is contextualized by—and in turn contextualizes—the making of other identities in linked histories of gender, race, and class. Indeed, the emergence of the concept "homosexual" marks a far-reaching and radical reorganization of the political economy of the body, for it is only in and through the body of this queer foil that a most improbable thing can happen: "The heterosexual," too, steps forth—as a person, as a way of life, as the Edenic paragon of normal, natural desire.[15]

This constellation of significations, this organization of "nature," remains surprisingly intact in recent "scientific" research, where, all euphemisms and self-deceptions aside, heteronormativity is a given and the homosexual is still conceived as a genetic mutation (the X chromosome marks the spot), a hormonal oddity (homosexuality as fetal stress syndrome), an intersexed monstrosity (still, after all these decades, a woman's brain in a man's body), in short: a freak of nature.[16] Given the genealogy of such a conception, the prejudices it distills, the dubious measurements it undertakes, and the pseudoscience it invokes, one should scarcely be surprised that whenever "the homosexual body," "the gay gene," or "the gay brain" are invoked, these phantasms also—simultaneously—raise specters of "selective" abortion, gene "therapy," and hormonal "treatments" for homosexuality. *How could they not conjure such associations?* Conceived as an "organic condition," homosexuality is imprisoned in a violent and exclusionary discourse on "nature." Reified as a "distinct morphology," the very idea of an innately homosexual body is lodged in a complex ideological force field defined by scientific racism, medical misogyny, and eugenics.

For these reasons, we scarcely need fear the day when a "gay gene" is identified, a reliable prenatal test is developed, or effective genetic therapies are marketed. The dream of a world without homosexuality belongs to the realm of fearful fantasies or wishful thinking, not careful, experimental science. Still, the history of medicine suggests that complacency is ill-advised. As Gail Vines has observed, where "homosexual disorder" or obscure "women's complaints" are involved, " 'cures' have invariably soon followed the diagnosis, and the fact that they did not work discouraged no one."[17] Keeping in mind how false ideas about sex nonetheless have had real applications—that is, real social and political consequences—we might well fear instead the implications of the genetic fetishism that is already fully with us and in which so many of our experiences are now, like fly to flypaper, caught.

18   The Biology of the Homosexual

Three studies, published close on each other's heels in the early 1990s, have been widely ballyhooed in the mass media as establishing the "organic seat," the "hormonal link," and the "genetic cause" of homosexual desire and gay identity: Simon LeVay's "gay brain" research, Michael J. Bailey and Richard Pillard's "gay twins" survey, and Dean Hamer's "gay gene" study. Major design flaws, problems with the definition and operationalization of terms, and alternative interpretations of the data were lost in the din of blaring headlines: "First Evidence of a Biological Cause for Homosexuality," "Genes Tied to Sexual Orientation; Study of Gay Men Bolsters Theory," "Study Shows Homosexuality Is Innate," "Genes Linked to Being Gay," "Report Suggests Homosexuality Is Linked to Genes," "Study Provides New Evidence of a 'Gay Gene'" . . . [1]

THE INTERSEXED HYPOTHALAMUS

Simon LeVay's much-cited "gay brain" study was published, with much fanfare, in 1991. The journal *Science* set the tone for press reportage, vigorously spinning LeVay's study to the media under its own press-release headline: "THE HOMOSEXUAL BRAIN: BIOLOGICAL BASIS FOR SEXUAL ORIENTATION?"[2]

LeVay found that the third interstitial nucleus of the hypothalamus (a neural structure at the base of the brain) is, on the average, smaller in gay men and straight women than in straight men.[3] (In theory, lesbians' hypothalami would resemble those of straight men—in other words, where gay men show a "feminized" pattern, lesbians would show a "masculinized" effect.) Because the hypothalamus affects certain endocrine functions and is

240

thought to influence "basic urges" such as hunger, thirst, and sexual arousal, LeVay's findings, which link the size of the third interstitial nucleus to sexual object choice, would seem to reiterate nineteenth-century understandings of sexual dimorphism. Disclaimers notwithstanding, LeVay's homosexuals appear as morphologically "intersexed" beings: female minds (or at any rate "female mating centers") trapped in outwardly male bodies, creatures who either mix or mediate male and female physical characteristics in one body.

The results of LeVay's research were widely disseminated in mass-media outlets, but LeVay's data are less impressive than the public was led to believe, and his study is plagued with methodological problems. LeVay's study examined the hypothalami of forty-one cadavers. While living, nineteen of the subjects were described in hospital records as "homosexual" (a figure that includes one "bisexual"). We do not actually know for how long, or with what degree of consistency, or for that matter even *whether* the "homosexual" subjects described themselves as gay. We know only what someone saw fit to observe (speculate?) in their hospital records. We also do not know how the other subjects described themselves when they were alive, nor do we know *anything* about *anyone's* sexual fantasies or sexual histories, but for purposes of LeVay's study, the sixteen other male subjects are presumed to have been "heterosexual," and all six women subjects are presumed to have been "heterosexual." Many critics have commented on the vagueness—indeed, the capriciousness—of the labels and classifications employed by LeVay.

Needless to say, important aspects of LeVay's research were not always given due weight in science journalism. Note, for instance, that the much-reported difference between "gay" and "straight" men in LeVay's sample is a statistical average, not an absolute difference. Individual measurements overlap: Some of the men in the "gay" sample had larger hypothalami than most of the men in the "straight" sample. Since many individuals did not fit the "average" picture, one could not thus predict who was what simply by looking at his hypothalamus. Such results in such a small sample mean that the resulting numbers lie close to the statistical margin of error—and that the reclassification of a small number of brains in the study would render LeVay's findings statistically insignificant.[4]

To make matters more complicated, LeVay talks as though identifying, delineating, and measuring the third interstitial nucleus were a simple matter. This is not the case.[5] The nucleus LeVay measured is a tiny structure by no means clearly differentiated from the similar neural tissue surrounding it. The fact that LeVay, rather than a colleague, performed the measurements, coupled with the absence of a "blind rater" to confirm his measurements independently, departs from the usual standards in research of this sort and does nothing to lend credibility to the findings.[6]

Worse yet, *all* of the "homosexual" men in LeVay's sample died from AIDS-related illnesses. Both AIDS and HIV medical treatments are known to affect a variety of brain structures. LeVay's inclusion of six (again, presumably) "heterosexual" men who died from AIDS scarcely addresses this problem.[7] Nor does the subsequent examination of the brain of one gay man who died from causes other than AIDS.[8]

In serious publications, LeVay rightly acknowledges that his results are open to a variety of interpretations. For instance, even if his results held—and to date his findings have not been replicated by a single subsequent study—it is by no means clear whether LeVay's average difference would measure biological "cause" or sociological "effect." As LeVay himself puts it, "It is not possible, purely on the basis of my observations, to say whether the structural differences were present at birth, and later influenced the men to become gay or straight, or whether they arose in adult life, perhaps as a result of the men's sexual behavior."[9] It is also not possible to say whether the average structural differences have anything to do with sexual object choice per se or with other aspects of life associated with sexual object choice. Certainly, extended anxieties, social stress, the experience of inequality, sexual activity and inactivity, and various other cumulative life experiences affect organic processes, brain structures, and hormonal systems in human beings. The sexual dimorphism claimed for the third interstitial nucleus, if verified, could be the organic *consequence* of male and female socialization patterns, not some "given" or "innate" biological difference between the sexes. And sexual orientation, with all that it entails, might logically be the *cause*, not the effect, of the same organic variation between gay and straight men.

LeVay has made far less cautious claims in public discussions of his study. LeVay's interpretation of his results, aggressively forwarded in a variety of media, is in no small part driven by his personal conviction that he was "born gay" and from his belief that the innatist scenario advances the social interests of gays and lesbians. LeVay thus favors a biologically reductive argument: The hypothalamus is the "seat" of sexual desire, and sexual object choice (or preference, or orientation) is physically *there*, in the third interstitial nucleus. As LeVay told *Newsweek*, "I felt if I didn't find any [differences between gay and straight men's hypothalami], I would give up a scientific career altogether."[10]

It might seem that the starting point for LeVay's inquiry—an association between sexually dimorphic structures in the hypothalamus, presupposing neural-hormonal links to sexual object choice—is sustained by other, earlier scientific studies. But in fact, the hypothalamic hypothesis is supported by a slippery mix of vague surmises, contested findings, biased interpretations, and outright misunderstandings.

First, it is not entirely clear which, if any, ill-defined portion of the hypothalamus displays the characteristics of "sexual dimorphism"—much less which portions might affect sexual arousal in humans. William Byne's review of the literature is instructive on this point, and I draw heavily on his arguments in the following reprise.[11]

In 1978, Roger Gorski identified a cell group in rats approximately eight times larger in males than in females. He called this nucleus the "sexually dimorphic nucleus of the preoptic area" (SDN-POA). Following up on Gorski's research, William Byne and Ruth Bleier then found that the SDN-POA was in fact "the midportion of a more extensive complex of sexually dimorphic nuclei" in rats. Although Byne and Bleier's findings have since been replicated in rats and certain other rodents, no structure comparable to the SDN-POA has ever been found in that other laboratory stand-in for human biology, mice. An obvious question, as Byne puts it, is: Why should a sexually dimorphic nucleus exist in humans, if not in mice? Notwithstanding this commonsensical caution, in 1985, Dick Swaab reported that what had been known since 1942 as the "intermediate nucleus" in humans is in fact the "sexually dimorphic nucleus," analogous to the SDN-POA in rats. Swaab also provided evidence that the size of the intermediate nucleus did *not* vary with sexual orientation. But two subsequent studies were unable to confirm Swaab's finding of sexual dimorphism in the intermediate nucleus, thus casting doubt on the claim that this portion of the hypothalamus is actually the human equivalent of the rats' SDN-POA. Gorski's UCLA laboratory then chartered three hitherto undiscovered nuclei—the second, third, and fourth interstitial nucleus—two of which (the second and third) were reported to be sexually dimorphic. (Gorski's group renamed Swaab's "sexually-dimorphic nucleus" the "first interstitial nucleus" for obvious enough reasons: If this portion of the hypothalamus is not "sexually dimorphic," then calling it the "sexually-dimorphic nucleus" would be a misnomer.) LeVay's subsequent study did not quite test or confirm Swaab's findings, but rather depicted the *third* interstitial nucleus as dimorphic according to both sex and sexual orientation. LeVay's findings thus rely on, but only partially confirm Gorski's, which diverge from Swaab's.[12]

As the foregoing might suggest, LeVay's study rests on an analogy between human and rat (but not mouse) brain structures. The rationale undergirding LeVay's study and the arguments he weaves around his findings both rely on presumably established connections between the hypothalamus and sex behavior in rats. But as Byne summarizes: "Aside from the fact that we now have two reputable scientists, LeVay and Swaab, suggesting that two different nuclei are the counterpart of the SDN-POA in the rat, it is crucial to note that the destruction of the rat's SDN-POA does not impair mounting or any other sexually dimorphic mating behavior or endocrinological function."[13] The actual connection between the hypothalamus and sex is this: The destruction of a portion of the hypothalamus in rats *above* the SDN-POA affects sexual *drive*. That is, it reduces the probability that a male will attempt to mount an estrous female.

In short, the problems are legion: with the antecedent studies LeVay cites as precedent, with the animal models he uses as analogies for human structures, and with his modeling of neural-hormonal-sexual links. It is not entirely clear which part of the human hypothalamus, if any, is analogous to the SDN-POA. In fact, this dimorphic structure in rats does not even appear to be directly linked to *sexual function*. And although a different portion of the hypothalamus has been linked to sexual motor functions in rats, it can scarcely be inferred that any of this is in any way linked to sexual "orientation" in humans.

The tradition treating the influence of prenatal and postnatal hormones on sex drive—the developmental point at which the hypothalamus supposedly "sets" sexual object choice—is even more problematic, although it is very much part of the flawed record backgrounding LeVay's study and informing its assumptions.

By manipulating the prenatal and postnatal hormonal levels of some rats, in effect chemically castrating and "feminizing" them, Gunter Dörner has produced what he calls "homosexual rats"—that is, male rats who routinely show a "feminine" lordosis response (squatting, allowing penetration) when other male rats mount them. Conversely, the East German scientist has also produced so-called lesbian rats—chemically masculinized female rats who mount untreated, squatting females (and/or chemically castrated males).[14] But such definitions of rat "sexualities" are decidedly odd—for who is displaying "homosexual behavior" in this scenario? The mounted? The mounting? Both? Or neither? If ordinary, untreated male rats will attempt to mount other male rats—chemically altered or unaltered—how could it ever be said that these ordinary rats are in any meaningful sense "heterosexual"? And if control group female rats will squat for chemically altered females, then it ought to be clear that the scientists are working from a set of almost unbelievably naive preconceptions about sex, temperament, and sexuality when they label one party "heterosexual" and the other "lesbian."[15]

All around the hypothalamic hypothesis swirls a shocking confusion of terms and practices: anthropomorphic projection, cultural preconception, ideological distortion . . . Not only does this confusion exist as a bad writing practice, but it undergirds the neurohormonal studies and sociobiological science fiction at deep methodological levels. The studies in question analogize between rats and humans by imbuing rats with human characteristics (desires, affects, intentions) and by depriving humans of even minimal human powers. Gay and straight, male and female, are portrayed as deeply, radically different from each other—but humans and rats are not. In view of its many gaps, LeVay's study provides "evidence" for genetic-hormonal causation only for someone who already wishes to believe that homosexuality is essentially a biological condition.

BROTHERHOOD

Only months later the same year LeVay's study appeared, Michael Bailey and Richard Pillard published the results of a survey they conducted among gay men and their brothers.[16] The researchers recruited respondents by placing ads in gay newspapers across the Midwest and Southwest, ultimately gathering information on 56 pairs of identical (monozygotic) twins, 54 pairs of fraternal (dizygotic) twins, 142 non-twin brothers, and 57 pairs of adoptive brothers. They found that the "concordance rate" of homosexual self-identification—that is, the percentage of pairs in which both brothers called themselves gay—was highest for identical twins (52 percent), next highest for fraternal twins (22 percent), and lowest for non-twin and adoptive brothers (roughly 10 percent each).

Once again, methodological concerns and alternative interpretations were ignored or brushed aside. And once again, headlines trumpeted "mounting evidence" of a genetic basis for homosexuality.

How one interprets this data is largely a matter of the perspective one takes. As Ruth Hubbard and Elijah Wald dryly observe: "The fact that fraternal twins of gay men were roughly twice as likely to be gay as other biological brothers shows that environmental factors are involved, since fraternal twins are no more similar biologically than are other biological brothers."[17] Indeed, genetically unrelated adoptive brothers show the same concordance rate as blood brothers, which would further seem to falsify the genetic hypothesis. And by definition, the monozygotic twins are genetically identical—yet *only half* of the pairs were sexually concordant. Given the conditions and assumptions of Bailey and Pillard's study, this figure could be viewed as surprisingly high or as revealingly low. It could even indicate that sexual orientation has *no* genetic basis whatsoever.

This is because twin studies normally use pairs of identical twins who were separated at birth. Such studies thus attempt to view the development of genetically identical individuals in (supposedly) different environments.[18] Since the identical twins in Bailey and Pillard's study in fact *shared* a family environment, it is a non sequitur to claim that the comparatively high (although theoretically low?) degree of concordance is genetically caused. It might just as easily result from the fact that the two occupy the same environment. As Hubbard and Wald put it: "If being a fraternal twin exerts an environmental influence, it does not seem surprising that this should be even truer for identical twins, who the world thinks of as 'the same' and treats accordingly, and who often share those feelings of sameness."[19] Gilbert Zicklin goes even further: "The intensely shared life of identical twins, including the phenomena of identification, mirroring, and imitation, might plausibly constitute fertile ground for the development of same-sex erotics."[20] Zicklin's suggestion is at least as plausible as the invocation of "genetic causation" to explain the 52 percent of identical-twin pairs who were concordant and "environmental factors" to account for the 48 percent who were discordant, an accounting that in no way follows from the data, but that dominated media presentations of the topic.

Consider the extraordinary anecdote related in *Newsweek*'s 1992 cover story, "Born or Bred: The Origins of Homosexuality."

> Until the age of twenty-eight, Doug Barnett (not his real name) was a practicing heterosexual. He was vaguely attracted to men, but with nurturing parents, a lively interest in sports and appropriate relations with women, he had little reason to question his proclivities. Then an astonishing thing happened: his identical twin brother "came out" to him, revealing he was gay. Barnett, who believed sexual orientation is genetic, was bewildered. He recalls thinking, "If this is inherited and we're identical twins—what's going on here?" To find out, he thought he should try sex with men. When he did, he says, "The bells went off, for the first time. Those homosexual encounters were more fulfilling." A year later both twins told their parents they were gay.[21]

The author of the *Newsweek* piece relates this tale as evidence of a fixed, clear-cut, and genetic basis for sexual orientation.[22] That is, no doubt, what the protagonist, "Doug Barnett," himself believes. But the tale could be read just as easily as a demonstration of the flux, ambiguity, and capriciousness—

indeed, the *suggestibility*—of sexual desire. The subject's description of his life as a "practicing heterosexual" is in no sense unusual. In various surveys, beginning with the Kinsey study, large percentages of men whose sexual activity is predominately or exclusively heterosexual agree, in principle, that everyone experiences "vague feelings" of "occasional attraction" toward members of their own sex.[23] Such findings are conveniently forgotten in the current rush to geneticize and typologize desire. The *Newsweek* anecdote could be understood as a particularly sharp example of the "twinning" behavior Zicklin invokes. Indeed, if taken seriously, it could even be understood from a constructionist perspective—why not?—as a gauge of the social force of reductionist theories in shaping personal life and identity formation.

In the end, even if we take Bailey and Pillard's figures as reliable ones, we simply do not know which had more of an effect on the identical twins' sexuality, shared genes or a shared environment, and we cannot even be sure whether we are monitoring a genetic tendency through degrees of sibling relatedness, a social tendency for twins—especially identical twins—to be alike, to mimic, mirror, and "twin" each other, or even a homoerotic tendency among identical twins.

So far, all of these interpretations lie within the realm of a generous reading of Bailey and Pillard's study—that is, within the assumption that their results are meaningful, that their numbers actually reflect real trends among siblings. But this is not necessarily the case.

The authors' sampling procedure almost guarantees a certain skewing. It is based on the self-selection of volunteers recruited through gay newspapers, rather than on a random sample of the general population. Given the stated aims of the study, which are clear enough in the ad, and given the cultural and political background of the question, which includes the active promotion of "innatist" scenarios in most gay newspapers, it is altogether possible that those who were most motivated to participate were those who already believed that sexual orientation is genetically determined. And it is altogether conceivable that those most likely to respond to the ad—to nominate themselves for study—would be concordant sets of identical twins.

These are not minor problems. They fatally undermine the study's reliability. As Zicklin elaborates:

> The overrepresentation of concordant MZ [monozygotic, identical] twins is quite possible, since gay MZ twins are likely to be more interested in studies that highlight the special meaning of close biological connections, and they might also have less trepidation about participat-

ing since there is a greater likelihood that they would be "out" with
one another than would any other pair of male siblings. Conversely,
some twins who perceive themselves as discordant on sexual orientation may be motivated to avoid studies wherein this difference may
be revealed. Thus Bailey and Pillard have a double problem: they attract
the kind of twins who fit their hypothesis and deter the ones who might
weaken it.[24]

Bailey and Pillard skirt the usual standards of twin studies, sampling procedures, and logical deduction. Again, only those already committed to the
notion that homosexuals have biologically marked bodies would be swayed
by this kind of evidence.

THE "GAY GENE"

The 1993 study by Dean Hamer and his associates is usually praised for being the most serious, sophisticated, and careful of the three major studies
purporting to substantiate a link between genes and male homosexuality.

Hamer's research team recruited an original group of 76 gay men for a
pedigree study. (A "pedigree study" is an attempt to determine how a trait
is distributed among members of a kin group.) One or more relatives from
26 of these men's families were also interviewed, for a total of 122 participants. Hamer's team found elevated levels of homosexuality among gay
men's maternal uncles and among their maternal cousins, linked by aunts,
as compared to their paternally linked relatives. Hypothesizing transmission of a homosexual gene through the X chromosome, the researchers then
recruited 38 pairs of gay brothers for a second pedigree study. These pairs
of gay brothers were specifically culled from families without known lesbians or paternally linked homosexuals in order to eliminate subjects likely
to display "nonmaternal" routes of "transmission." The second pedigree
study found a somewhat more pronounced maternal pattern. Finally, the
Hamer team performed DNA linkage analysis on the 38 pairs of gay brothers from the second pedigree study, plus two pairs of gay brothers from the
first study. Hamer et al. reported that 33 of 40 pairs (or 82 percent) shared
a DNA marker, Xq28, located on the tip of the X chromosome. (The term
"DNA marker" denotes a strip of DNA that is usually transmitted "whole"
from parent to offspring; it thus allows geneticists to work with units of a
few million base pairs of DNA, rather than trying to sort out individual genes
from among several billion base pairs. Xq28, as the authors note, is large

enough to contain several hundred genes.)[25] Hamer et al. conclude: "We have now produced evidence that one form of male homosexuality is preferentially transmitted through the maternal side and is genetically linked to chromosomal region Xq28." The authors suggest that a thorough mapping of the region will eventually yield a gene involved in homosexual expression, but they also suggest that more than one gene might contribute to sexual orientation, and that environmental factors also play a role.[26] ("Hey, Mom, Thanks for the Genes!" is the message that with minor variations appeared on gay T-shirts across the country—a line that proved even more popular than the camp come-on "How Big Is Your Hypothalamus?")

In his scientific (as opposed to journalistic or popularizing) publications, Hamer has been careful to avoid extreme variants of biological determinist arguments. Indeed, Hamer himself often points out that a "link" is not the same as a "cause." He distinguishes between "genetic influences" and "genetic destiny," and even while in search of a "gay gene," he often puts the term inside eyebrow-raising quotation marks.[27] Still, there is something less than fully congruous about searching for a "gay gene" while claiming that one does not exist, and the problems with Hamer's study are quite serious.

The pedigree studies invite certain preliminary observations. First, not all of the families in Hamer's samples exhibit the "maternal pattern" highlighted in the subsequent genetic study of gay brothers. The results suggest a "significant" but not dramatic elevation of homosexuality among the maternally linked relatives of gay men.

Next, some of the raw numbers supporting the idea of a maternal linkage are in fact quite low. In the first pedigree study, 7 of 96 gay men (7.3 percent) reported having a gay maternal uncle, as opposed to only 2 of 119 (1.7 percent) who reported a gay paternal uncle. But there is little difference between the 4 of 52 (7.7 percent) who reported a gay maternal cousin on their aunt's side and the 3 of 56 (5.4 percent) who reported a gay paternal cousin on their uncle's side.

In consequence, the difference between rates of homosexuality among maternal and paternal kin is statistically significant *only* if one assumes a (relatively low) 2 percent "base rate" of male homosexuality. As Edward Stein and others have pointed out, the difference becomes statistically insignificant if one assumes a (more plausible) base rate of 4 percent.[28]

Finally, given such small raw numbers, Hamer's pedigree analysis is open to charges that it fails to account for even the most obvious relevant effects of gender and family relations in American society. Women—mothers—play a much greater role than men in negotiating and cementing family ties, a tendency that is well established in the sociological and anthropological literature.[29] As a result, Americans tend to be closer to and to know more about their maternal relatives than their paternal ones. This sociological effect is likely to be even more pronounced in the case of gay men than in society at large. Given the role of fathers in perpetuating cultural expectations of masculinity, given the cultural anxieties that a gay son reflects upon his father, and given the nature of the idealized maternal role (nurturing caregiver), it is certainly conceivable that on the average, gay men tend to be closer to their mothers and to know more about their maternal, consanguineal kin than they are to their fathers, about whose blood relations they know correspondingly less.[30]

Hamer's team did attempt to apply a reasonable check on information provided by the gay men. They also interviewed at least one relative each for twenty-six participants (for a total of forty-six relatives interviewed). On this basis, Hamer concluded that the information provided by the seventy-six total participants was reliable. One might suggest, instead, that the claims were merely *consistent:* that one relative tended to think pretty much what another relative thought. Since extensive networks of the gay men's relatives were not systematically interviewed, either or both of the above sociological factors could fully account for the maternally skewed results of Hamer's pedigree study.[31]

At this point in a review of Hamer's study, it is usually conceded: "Yes, but Hamer's group nonetheless found *something*—a genetic marker—shared by gay brothers, and that is in itself significant." And after all, Hamer's group claims only to have established a genetic link for "one form" of male homosexuality—presumably the kind genetically transmitted along maternal lines. Still, there is considerably less significance here than one could glean from media reports, which took Hamer's study as the charmed third to seal the argument.

It is important to specify first what has *not* been shown by Hamer's group. First, no "gay gene" has been identified. Nor can we safely conclude that one is there, in Xq28, like a needle in the proverbial haystack, awaiting discovery. All kinds of traits "run in families" without having a genetic basis. And because human populations are quite variable, when a trait "runs in families," a "DNA sequence that is a marker for a particular trait in one family may not be associated with that trait in another."[32] The complexity of the relationships between genes, heredity, and even relatively simple phenotypic and behavioral characteristics has frustrated the search for genes "for" all manner of things that would appear far more straightforward than sexual desire.

There also is no genetic on/off switch for homosexuality. Even after a deliberate screening and selection process designed to produce a "maternal pattern" of "linkages," if not "transmission," not all the pairs of gay brothers whose X chromosomes were examined shared DNA markers for Xq28. A subsequent study by the Hamer group reported a somewhat lower percentage of Xq28 concordance among gay brothers.[33] Even a generous interpretation of these results along the lines laid out by the authors clearly does not indicate a simple or direct genetic "cause" for homosexuality.

There is no conceivable genetic "test" for homosexuality. Specifically, it has been reported that a percentage of pairs of self-identified gay brothers, culled from certain highly selected samples, share a genetic marker. Note that this selectively culled group of gay brothers share that marker *with each other*, not with unrelated gay men. Thus, even if Hamer's results hold, no one can take a blood sample and look at this genetic marker to determine whether a person is gay or straight.

Finally, in larger terms, the search for an "organic seat" or "biological cause" of homosexuality remains an undemonstrated conceit—a mishmash of selective citation from the animal kingdom and speculative parallels to poorly understood human processes. Although various commentators have speculated that some gene in Xq28 *might* play a role in sexual orientation by way of neurohormonal links with the hypothalamus, no one has specified exactly how this might happen, much less tested a coherent hypothesis. In view of the aforementioned problems with LeVay's hypothalamus work, it is unlikely that they will.

If a generous but dispassionate look at Hamer's study scarcely gives cause for panic that homosexuals are fast becoming an "endangered species," there are moreover very good reasons to be skeptical of the Hamer group's results and its limited conclusions. As Edward Stein points out, the use of genetic markers in genetic linkage analysis "is best suited for discovering the genetic basis of traits that are controlled in a genetically simple manner rather than traits that are controlled by several genes working in concert." Mistakes and "false positives" are especially likely "in the case of genetically complex (that is, polygenic) traits or traits that are strongly affected by environmental factors." Stein adds that "linkage analysis has not been a fruitful method of study for behavioral traits" and goes on to note dryly that since sexual orientation is a cognitively mediated behavioral trait, "linkage analysis does not seem an especially promising technique with which to study" it.[34]

The questions raised about the reliability of Hamer's pedigree studies are crucial. Because the pedigree study is based on such poor design for sociological research, the likelihood is increased that the Xq28 concordance rates are the equivalent of "false positive" readings, results that appear to be significant but that are not replicated in subsequent research. (This kind of result happens all the time, even in unimpeachable, well-designed research.)

Notably, the Hamer group did not try to determine how many *nongay brothers* share this region of the chromosome with their gay brothers, much less whether pairs of straight brothers exhibit high rates of Xq28 concordance among themselves. This is not a trivial matter, because unless we know the Xq28 concordance rates for gay men with their heterosexual brothers, we have no way of interpreting the meaning of the 82 percent rate among gay brothers reported in the first study or the 67 percent rate reported in the Hamer group's follow-up study. Hamer's conclusions—that the gay men received a maternal chromosome for homosexuality and that Xq28 is *a* (or even *the*) genetic site involved in sexual orientation—depend on a viable control group that has never been established. The absence of such a control group renders Hamer's first study's results virtually meaningless.

The follow-up study, which found a lower rate of Xq28 consonance between gay brothers, did report a *very* small sample of eleven families in which two gay brothers shared the Xq28 marker and also had a nongay brother. It is reported that nine of the nongay brothers did not share the marker with their two gay brothers and that two did—but these numbers are very small indeed, scarcely adequate for a viable control group.

Perhaps most significantly, the Xq28 concordance rate for gay brothers fell from 83 percent in the first study to 67 percent in the second study. As Jonathan Marks makes clear in his discussion of inflated claims in the field of "behavioral genetics," the design of Hamer's study makes it extremely sensitive to a small number of families matching or not. The real question is not "Is there a gene for homosexuality?" but rather, "Is the 82 percent concordance result sufficiently different from the 50 percent rate that would occur by chance to be meaningful?" The concordance rate in the second study lies considerably closer to the 50 percent rate that would presumably occur in a DNA linkage analysis of pairs of brothers chosen entirely at random.[35]

More significant than any technical problems with Hamer's research design, however, are fundamental problems with the conception of the research and with the untested and untestable assumptions embedded therein. As is frequently the case with such research, the Hamer study implicitly understands phenotype (the aggregate physical and behavioral characteristics of an organism, usually understood as the product of a dynamic interaction between genes and environment) as the more or less direct expression of genotype (the state of the organism's genes, or the inherited genetic "givens" that are brought to the interaction), thus demoting "environmental" factors to an order of secondary importance. Whereas genes "for" this or that trait are conceived as playing a stable and "active" role in constructing the person, the environment serves as a backdrop and plays an essentially "passive" role, either speeding along the pre-given results or posing obstacles for the normal course of their expression.

This conception has the effect of obscuring the peculiar environment established by the study itself. Note the selection process that produced the sib-pair sample: There are always two gay brothers, maternally linked to other homosexual kin. We do not know how to compare this very specific sample with gay men who do *not* have gay brothers or other gay kin. This is no minor quibble, for the sampling procedure makes it impossible to distinguish environmental and social factors from genetic ones. There might well be major *social* differences between the development and experience of sexuality where a gay sibling is present, as opposed to sexual development and experience in other kinds of settings. Hamer has presumably accounted for this objection by claiming that he has identified "one form" of male homosexuality—presumably the kind genetically "transmitted" from mother to son. But this does not necessarily follow, and there are no compelling grounds for concluding it, unless one assumes that Xq28 in fact hides a "gay gene," which has not been demonstrated.

An alternative hypothesis, then: If older and successfully homosexual relatives serve as role models, fostering a sense of esteem for the homosexual feelings of younger relatives during crucial periods, then the "trait" in question might actually be "transmitted" *socially*, from uncle to nephew, from cousin to cousin, from brother to brother. . . . And the "form" of homosexuality identified here might mean only that there is an environmental difference—in that having a gay brother constitutes a different environment than not having a gay brother.

In this context, consider the most generous possible reading of Hamer's results, on their own terms—including the assumption that there must be some kind of "linkage" between genes and sexual object choice. Even assuming that Hamer's data, in toto, are reliable, there is no way of specifying exactly *what* is shared by gay brothers in Xq28: some gene *directly* related to sexuality and sexual orientation? Or some gene that has nothing to do with sexuality *directly,* but that can become linked, *indirectly* and under certain circumstances, to sexuality? In other words, the question of cause versus effect—indeed, of multiple causes and effects—has not been settled. Are consonant sibling pairs simply expressing a genetic predisposition toward homosexuality? Or are they being subtly socialized into homosexuality based on some other characteristic or set of traits? Or are they indirectly prodded toward the resolution of various conflicts through a homosexual outcome? Or even, yet again: Are they discovering and/or coming to emphasize a homosexual potential by way of some other characteristic, or by way of some other affinity with close kin?

Media reportage of genetic research like Hamer's invariably traffics in oversimplified, folkloric understandings of genetics and heritability. Headlines tell us that biologists have unearthed the "roots" of sexual orientation, or that geneticists have identified the gene "for" thrill seeking or a love of novelty. . . . [36] Such reportage, directed at the lay public, inevitably glosses complex technical questions. But it is not always clear that the research itself, considered apart from its splashy publicity, maintains a properly scientific approach to the question of heritability or the role of genetics in biological processes.

In the vernacular, heredity denotes what is "given," what is "in" the "blood": It is the part of human variation that is "caused" by genetic "nature," rather than by environmental "nurture." The folk conception of heredity also implies "immutability": The leopard cannot change his spots, and short of wearing colored contact lenses, human beings cannot change the color of their eyes.[37]

The biological conception of heritability is more precise and less deterministic. In biological terms, heritability is a measure of the likelihood that a trait present in one generation will recur in subsequent generations sharing a common gene pool in the same environment. Expressed as an equation, heredity includes both a numerator and a denominator. Richard Lewontin, Steven Rose, and Leon Kamin give that equation this way:

$$\text{Heritability} = H = \frac{\text{genetic variance}}{\text{genetic variance} + \text{environmental variance}}$$

where "genetic variance" refers to "the average performance of different genotypes" and "environmental variance" refers to the "variation among individuals of the same genotype."[38]

Two important qualifications follow from this formula. First, scientists attempting to determine the heritability of a trait assess average genetic variation. They do not measure genetic "causes." Second, environmental variation is part of the denominator—a basic point that is often forgotten in genetic research on complex human behaviors.

Note what limited arguments a properly biological conception of heritability and genetic factors actually permits. To say that a trait is "highly heritable"—that a high percentage of phenotypic variation is correlated to genetic variance—does not preclude saying that the trait also responds dramatically to environmental conditions. For example, if we say that height among a group of human beings has a heritability factor of about .9, or 90 percent, what this implies is that children in that group tend to be about the same height as their parents, all other things being equal.[39] But height also responds, impressively, to environmental factors, especially to childhood nutrition. Drought in the Sahel and famine in North Korea produce children whose height is substantially less than that of their parents, as is their body weight, among other things. In much of Asia, a shift away from traditional rice-and-fish staples to a cuisine more closely resembling the Western diet, with its emphasis on red meat, has dramatically raised the average height—along with body weight, average cholesterol levels, cardiovascular ailments, and the like. Heritability, then—even an extremely high measure of heritability—does not imply inevitability, immutability, or even genetic "causation." To say that a trait is "highly heritable" for a given population means only that the trait in question recurs at a certain rate among genetically related kin reproducing in a shared and relatively stable environment. It also implies a number of very substantial contingency clauses. If the environment changes, whether by accident, by migration, or as a result of changes introduced by the activity of the population itself, then the trait in question could also change dramatically.

To make matters yet more complicated, the heritability of a given trait can vary from group to group and place to place: "Some populations may have a lot of genetic variance for a character[istic], some only a little. Some environments are more variable than others."[40] For certain complex traits correlated to polygenic factors, environmental changes could signal the appearance of the trait in families where it was previously absent—or its elimination from lines where it had previously occurred. Finally, some simple, highly heritable traits in some species respond dramatically to changes in the environment—but not in any linear or straightforward way.[41] Famous experiments by Jens Clausen, David Keck, and William Heisey elegantly illustrate this principle.[42] The scientists took three clippings each from several different individual plants of the species *Achillea millefolium*. Such clippings will produce "clone" plants genetically identical to their parent plant and to each other. The scientists planted the clippings from each of the different plants in three different environments to observe how they grew under different conditions: one each at low, medium, and high elevations. The genetically identical plants grew to different heights at different elevations, but some were "tall" at low elevations, "short" at medium elevations, and "tall" again at high elevations. Others exhibited the opposite relationship: "short," "tall," and "short" from low to high elevation. Some showed a wide range of variation in different climates, others a narrow range. Although it was clear that the plants' heights were affected by elevation, it proved impossible to predict just how individuals would actually respond to different environments.[43]

Let us imagine, then, that homosexuality *has* a heritability factor, and that Hamer and his team are on to something. Even if one takes the Hamer results at face value—and I have tried to indicate some of what might be wrong with the research itself—and even if the findings withstand subsequent restudies, which is already very doubtful, the correlation of some form of sexual variation with some kind of genetic variation has many fewer implications than the lay public (or for that matter much of the science establishment) seems to think. Even a relatively high correlation—a high heritability factor (the worst-case scenario for partisans of a constructionist perspective)—could not preclude dramatic or unpredictable environmental effects on sexual orientation. Nor could it preclude the possibility that, under other circumstances, the "trait" in question could manifest itself differently or among altogether different kin groups.

Genetic research like Hamer's almost never announces itself with anything resembling the range of caveats appropriate for properly restrained biogenetic research. More often than not, it lapses into an essentially folkloric understanding of heritability: the search for "the" "gay" "gene," the confusion of "genetic correlation" with "genetic causation." That is because biologists, as a group, tend to be committed to an ideology of biological reductionism, with its reification of practices into things, even when such reduction runs contrary to their own best methods.[44] They also tend to reject the notion that science cannot answer every question.[45]

Readers will no doubt see where I stand. I do not believe that homosexuality is really susceptible to even "good" biological research. As a complex, meaningful, and motivated human activity, same-sex desire is simply not comparable to questions like eye color, hair color, or height. I am not even convinced that "desire" can be definitively identified, isolated from other human feelings, objectively classified, gauged, or compared. For how are we to measure the "occurrence" (or non-occurrence) of a "trait" that is itself relational, subtle, and subject to varied modalities and modulations? And how are we to measure environmental constancy across generations on a subject defined by contestation, volatility, and change?

19 Desire Is Not a "Thing"

Arguments about the "nature" of desire invariably turn on broader claims about human nature. Now, as in the past, innatist explanations for homosexuality lean on essentialist theories of gender—on heteronormative (and heteronormalizing) models of how biology "works" and what sex is "for." Dean Hamer, for instance, devotes a crucial chapter of *The Science of Desire* to an entirely speculative sociobiological discussion of the "gay gene" in human evolution. The chapter concludes with an astonishingly reductionist (and reproductivist) argument about the basic human design: "Since evolution selects genes that allow people to reproduce, there must be genes that encourage men and women to have sex and raise children." (The "gay" gene, it would appear, is but a variation on the "straight" gene. A subsequent book by the same author, *Living with Our Genes*, proposes genetic explanations for all manner of complex human behaviors.)[1] The trajectory of LeVay's *The Sexual Brain* is also illustrative: It traces the biological roots of human sexual behavior back to the familiar evolutionary just-so stories about the eager, aggressive sperm and the coy, choosy egg: default modes from which the homosexual differs (and which the homosexual type in some way mediates).[2]

Hamer, LeVay, and others seem serenely unaware of the lessons drawn by sophisticated historiography and serious anthropology in recent years about desire as a social phenomenon, contingent, variable, and varied. What they depict as universal human norms are in no way verified by cross-cultural evidence. And what their studies identify as the phenomenon to be demarcated, measured, and reduced to its biological basis—a homosexual identity, typified by lifelong, exclusive homosexual behavior—is largely a modern, Western, even Northern, phenomenon.

It is not clear what the Band of Thebes, bound to each other as "an army of lovers," would make of our notions, nor can we know what the aristocrats of classical Athens—that great pedophile ring of Western antiquity—would make of our conventions.[3] The medieval followers of Saint Paul might have difficulty understanding why we believe that only men who have sex with men (or women who have sex with women) commit the sin of "sodomy." And if the Sambia or the Etoro of Papua New Guinea were to contemplate notions as improbable as the scientists', they might ask how it comes to pass that a boy grows into a man in the United States—for, according to *their* theories of human nature, although manhood is an essential quality of semen, that semen is acquired from without, rather than produced "naturally" from within. A boy thus grows into a man by regularly ingesting the semen of senior boys and men over a period of several years, after which he becomes a semen donor who in turn "grows" junior boys into men. The Sambia "norm," if they were to name it as such, would turn on the properly directional circulation of semen among affinally linked males of different age cohorts.[4]

But it is not just peoples from ancient civilizations or exotic cultures who might be perplexed by our society's emphatic distinction between homo and hetero, with its concomitant lack of attention to other kinds of distinctions that might be drawn by sexual intercourse. Even people from modern, industrialized cultures close at hand—from parts of Latin America or the Mediterranean—understand and demarcate "the homosexual minority" according to rules and assumptions dramatically different from the cultural notions that guide the bulk of scientific sex research in Northern countries. In these settings, it is not simply the man who has sex with men, but specifically the anally receptive (rather than the phallicly active) party who is understood as "homosexual." Provided he follows established conventions, the "active," phallic, penetrating partner receives no special label or stigma. "And what real man would not become aroused at the idea of penetrating an attractive young man?" opines a very large sampling of contemporary societies.[5] "Homosexual desire," singled out and specified as such, is in no sense a universal or transcultural category. And the "homosexual body," posed as foil to a "normal" body, is a historical product, not a "natural fact."

Scientific attempts to establish the supposed "organic seat" of or the "genetic basis" for homosexuality make no allowance for such dramatically different cultural forms of sexuality, different sets of rules and prohibitions, different systems of sexual classification. They simply argue that homosexual identity, as it manifests itself in late modern Northern culture, is the phenotype expressed by a given genotype. One thinks of the coming-out scene in the movie *In and Out*, in which the naive protagonist's obsessions with Barbra Streisand, his femme taste, and his love of disco all point him irrevocably to the conclusion that, despite all doubts and denials, he really *is* gay—a gay identity he seems to discover even before he discovers lust for men. The movie image distills a certain picture of identity: Like phenotype to genotype, the outer signs all vouch that the character is, in his innermost essence, gay. Kevin Klein's character is convinced—and the audience is convinced—by "outer signs" that swirl among wholly modern consumer objects and media images. We thus read the "outer signs" as evidence of an "inner destiny" in a milieu where both are swayed, not by eternal nature, but under the changing firmament of history.

Given the limitations of this cultural and historical horizon, the assumptions it contains and the conventions it invokes, it is not enough to say that "environmental factors" like family life or social attitudes might accelerate, impede, or mask the expression of an innate homosexual tendency.[6] For the "tendency" itself is by no means independent of the social and cultural conditions that contextualize it. The identity in question occurs only at the conjuncture of certain institutions, practices, and meanings. The very idea of a distinctive homosexual tendency, demarcated from a heterosexual norm and supporting a distinct homosexual identity, is simply not sustained in sophisticated historical or anthropological research: Is this desire the love of men, or the love of boys—or is it perhaps not love at all? And is the identity in question defined by this act, or by that one, or by both acts—or neither?

Even notions as apparently straightforward as "same-sex desire" or "same-sex intercourse" require considerable translation to provide a pivot for cross-cultural or historical research. I have sometimes resorted to the circumlocution "what we would call same-sex desire or intercourse"—for precisely what "counts" as "the same" is by no means self-evident. Is the Greek youth "the same" as the adult male? The senior inseminator "the same" as the junior initiate? Is the *macho* "the same" as the *loca* he beds? These social actors are counted as "the same" only within the horizon of a certain cultural way of accounting. Other cultures make other accountings, draw other distinctions.

This is not to say simply that sex, its meanings and practices, varies from culture to culture according to certain more or less intelligible principles. It is equally necessary to acknowledge that variations in the warp and kink of the social fabric complicate matters—exceptions and niches that ever lend themselves to anthropologizing tropes on, say, the milieu of truck-stop rest rooms, or, for instance, English boarding schools: "as curious a culture as anything you'd encounter up the Amazon," the site of "bewildering erotic customs whereby any boy might have sex with another boy but only certain boys are branded 'queer.'"[7]

To draw a reasonably complex picture of sexuality, one must admit, moreover, the salient facts of individual variations within the heterosexual or homosexual types. I think here of a widespread alternative system of folk classifications from the straight world: "leg men," "breast men," men who are into redheads, and the like. . . . And in truth, I do not know very many "gay men" per se. On the ground of sexual practices in late modernity, desire is often quite specific. It tends to seek distinctive markets or particular niches. I know young men who are attracted to men who are older than they are, as well as men who lust only after younger men. I know men who are attracted to men of a specific age range, and I know men whose objects of attraction vary according to their own age. Some are avid consumers of pornography. Others express not the least interest in paper or celluloid sex. Some men are into blondes, others are into brunettes, yet others singularly pursue redheads, Asians, blacks, whites, Latinos, men of a certain stature or girth, muscle boys, butches, drag queens, and so on. I know Asians who sleep only with Latinos. I know white blondes who sleep only with other white blondes. Among my acquaintances, I know men who are, as they say, "versatile," but I also know men who are exclusively "tops," "bottoms," cocksuckers, fistfuckers, leather queens, S&Mers, masturbators, and men who only get blown. I know a man who is very enthusiastic about being on the receiving end of anal intercourse—but only while he fucks his wife. I am sometimes struck by the odd rules people construe in a narrow circle around their sex acts: I know one man who likes to get fucked by his lover, but who insists on fucking those men with whom he has more casual relationships. I know of another man who calls himself "straight" because he fucks only women—his penis, he will tell you, is "for" the vagina. His ass, however, falls under a different set of self-made rules, so that no number of rear entries suffice to make him, in his mind, "homosexual" or even "bisexual."

It has always seemed to me that such pronounced preferences and practices have at least as much salience as whether one's "object choice" is hetero or homo—that they say as much about one's desires, life experiences, and emotional investments as whether one is "gay" or "straight." But neither the heteronormative scientific apparatus nor the advocates of homosexual exceptionalism really wish to get down to the specifics of sexual desires and sex acts.

Michael Warner gives the following striking example, in which biologists inadvertently illustrated how history and caprice come together in individual sexual orientations: "One of the genetic studies . . . tracked the sexual preference of identical twins reared apart, hoping to see whether genetic and individual factors could be distinguished. The researchers found a very striking case of male twins, separated from early childhood, both of whom shared the same sexual preference: masturbating over photos of construction workers."[8] To my knowledge, no one has ever seriously proposed reducing such highly specific preferences to an organic seat, to a genetic basis, or to origins in the dim evolutionary past.[9] As Warner points out, key features of the twins' desires could not possibly be genetic, since they are clearly contingent on cultural history: on the existence of photography and on the existence of the profession "construction worker." Even if one takes the presuppositions of genetic research into sexuality seriously, it would be difficult to say just what part of this constellation was "mutable" and what part was "immutable."[10] And if you asked the twins themselves, they'd probably tell you that they felt not the slightest twinge of desire to masturbate over, say, Impressionist paintings of farmers or lifelike drawings of businessmen.

It might be objected that preferences of this sort are "fetishes"—derived or secondary desires, to be distinguished from whether one is heterosexual or homosexual.[11] But such a distinction is convincing only within the framework of a folk culture whose primary obsession (fetish) is over the sex of the actors rather than the sex acts of the actors. Indeed, sustained reflection on "fetishisms" tends to dissolve one-dimensional notions of a homogeneous heterosexuality or unitary homosexuality.

But the problem of desire is yet more profound. At bottom, it is not simply a question of cultural, subcultural, or individual variation, but of phenomenological complexity. At issue is what counts as "sex" and how to identify "desire" at all—an issue complicated by a heterocentric cultural apparatus.

To take a page from literary queer theory: Henry James's short story "The Beast in the Jungle" is usually read as a heterosexual love story in which the protagonist, John Marcher, is tragically unable to reciprocate the love of May Bartram—not, that is, until it is "too late." According to the conventional reading, this is a tale of lost chances: after Bartram dies, Marcher breaks down in the cemetery when he realizes the true extent of his loss and emptiness:

> *She* was what he had missed. . . . The fate he had been marked for he had met with a vengeance—he had emptied the cup to the lees; he had been the man of his time, *the* man, to whom nothing on earth was to have happened. . . . [A]nd she had then offered him the chance to escape his doom. . . . The escape would have been to love her; then, *then* he would have lived.[12]

Figure 26. Farley Granger and Robert Walker in Alfred Hitchcock's *Strangers on a Train*. Warner Brothers, 1951. Courtesy of Photofest, New York, N.Y.

Eve Kosofsky Sedgwick interprets James's story differently. In "The Beast in the Closet," she suggests that this is the tale of a secretly gay man who is unable to acknowledge the truth of his homosexual desire.[13] May Bartram sees and understands Marcher's much-invoked "secret" and attempts—unsuccessfully—to coax her companion into an acknowledgment of his homosexuality. In the final scene, Marcher confronts a passing male figure in the cemetery:

> His pace was slow, so that—and all the more as there was a kind of hunger in his look—the two men were for a minute directly confronted. Marcher knew him at once for the deeply stricken.... What Marcher was at all events conscious of was in the first place that the image of scarred passion presented to him was conscious too—of something that profanes the air; and in the second that, roused, startled, shocked, he was yet the next moment looking after it, as it went, with envy.[14]

According to Sedgwick, this scene embodies the concept of "homosexual panic." Properly understood, it gives a very different context to the frantic passage on how "*she* was what he had missed" and how he was a man struggling to escape his fate ("*the* man, to whom nothing on earth was to have happened"). Marcher becomes not, as others have claimed, "the finally self-knowing man who is capable of heterosexual love, but the irredeemably self-ignorant man who embodies and enforces heterosexual compulsion."[15] In these passages, Marcher turns his back against the passing stranger's "hungry look," the "arousal" it provokes, the self-recognition it threatens to effect: "He saw the Jungle of his life and saw the lurking Beast; then, while he looked, perceived it, as by a stir of the air, rise huge and hideous, for the leap that was to settle him. His eyes darkened—it was close; and, instinctively turning, in his hallucination, to avoid it, he flung himself, face down, on the tomb."[16] In James's coded text, the beast that lurks within is closeted homosexual desire.

Perhaps by pursuing this tale of the Jungle and the Beast within—by triangulating James's story with both conventional and queer interpretations, and by holding the entire constellation up against those dreary, heteronormalizing narratives on the mist-enveloped time of evolution, populated by mythic creatures and other kinds of beasts—we might come to a very different understanding of desire, its "nature," its expression, an understanding expressly urged by a current of queer theory since at least the 1980s, but suspected not in the least by the oh so earnest scientists. Sexual desire in the James story, like a murderous rage in Stendahl, "is not in the words at all. It is between them, in the hollows of space, time, and meaning they mark out."[17]

So, too, in all our stories about sex. In the intralinear version of the story, it is not always so clear in which direction the arrow of desire points. "Male bonding," that consummate expression of masculine heterosexual identity formation, obviously enough trades in a homoerotic currency. This was clear enough to Lionel Tiger when he first proposed the term, and the irony has been milked as comedy shtick for years on prime-time TV.[18] But even heterosexual courtship and marriage might be understood as forms of homosocial rather than heterosexual bonding, a circulation of desires and meanings that was by no means invisible before they were given a sustained analysis by Sedgwick.[19]

It is not altogether clear what "The Beast" might have meant to James or how he might have intended it to be understood—or even whether the author's relationship to the text ought to affect our reading of it. We know only that meanings resonate in and around the text. Sedgwick's interpretation of how homosexual desire lurks in the shadows of James's text is a convincing one, but it does not necessarily rule out other viable interpretations, for I can think of no claim about desire that is immune from doubt, second guesses, or ambiguity.

This is because desire is not a clearly demarcated thing, despite efforts

Figure 27. Harry Holland, *Two Men.* 38 inches × 36 inches, oil on canvas, 1996. Courtesy of Jill George Gallery, London.

to reify it, but instead is present in human life "like an atmosphere," in Merleau-Ponty's memorable phrase.[20] "Diffused in images," it "spreads forth like an odour or like a sound." Desire clings to our actions and perceptions, "coextensive with life." For this very reason, it can never quite be decided whether Marcher (like all the real-life Marchers he represents) is motivated by his desire and his denial of it, or whether he moves in a sexual medium attributed to his life by the reader, the observer, the critic. In sexuality, as everywhere, "ambiguity is of the essence of human existence, and everything we live or think always has several meanings," such that in the end, it is "impossible to label a decision or act 'sexual' or 'nonsexual.'"[21] It is not only that in sex the one thing might serve as a screen or substitute for the other, but that we find ourselves "caught in a secret history, in a forest of symbols," where "everything symbolizes everything else."[22] In this forest, as everywhere, the difference between perceiver and percept, between desire and its object, is difficult to measure—and can never be measured precisely.

Desire, then, is not there to the one side or the other of cultural meanings. Nor can it be bracketed off, made to stand as a stationary, autonomous "thing" against a backdrop of flickering signs and changing social relations. In real life, the one thing suffuses the other: Desire and significance saturate each other such that it is always arbitrary to mark boundaries or draw distinctions between them.

In the end, I distrust anyone who says too emphatically that he knows what he wants. The woman, or the slash of red lipstick signifying a certain kind of femininity? The man, or the variant of manhood evoked in the humid musk of his cheap aftershave? The cut of his body, the cut of his clothes, or the way the one somehow seems implicit in the other? Desire, insofar as it makes distinctions, can take shape as desire only when it conveys or blocks cultural meanings in a social context. But it is not even right to say that I want men as their manliness is shaped or rebelled against, as their bodies are conditioned, experienced, or transformed—which is to say: I want men of my time and era, and no other. I might, after all, desire beyond my own time and culture.

Having put desire squarely within a social purview, I hope it is not understood that I wish thereby to reduce it to a given semiotic code or a definite discursive formation. This move, common enough in academic theory, is but a mirror image of the geneticist fallacy that reduces desire to a genetic code or morphological function. Desire is not an inert precipitate of discourse, but a living process of communication. It is not the expression of a code, or even a construct modeled after a blueprint—images that, like those tales of the geneticist vulgate, express a desire for closure and finality—but a kind of creativity.

Desire is an opening, not a closure, a relationship, not a thing. It is not within us, but without. It encloses us. We cannot quite *see* our desires, because we are immersed within them: They are the very medium of our actions and thoughts, the perspective from which and through which we see and feel. This desire is on the side of poetry, in the original and literal sense of the word: *poiesis,* "production," as in the making of things and the world. Not an object at all, desire is what makes objects possible. In Kojève's reading of Hegel—a poetic meditation on the power of desire in the making and unmaking of the human world—subject and object, the identity of the person and the accessibility of the world, are given as moments in the dialectical motion of desire:

> Desire is what transforms Being, revealed to itself by itself in (true) knowledge, into an "object" revealed to a "subject" by a subject different from the object and "opposed" to it. It is in and by—or better

still, as—"his" Desire that man is formed and is revealed—to himself and to others—as an I, as the I that is essentially different from, and radically opposed to, the non-I. The (human) I is the I of a Desire or of Desire.[23]

Desire, then, like nature, has no "nature" apart from the one we give it—even if we don't always know just how we give it what we give. Not an essential aspect of identity, desire is the realm of practice and perception through which identity takes shape or is dissolved. Not an object, it is what imbues the object in a flash with its objectness and the subject with a corresponding subjectivity. We are never outside of our desires. We never meet our desires face to face. By understanding desire as a fixed, particulate object, imbued with essential and unchanging properties, the scientists have not understood it at all.

20 Familiar Patterns, Dangerous Liaisons

Every critic who has written on the subject has noted a recurrent pattern, evident for several decades in the biological research on homosexuality. A scientist working with a small or nonrepresentative sample claims to have discovered the hormonal source, anatomical seat, or genetic cause of homosexuality. His numbers lie close to the margin of error, but his findings invariably conform to the prevailing stereotypes of gender and sexuality. The media balloons the study into a major event—not because of any secret agreement or institutional conspiracy, but because the media, no less than the science establishment it reports, is part of the culture in which it operates. Over time, subsequent studies more carefully executed than the first fail to reproduce the results in question. But because these restudies do not confirm the prevailing paradigm, they are not extensively reported. In consequence, a sector of the public is left with the impression that homosexuality really is a distinct biological condition, induced by genetic, hormonal, or anatomical anomalies. And the no less credulous scientific establishment concludes, not that there is no causal connection between biology and homosexuality, but only that the *real* cause has not yet been uncovered—in the relevant part of the brain, in the proper gonadal tissue, in the developmental intricacies of the endocrine system. So, after a brief pause, a new study appears, and the cycle begins again: A scientist working with a small or nonrepresentative sample claims to have discovered the hormonal cause, anatomical seat, or genetic source of homosexuality. . . . The social and political effects of this cycle are disastrous for straights and gays alike.

ALL THE DATA THAT FIT?

The three studies of the early 1990s already bend to the arc of this pattern. Dick Swaab's 1995 study of sexual dimorphism in a different part of the hypothalamus failed to replicate the general pattern of LeVay's results. Although Swaab's laboratory found male/female dimorphism in the central subdivision of the bed nucleus of the stria terminalis (the BSTc), it found no substantial difference between gay and straight men. Rather, Swaab's study reported an average difference between male-to-female transsexuals and non-transsexual men. Although this study has been cited as establishing for transsexuals what LeVay's study supposedly established for gay men—an organic seat of orientation—its findings, too, are open to criticism. First, as in LeVay's study, the total number of subjects is quite small—six transsexuals in all. Second, the reported variation between male-to-female transsexuals and non-transsexual men may even be wholly explicable in terms of the surgical and hormonal regimens transsexuals undergo.[1]

More to the point, William Byne reports that he has now concluded a replication of LeVay's "gay brain" study. Owing to considerable difficulties in obtaining, classifying, delineating, and measuring legitimately comparable hypothalami, Byne's restudy took several years (a good deal longer, one might note, than LeVay took with his original study). Byne reports his results as "inconclusive" while suggesting that a careful, methodical study of the question using suitable brain specimens would require considerable funding.[2]

In a study of Australian twins, J. Michael Bailey, Michael P. Dunne, and Nicholas G. Martin have reported homosexual concordance rates among siblings dramatically lower than Bailey and Pillard's earlier findings—with the same interpretive dilemmas, since siblings share both genes *and* an environment.[3]

Hamer's study, once the most respected of the suite, has perhaps faired the most poorly in retrospect. A member of Hamer's own research team has challenged "unspecified methods of data collection," according to a report on the front page of the *Chicago Tribune*.[4] An investigation of experimental irregularities by the Office of Research Integrity of the U.S. Department of Health and Human Resources ended in December 1996, "in effect absolving its subject, Dean Hamer."[5] But by this time, George Ebers, a professor of neurology at the University of Western Ontario, had been through more than four hundred pedigrees and one linkage analysis. He reports that his extensive (and at the time still unpublished) research corroborates neither Hamer's pedigree study nor the Hamer group's findings on Xq28 concordance.[6] To date, then, Hamer's is the only laboratory to link male homosexuality with a genetic marker on the X chromosome.

It was in a story buried on page 19 that the *New York Times* summarized the results of research conducted by George Ebers and George Rice at the

University of Western Ontario, as reported in the journal *Science:* "Study Questions Gene Influence on Male Homosexuality." Using families "not very different" from those studied by Hamer, Rice's team was unable to duplicate the results of the earlier "gay gene" study. Still, *Times* reporter Erica Goode did her best to spin these negative results in favor of a positive geneticist accounting. The first sentence of the story suggests that Rice's study is a minor setback, underscoring "the difficulty scientists face in finding genes that underlie complex human behaviors." The first line of the second paragraph accentuates the point: "Experts in behavioral genetics say the new report . . . does not invalidate the notion that genes influence sexual orientation, which many scientists believe is the case, on the basis of other types of studies." "But," the third paragraph sighs, "the failure to replicate the earlier study is typical of the zigzagging course of research efforts to ferret out genes that influence complicated traits, or that contribute to illnesses with a behavioral component, like schizophrenia, manic depression, alcoholism and hypertension." The article gives Hamer five paragraphs of rebuttal space in a nineteen-paragraph story.[7]

A month later, the media gave even slighter coverage to experiments by John Crabbe that demonstrate just how difficult it is to establish a "genetic basis" for "complex behaviors," even in laboratory mice observed under closely controlled conditions. Under Crabbe's direction, three laboratories conducted a series of carefully standardized tests, each using 128 seventy-seven-day-old mice from the same eight strains. Although the tests were carefully synchronized to occur at the same time, and although food, the light-dark cycle, and other conditions in the three laboratories were meticulously equated, genetically identical mice behaved differently on a battery of tests depending on whether they were tested in Portland, Albany, or Edmonton. Indeed, Crabbe's famous "alcoholic mouse"—a strain that was once claimed to establish the importance of serotonin pathways in addictive behaviors— proved no fonder of intoxication than control-group mice in all three settings. Crabbe's team cited a number of possible factors that might affect outcomes in different laboratories, including the chemical composition of drinking water and the smells and appearances of laboratory technicians.[8]

One might say that Crabbe's findings show how difficult it is to pin down the genetic component of behavior. Or one might say that what these experiments show is that even extremely subtle environmental cues sway the course of even basic organic processes, often in unpredictable ways.

Meanwhile, philosopher Edward Stein published *The Mismeasure of Desire: The Science, Theory, and Ethics of Sexual Orientation.* Stein's methodical, meticulous, and highly readable critique of scientific research on sexual orientation became available at about the time I was working on revisions for the present text. It covers many of the same problems with research by LeVay, Hamer, and Bailey and Pillard that I have indicated. It also examines and critiques a much wider range of scientific studies of sexual orientation, both studies that assert a biological basis for sexual orientation and studies claiming that sexual orientation develops from early childhood experiences.[9] Stein's important and cautionary text was given less attention than it deserved in the gay press, and it was hardly noticed at all in the mainstream media, a measure of the privileged place occupied by "nature" in the serious public sphere.

PERMUTATIONS

Meanwhile, notwithstanding so much countervailing evidence and so many difficulties, efforts to reify continue apace, the media frenzy for geneticization continues, and the contradictions multiply. On the lecture, conference, and interview circuit, Hamer, who first spoke as a straight advocate of "what gays have been telling us all along"—that they were "born gay"—has subsequently come out as a gay man, questioning the rights of straights to comment on technical difficulties with his research.[10] Simon LeVay, too, remains a featured media commentator on questions of homosexuality and biology. LeVay writes regular columns, syndicated in some gay newspapers, to popularize what he calls "Queer Science." And gay political culture in general has banked heavily on the scenario that gays are, like left-handed people, a genetically determined (or perhaps neurohormonally determined) variation on human nature.

As new discourses on "nature" expand the space of identity politics, they also come to fill unexpected niches, to serve unanticipated political ends. Gabriel Rotello has argued that the freewheeling promiscuity of the gay subculture is "not ecologically sustainable"—a thinly veiled euphemism for calling it an "unnatural practice." Rotello proposes the use of carrots and sticks to construct a "viable," "sustainable" culture for gay men, a culture where men's promiscuous natures would be reined in and made to harmonize with natural law, a culture where gays would be compelled to act more like straight men (who at least have the civilizing influence of women) and would be induced to practice monogamy (or some semblance of it).[11] As though on cue, Larry Kramer took off from Rotello's book with an angry diatribe, fronted on the *Advocate*'s cover as "AIDS:

We Asked for It"—a tale of nature's retribution against those who flout its laws.[12]

When Ellen DeGeneres's eponymous sitcom character came out of the closet to skyrocketing ratings on prime-time TV—along with the actress herself and her new girlfriend, the actress Ann Heche, a woman who says she had never felt even an twinge of lesbian desire until she met DeGeneres (and who subsequently left DeGeneres to marry a man)—talk shows and some journalists in the gay press used the opportunity to weigh, once again, the premise that people are "born gay." Events around this highly concocted media event were sometimes read as weighing in favor of an "innatist" scenario—despite the absence of any sense to this interpretation and despite the fact that virtually no one had ever claimed to find any hard evidence that lesbianism (as opposed to male homosexuality) is "inborn."[13]

Only a few years ago, lesbian feminists asserted that same-sex affective and erotic relationships were not only every woman's potential, but also a common front in the battle against a heterosexist patriarchy. The early voices of gay liberation exulted in polymorphous perversity—the idea that everyone has a bisexual potential.[14] By way of contrast, consider the advice given by Justin Richardson, a gay psychiatrist and consultant "who helps private schools deal with gay issues." According to the *New York Times,* Richardson was called in to various Manhattan single-sex schools to "hand-hold and lead discussion" after an outbreak of "bisexual vogue" among students.

> Parents often ask Dr. Richardson how they should react to same-sex experimentation. Their No. 1 question, he says, is whether the experimentation will affect a child's later sexual orientation.
> "The answer is no," he said. "In fact, if this is a girl who has the genetic predisposition and early experience to grow up to be a heterosexual, then bisexual experimentation will probably only help her clarify that she is more attracted to males than to females." On the other hand, he says that if "she started life on the path to being a lesbian, teen-age experimentation might help her to develop her lesbianism in a healthier way than if she were forced to ignore her true desires until adulthood."[15]

The psychiatrist avoids the worst excesses of reductionism by claiming that sexual orientation is somehow the outcome of a "genetic predisposition" *and* environmental conditions ("early experience"). Still, his statement represents a remarkable extrapolation: He posits a genetic influence (on lesbianism) no one has ever seriously claimed to document. And this "gay gene," now extended to lesbians, serves as such an effective border guard that even sex with one's own sex is of no real consequence. The truth of desire is indelibly stamped in the gene or, perhaps, in some equally reductive imaginings about "early experience." Everything else is epiphenomenal, even illusory.

In yet another permutation of the geneticist argument, scattered gay anti-abortion groups cite the impending development of a prenatal test for homosexuality, beckoning other gays to join them in the anti-abortion camp. Their positions are periodically aired as letters or columns in gay newspapers—and their supporters occasionally show up at anti-abortion or gay rights rallies carrying pictures of "gay fetuses." Of course, support on such grounds for the anti-abortion cause is hardly welcome on the "pro-family" religious Right—which has in turn joined the fray over "nature" by distributing the results of research indicating that homosexuality is "learned," not "inborn," and can thus be "overcome" with prayer, Christian counseling, and "reparative therapy."

And so in the midst of all this confusion and turmoil, at the end of the *Blade* piece that summarizes James Watson's fantasies of a eugenic future in which selective abortions cleanse the species of homosexual deformity, we encounter Hamer—arguably the very figure most responsible for such confused scenarios—exclaiming that sexual orientation is "just much more complicated than that" and "there never will be a definitive test [to identify a gay gene]—never ever ever."[16]

BACKTALK BACKLASH

This is how Foucault summarizes the volatile and unanticipated results of a past era's specious science and dubious anthropometrics:

> There is no question that the appearance in nineteenth-century psychiatry, jurisprudence, and literature of a whole series of discourses on the species and subspecies of homosexuality, inversion, pederasty, and "psychic hermaphrodism" made possible a strong advance of social controls into this area of "perversity"; but it also made possible the formation of a "reverse" discourse: homosexuality began to speak in its own behalf, to demand that its legitimacy or "naturality" be acknowledged, often in the same vocabulary, using the same categories by which it was medically disqualified.[17]

Such are the spirals of discourse and power, representation and its effects. The calibrations designed to "diagnose" and "treat" perversions, the system of classification that gave birth to "the homosexual" as an object of administration also became the bases for new forms of subjectivity, new forms of resistance. . . . In no time, the homosexual was talking back, speaking in his own given name.

And what now, only a little more than a century after the invention of the homosexual? Mindful of how the concept of "nature" has twisted and turned in contested histories and wary of how discourse lends itself to abrupt and ironic—but necessary—reversals, we might ponder a series of timely questions: In whose language will we speak today? How will the identities we construct also construct us? What will the discourse on "nature" disclose? What possibilities will it foreclose? Who will it enclose?

REPRESENTATIONAL WHIPLASH

A segment of the gay/lesbian movement has always viewed homosexuals as a distinct, definable minority. This segment has often argued that same-sex desires are inborn, congenital, fixed, immutable, and invariant. The social and political strategy that follows from these propositions is familiar and well known. A process of self-discovery, self-acceptance, and self-disclosure is to be followed by space claiming, community building, and advocacy for gay rights. Anything short of this is a failure to embrace the truth of one's own self and interests.

This "minoritizing" or "rights-oriented" approach has usually coexisted in a productive tension with a very different set of "universalist" or "liberation-oriented" approaches. These latter approaches understand same-sex desires as a widely distributed human capacity whose liberation implies broad social changes, not just minority rights. An old gay-liberation maxim played off a line by Allen Ginsberg to good effect: "Everyone's just a little bit homosexual, even if they sometimes forget." The early works of gay theory—classics like Dennis Altman's *Homosexual Oppression and Liberation* and Guy Hocquenghem's *Homosexual Desire*—argue from this universalist premise. Instead of claiming an airtight, unimpeachable identity or seeking accommodation for a well-defined homosexual minority, liberationist politics have tended to emphasize grander strategies: the displacement of sexual intercourse from the discourse on "nature," the subversion of gender roles and sexual rules in society at large, an emphasis on the tentative, even contrived, nature of all identities, and the cultivation of alliances with other social movements incubated in linked histories.[18]

If a liberationist sentiment to "change the world" animates moments of radical upsurge, the pragmatics of day-to-day work tends to favor the rights approach as the "steady state" of gay politics. Certainly, gay institutions—gay advocacy organizations and self-help groups, gay neighborhoods and real estate, gay social clubs and businesses—foster the sense of a well-defined, territorially demarcated, essential identity. In recent years, therefore, a minoritizing, rights-oriented, and innatist perspective has prevailed in gay/lesbian political culture—in part because "innatism" better sustains "community building," and is better sustained by the gay subculture itself, in part because arguments that "I was born that way" and "I can't change the way I am" offer ready, economical refutations to conservative and Christian fundamentalist demands that gays abandon their "unnatural" choices through psychological counseling, religious conversion, or sheer willpower. The U.S. legal system, like the culture in which it is immersed, would likewise seem to promote the saliency of certain rights-oriented arguments. Following guidelines established by the bulk of civil rights law, which prohibits discrimination against people on the basis of "immutable" characteristics (e.g., race, sex, national origin), major gay political organizations have increasingly come to predicate their claims to legal redress in terms of that ultimate underpinning of a fixed gay identity, a gene for homosexuality.[19]

So in 1993, the Human Rights Campaign, the nation's largest gay rights lobby, distributed free copies of conservative journalist Chandler Burr's *Atlantic* piece, "Homosexuality and Biology," on Capitol Hill.[20] The HRC hoped to convince Congress that sexual orientation is innate and immutable, and in so doing, to sway federal policy toward grudging support of gay rights. Exposing this non sequitur, Burr's subsequent book, *A Separate Creation*, foresees instead the development of genetic potions that will turn gays straight, chemical keys inserted into genetic locks that, by dint of the most subtle transformations, will steer the course of sexual desire.

DESIRE IN A TEST TUBE

Here is how Burr imagines the not too distant future: First, scientists discover the genetic variation that produces homosexual men. Then they discover what that gene "does." The imaginary gene, "GAY-1," makes a protein, which in turn helps make a specific receptor molecule, which in turn allows MIH (Mullerian Inhibiting Hormone) into the third interstitial nucleus of the hypothalamus (INAH-3). MIH is the hormone that suppresses female genital characteristics in males in utero. It allows testosterone to "masculinize" the developing fetal body, despite the preponderance of "female" hormones in the uterine environment. Burr's scenario rests on the assumption, widespread in biological research on homosexuality, that MIH plays a far wider role, also "defeminizing" (or allowing the "masculinization" of) the male brain. In Burr's scenario, gay men are homosexual because of a genetic variation that renders "impotent" their MIH receptors on the part of the brain that controls sexual orientation so that they develop "feminized" as opposed to "masculinized" brains. They thus develop the "natural" sexual desires of women for biological men rather than those of "normal" men, whose brains are preprogrammed to seek out sexual pleasure with women.

Finally, scientists figure out how to perform "genetic surgery" by delivering the dominant variant of GAY-1 to the appropriate site, using viral agents as hosts.

> Dr. Link holds a little syringe in his hand. It contains a solution filled with engineered viruses, each one loaded with the dominant and by far most common allele of GAY-1, an allele that makes a functional MIH receptor in the brain. Link sticks the needle into the man's arm and injects him with millions of gene-carrying viruses. They spill into his bloodstream, making their way through his system. As they drift past the hypothalamus in the current of blood and fluid, they brush up against the non-functioning MIH receptors, prepare their stealthy incursion, bind to the cells, and invade. The viruses deliver their genes and then, as programmed, break apart and self-destruct.
>
> Dr. Link shakes his patient's hand and says, "I'll see you in six months," and he knows that when the man returns, the gene will have reconfigured a tiny portion of his cellular machinery. After a few days it will begin pumping out a protein that will make the MIH receptor functional, allowing Mullerian Inhibiting Hormone into the brain cells for the first time, changing its neural architecture after a few months and coincidentally making it grow by almost two times over the next half year. [Presumably Burr means that the INAH-3, not the brain, will double in size.] And gradually there will be a change in the man's internal erotic and emotional responses. From as early as he could remember, he was attracted to boys, involuntarily and instinctively, but now, almost without noticing it, he finds himself getting out of a taxi or stepping into a room and noticing women in a way he never has before. Involuntarily, he realizes he is looking, and for the first time in his life he is sexually aroused. He wonders, after a season passes, at the fact that he was ever attracted to men, an attraction that seems not offensive now but simply alien to him, simply unfathomable, the memory of a vivid dream.[21]

Reading this new version of *Fantastic Voyage*'s trip through the bloodstream, I cannot help but think of Jeffrey Dahmer—poor, mad Dahmer—who attempted frontal lobotomies on his lovers: He wanted to steal their

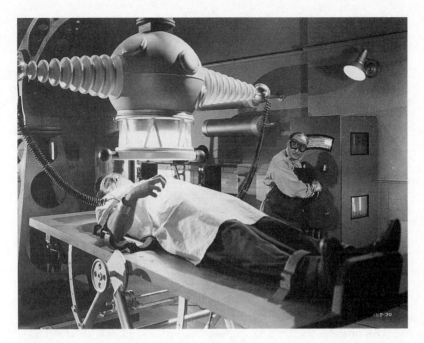

Figure 28.   Lon Chaney and Lionel Atwill in *Man Made Monster*. Universal Studios, 1941. Courtesy of Photofest, New York, N.Y.

wills, to turn them into zombie slaves at his command. I think of sci-fi horror movies, their dimly lit dungeon laboratories populated by maimed creatures, the outcome of evil experiments by mad doctors. I marvel at the oscillating course of reification in a world of fantasies and fetishisms: Human volition, mercurial and evanescent, is distilled into an objectified, particulate substance, which then exercises some malevolent, magical will of its own over the human subjects from whom it was alienated. I wonder whether Burr has ever had "a vivid dream" and if he remembers just how *sensuous* its memories can be, how much carnal reality they assume, when they play across the waking body in the light of day. Indeed, it is difficult for me to imagine that the author of these fantasies has ever felt desire—or even that he has ever known anyone who has.

MAGICAL SCIENCE, WISHFUL THINKING

Whatever the dubious logical, empirical, or scientific grounds for the claims now in play, it is conceivable that new "technologies of the self" are being invented, that new subjectivities are being fostered, that new conceptions of agency are being deployed—now, as in the nineteenth century.[22] In certain scientific circles, and in the mass media outlets that ventilate their views, sexuality is increasingly conceived as the active expression of personified genes in a depopulated and passive environment. Desire, in these paradigms, appears well demarcated, self-enclosed, solid, and particulate, an object or a thing detached from the person who feels it and the meanings that support it. A curious symptom afflicts "desire," imagined this way: the more volition, activity, and creativity are attributed to genes, hormones, and brain structures, the less they are seen as inhering in people's activity, in their relations with others, in their bond with the world. A perverse political symptom follows, too. As long as one seeks categorical confirmation of an essential identity in the self-evident light of nature, our own desires, acts, and ways of being appear to us as alien and inert things, as finished products, already here before we arrived. This strange and estranging conception of desire has implications far beyond the question of carnal relations, per se, for in these fantastic representations of "nature" and of humankind's place in it, the play of history and the sway of culture have likewise been drastically reduced. Not that unstable theater where intentional actions collide with unintended consequences, cultural history is now conceived as the slowest and most long-term effect of chemical changes.

This kind of magical thinking at least makes sense for cultural revanchists: Faced with change and disorder, they construct a hypothetical world of order and stability. They represent desire, uncontrollable and volatile desire, as a gene so as to classify, to know, and in some sense to control it. But genetic fetishism and nostalgia for nature are dangerous indeed for those creatures whose existence lies at the margins of the naturalistic imaginary.

It would be premature to claim to know in advance all the effects of today's minoritizing discourse, the magical "science" it embraces, and the genetic fetishism it purveys—or even to guess whether such representations will "take" against the flux of other, very different, contending representations. What will be the thoughts and introspections of future adolescents, as they connect wayward desires with images of genetic blueprints? Perhaps "vague feelings" and "occasional attractions" like those reported by the *Newsweek* twin are already convincing legions that they are "gay by nature." Or perhaps a generation of youth will engage in sexual experimentation of every imaginable sort, in every conceivable combination, reassured by the belief that none of it will have any implications for them unless they carry "that gene." ("LUG," or "Lesbian until Graduation," has already become a badge of identification on college campuses.) And of course, if they do carry "that gene," they might as well enjoy it. Or perhaps youth, being skeptical, already takes "the gay gene" with a grain of salt—understands it for its camp value from the start.[23]

Or perhaps we stand on the threshold of a strange but eerily familiar landscape where new vistas of magical technologies await us—a brave new world where expecting mothers will visit high-tech witch doctors to divine the shape of the future, where mutant outcasts confront all the superstitions of a latter-day scientific racism, where instead of kneeling at the prayer bench, clutching a rosary, or confessing to a psychiatrist, the tormented faithful will take "straight pills," which, the doctor confides, "will only work if you really, really want them to," and, no doubt, where sexual reprobates take black-market elixirs in pursuit of scientifically enhanced perverted pleasures. . . . [24]

VARIATION AND DEVIATION

For the time being, however, those who would claim their human rights based on the idea that people are "made homosexual by nature" play a dangerous game indeed. LeVay, Hamer, and other innatists fully assent to the prevailing biomythology, with its belief that heterosexuality of the modern era is the biological "default mode" for human beings—but they then assert that homosexuality is a benevolent variation on, rather than an unhealthy deviation from, this essentially heterosexual "nature."[25] In making this move, today's innatists are attempting, one more time, to both play off and reverse the thrust of a heterocentric sexual science. This is an old move with an established legacy, but it is also a dangerous move, like a dance with death. Once it is alleged that men are by nature this way and women are naturally that way, nothing very clear distinguishes a good "variation" from a bad "deviation." The new innatists have made no new discoveries, nor have they developed any new insights or theories. They are simply making a claim and hoping that others will agree with it.

In fact, it is difficult to believe that that bastion of social conservatism, the medical establishment, will simply go along with the idea that homosexuality is benign—that it will come to view homosexuality the way modern medicine tends to view left-handedness, as a more or less benevolent variation on human neural wiring, rather than, say, the way it understands color blindness (a defect) or even lupus (a genetic disease). When homosexuality is considered as having a genetic basis, as in the *Times* reportage quoted earlier in this chapter, it invariably appears alongside "illnesses with a behavioral component, like schizophrenia, manic depression, alcoholism and hypertension."[26] As such associations suggest, no amount of sober talk about "medical ethics" will prevent the remedicalization of homosexuality as long as it is viewed as a genetic or hormonal condition at variance with the "normal" workings of male and female genes, hormones, and neurobiology in the default mode—a perspective that logically would seem to place homosexuality squarely within the medical purview.

At best, then, the new innatist claims carve out a protected niche for homosexual exceptionalism. At worst, they reify the prevailing logic of heterosexual metaphysics and thus actively contribute to the reproduction of an exclusionary homophobic—and sexist—environment. For gays can only be gay "by nature" in a "nature" that already discloses men and women whose deepest instincts and desires are also different "by nature." In the resulting sexual imaginary, biologically engineered "real" men are always in hot pursuit of "real" women, who always play coy. In such a paradigm, every conventional gender norm, down to the last stereotype, is attributed to a fixed, immutable biology. Men do better at math and science because of that thing in the brain. Women are better at housework and child care because of their hormones. Men are aggressive and women are nurturing because we are hunters and gatherers in our heart of hearts. And gay men are gay because they inherited a genetic defect, which caused something to go wrong in that thing in the brain. It's normal. It's natural. It's just the way men are and women are.

Norms reified; men and women trapped in their "natures"; a radical division of gay people from straight people, of queer sex from normal sex, of our experiences from theirs. . . . One scarcely has to imagine extreme scenarios to see that this is not good for gay people. Or for straight people, either.

It was sweet irony that the homosexual ever "talked back" in his given name. But it would be a cruel contradiction if we lost our desires by taking them for things and sacrificed freedom in the pursuit of identity.[27] By turning their backs definitively on liberationist dreams and by philosophically retreating to the confines of a small and well-demarcated homosexual enclave, the new innatists in the gay rights movement also abandon society at large to the rule of an absolutist heterosexual normalcy.

This heteronormativity, so necessary to the current biomythologies, will not stay on the other side of the divide its rule creates. Today's sex scientists are simply not being honest about the implications of their models. When Hamer and others entertain speculation (subsequently disconfirmed) that "insensitivity of androgen-receptors" is involved in male homosexual orientation, when Hall and Kimura look for asymmetries in the number of fingerprint ridges on gay men's thumbs and little fingers, when so many others still cling to the outdated "fetal stress" model, which holds that gay men were hormonally poisoned in utero by anxious mothers, they are engaging in the occult science of a latter-day phrenology.[28] They are also diagnosing causes and implying treatments. The implications are there, to be drawn by anyone who understands the arguments.

Syndicated radio talk-show personality "Dr. Laura" Schlessinger gets the implications of the arguments—perhaps better than some of their advocates. In response to news reports that lesbian students at Barnard College had successfully lobbied the school to revise its recruitment brochures—which had for years promoted Barnard as a place to meet eligible bachelors, claiming that alumni of women's colleges were more likely to marry and have children than women who attended coed schools—"Dr. Laura" lashed out against gays and lesbians: "I am sick and tired and fed up with the tyrannical power of gay and lesbian activists. . . . I'm sorry, hear it one more time perfectly clear: If you're gay or lesbian, it's a biological error that inhibits you from relating normally to the opposite sex." "Dr. Laura" went on to chide that gay activists should stop "interfering with the freedom of normal, regular heterosexual people."[29] (One might have imagined that even normal, four-square heterosexual women would want to be evaluated on the basis of what they know and who they are rather than whether they're married, or that colleges like Barnard might logically want to tout their academic standards rather than their mixers with eligible Columbia lads from across Broadway.)

Meanwhile, William Turner, following Hamer's lead on "maternal transmission," links homosexuality with elevated maternal-side suicide, infertility, and "fetal wastage" (a tendency to miscarry male fetuses), thus associating homosexuality with a family of very serious genetic defects.[30] With such dark talk in the air, we are fully in the thick of a new eugenics already.

The Ends of Nature
The Weird Antinomies of Postmodern Mass Culture

The idea of nature contains, though often unnoticed, an extraordinary amount of human history.

RAYMOND WILLIAMS, *Problems in Materialism and Culture*

21 "Nature" in Quotation Marks

Representations of nature both reflect social preoccupations *and* provide the means whereby new social relations come into being. Contentions over the meanings of nature or the implications of its design are thus, by definition, political contestations: struggles over the slippery relationship of "is" to "ought." A recurring front in these simultaneously material and ideological conflicts is the question of where to draw the line between the one thing and the other: between the inborn and the cultivated, the given and the made, the authentic and the artificial, the original and the derived, the real and the contrived; between the everlasting immutable and the oh so changeable aspects of human existence . . . It has never really been an easy matter just where to draw this line. This difficulty is of a special sort today.

I want to underscore this difficulty, its peculiar relationship both to what Foucault called "bio-power" as it circulates today and to current fantasies about natural origins and their implications.[1] Notwithstanding the sense of certainty and reassurance they convey about states of nature and meanings of life, today's tales about the evolutionary design of timeless human norms circulate at a decidedly odd moment in the history of nature, a moment that augurs not just the end of "natural sex" but also the end of "natural reproduction" as people living in Northern industrialized countries have understood these concepts for much of the twentieth century. That is to say, scientists, journalists, and policy makers invoke assorted "facts" of reproduction to stage counterfactual claims about the nature of desire at a moment when not just social relations, but also material means of reproduction, are in transition and turmoil.[2]

(RE)PRODUCING NATURE

In 1978, in-vitro fertilization definitively separated the sex act from reproduction while blurring the boundary between biology and technology. At the time, reasonable people experienced some puzzlement over the implications of this new procedure: Would the resulting human beings, deprived of a natural environment during conception, be adversely affected in some subtle or dramatic way? How, some queried, would children feel upon discovering that their "mother" was a test tube? Some conjectured that the application of new technological means to the mysteries of conception would have the effect of cheapening the meaning of human life. Others wondered aloud just how the soul would be able to enter the body under such unnatural conditions.

In the twenty-some years that have elapsed since the birth of the world's first test-tube baby, new reproductive technologies have multiplied, spawning logical conundrums and unprecedented possibilities. *New York Times* reporter Gina Kolata elaborates how the use of donated embryos and donated eggs challenges traditional conceptions of parenthood, reproduction, kinship, and identity: "It is now possible for a woman to give birth to her own grandchild—and some women have." I note that it is no less possible for a woman to give birth to a child genetically unrelated to her—and some women have. "It is possible for women to have babies after menopause—and some have." Using existing techniques for cloning embryos outside the womb, it is now feasible "for a couple to have identical twins born years apart. It may also become feasible for a woman to have an ovary transplanted from an aborted fetus, making the fetus the biological mother of a child." Thus, "women who donated their fetus's ovaries might become grandmothers without ever being mothers."[3] As Faye Ginsburg and Rayna Rapp note in their 1995 collection *Conceiving the New World Order: The Global Politics of Reproduction*, "The creation and birth of a baby may involve as many as five adults contributing everything from genetic material to gestational nurturance to the social life of a family, not to mention a cast of thousands in medical laboratories, legal offices, and insurance agencies."[4] More recently, fertility clinic scientists have produced the world's first genetically altered human beings, babies whose DNA derives from three, not two, "parents."[5]

At the door of the fertility clinic, the modern birth mother and her entourage might think themselves fully arrived at the threshold of some brave new world of cybernetic reproduction, a world far removed from the heterocentric hold of nature. In practice, if not in theory, modern reproduction has traveled a long, long distance from that recurring figure in fables of nature, Adam and Eve, the heroic heterosexual pair, bonded for life, whose mutual desire is all that is necessary to conceive the world. Matters promise to grow yet more complicated in the wake of recent breakthroughs in embryology, genetic engineering, and cloning.[6] Soon, human reproduction will not only

Figure 29.   *Fetal Attraction.* RALL © 1998 Ted Rall. Reprinted with permission of Universal Press Syndicate. All rights reserved.

not require heterosexual intercourse, it may not even require male and female, egg and sperm. It is by no means unthinkable that over the next century, large numbers of people in affluent, developed countries will opt to reproduce—or replicate—entirely outside of the human body. Yet in news story after news story, when doctors and patients are questioned about the logical and ethical complexities of the new reproductive technologies, they tend to give the standard justification for these and other extraordinary interventions: People's willingness to do almost anything to have children—especially children who look like them—is "only natural."

THE MOST NATURAL THING IN THE WORLD

With Viagra, not just reproduction, but desire itself undergoes something of a technological transformation—but still, there is nature, as natural as you please, a spectral presence haunting biochemical reactions:

> "When you're diagnosed with this kind of disease [prostate cancer], you're absolutely devastated," Mr. Bierly said, and so, although he

knew that impotence might result from the cancer surgery, it hardly peeked over the horizon of his concerns.

Later, when he discovered that he could no longer have intercourse, he felt desolated and frightened. But he kept his suffering to himself. "It wasn't a problem that guys discussed," he said.

Eventually Mr. Bierly tried injections of chemicals that can force a flaccid penis to become erect.

"It simply didn't work," he said. "It was cumbersome and painful. I was clumsy—you have to mix things up, and that took me 10 or 15 minutes. By the time I get done, she's asleep."

One time his erection lasted five hours despite intercourse during the interim. "I would have gone to the emergency room," he said, "but I didn't know what I would tell them."

Mr. Bierly participated in Pfizer's clinical tests of Viagra, and the drug worked for him.

"It is the most natural thing in the world," he said.[7]

LOOK AT MOTHER NATURE ON THE RUN

Along with "the waning of affect," "ironic detachment," aesthetic "depthlessness," and a renunciation of the idea of authenticity, "the disappearance of nature" is often listed as a signature trait of postmodernity.[8] But on the contrary, even a cursory trip through any province of popular culture shows that "nature" is abundant in late capitalist representational and discursive formations.

As every medium reminds us, our flesh, beneath its pores and down to its very atoms, begins in "nature" and is given by that same "nature." Cybernetic inventions are said to express the principles of natural selection— indeed, the wired economy in toto is said to be a "rainforest": Too complex for anyone to control, it supposedly provides a "habitat" within which a variety of forms might happily evolve.[9] In biomythologies both vernacular and scientific, that heterosexual duo, Adam and Eve, begin in the garden of "nature" whose transparent design is stamped in the self-evidence of these, our terrestrial bodies. And it is in the name of this ghostly, immaterial "nature" that high-tech Adams and Eves—*and* Adams and Steves—effect hitherto unimaginable transformations of body, desire, and reproduction. Our social practices either express "nature" or compensate for it—take your pick and make your argument.

Like feral eyes gazing through thick undergrowths in Rousseau paintings, "nature" peers out at its quarry from newspapers, newsweeklies, infotainment programs, and sitcoms—not to mention TV "nature specials" and "natural disaster" movies. Twisting, recombinant "nature" snakes its way across strands of DNA in petri dishes, annexing that most unnatural of spaces, the modern laboratory. "Nature" is the chatter of jungle noises at night, the trill of sinister characters treating "the human condition" on late-night radio talk shows as well as the utopian ululations of rainforest conservationists, who sometimes blur the distinction between cultural and biological diversity. Far from vanishing, "nature" presides over central arenas of postmodern culture. It plays the part of the teacher, but also that of the trickster. It is the compressed image of those other supposedly waning values, affect, immediacy, depth, and authenticity, all in one. Or, as Anne Fausto-Sterling quotes from the poetry of Marilou Awiakta: "Nature is the human heart made tangible."[10] Postmodern economists, dreaming of a deferred genomic revolution, boast that modern industry, led by biotechnology, is on the verge of achieving a "new relationship to 'Nature'"—then cast a furtive, half-superstitious glance over their shoulders, as if to see whether the *old* "Nature," with whom they entertained a love-hate relationship, might punish such impudence. And further out, on the margins of mass culture, nature's modern acolytes—prophets and prophetesses of the new religion—include a motley and inconsonant assortment of chaos theorists, Gaia thesis theologians, Trekkie philosophers, wacked-out astronomers, Druid astrologers, weirded-out hackers, amateur sociobiologists, black supremacists, white racists, eco-feminists, men's-movement enthusiasts, queer majik faeries, wilderness conservationists, neopagan nature worshippers, and many others besides.

Future historians might well define our era by its insatiable obsession with and over "nature." Postmodern culture expresses contradictory desires to learn from "nature," to emulate it, to visualize it, to listen to it, to capture it, to clone it, to use it, to live it, to achieve a new relationship to it or with it, and to get beyond it—all at the same time. Thus, it seems to me that when Marilyn Strathern discusses how the new reproductive technologies "destabilize" the received nature-culture binary pair, she is half-right.[11] The technologicization of reproduction does indeed logically compromise what (for want of a better phrase) might be called the "naturalness" of nature, at least for those who are prepared to contemplate the problem logically. Strathern correctly identifies a key contradiction between modern technological means and received social relations of reproduction. But since the whole point of technoscience is to *get at* nature, any biotechnological intervention, including the new reproductive technologies, can also *reproduce*, even magnify, received ideas about "nature" as a cultural category. The "special difficulty" for discerning the boundary between the "authentic" and the "artificial" becomes the enabling condition for a runaway naturalization of everything in sight.

Echoing cadences from Raymond Williams, Fredric Jameson, and Donna Haraway, then, it might be better to put matters this way: In inverse proportion to the shrinkage of those islands of "wild"—unexploited, unmanaged—nature (and much like the return of other things repressed), "nature" spreads, lush and verdant, colonizing new habitats in contemporary culture.[12] Analogously, and by way of some conceptual somersault, the more extraordinary the artificial human interventions into basic biological processes (death, reproduction, cell replication)—and by the very measure of their drama—the more forcefully the image of "nature" inscribes itself in our thinking about culture, humanity, existence. The more paltry, puny, and reduced the depictions of nature on the technocratic side of the equation—the same side that gives birth to evolutionary psychology's just-so tales and to mathematical formulations of beauty—the wilder, the greener, and more extravagant the ideas that sprout on the other side. It is as though someone has tripped that "Wild Plant Growth" lever depicted on the record sleeve of Jefferson Starship's *Blows against the Empire*. Nature, as dense and hallucinatory as 1960s psychedelia, keeps flashing back.

This is no mere paradox, but a systematic contradiction implicit in modern means of production and reproduction. Through discourse, science represents a world beyond discourse.[13] Technology—the apparatus informed by these models—is designed to leverage that world "beyond" (or "beneath") culture, but also made possible by it. "Nature" is thus made more powerful, more awe-inspiring, precisely by its supposed opposite, technoscience, whose role it is to harness possibilities of the physical world, but which also augments them, distilling them into . . . "nature."

THE PREDICAMENT OF NATURE

At the interstices of productive science and fecund matter, then, "nature" explodes: magical, mythic, and omnipresent, but also oppressed, imperiled, and solicitous, a support for technology and the possibility of a wholly technocratic world, but also object of a mad desire for unspoilt authenticity. "Nature" embodies an order, full of meaning, but because its body is prepersonal and nonrational, its meanings turn away from us. Boundless and regenerative, "nature" nonetheless stands as a limit. Self-moving, life-giving, and life-sustaining, this "nature" is also in crisis, because every human crisis passes through it.

Contemporary discourses are perpetually quoting "nature" because modern technoscience is always citing, resourcing, refining, and producing it. I put this "nature" in quotation marks not to dismiss it, undialectically, as a wholly linguistic construct—Evelyn Fox Keller has suggested that (contrary to extreme constructionist positions) technoscience would scarcely be effective if it only intervened in its own "representations"[14]—but to relieve it of its inertia and objectivity, to problematize its usual conception as a purely physical world estranged from human involvement, and to thus highlight its ongoing "enmeshedness" in and with the practical projects of the human social world.

Understood this way, "nature" can be discovered, brought into existence, made intelligible, and acted upon only *historically*, by social, perceptive beings. As David Rothenberg puts it at the beginning of his book on technology and nature: "We know no definition of the world outside our definite, practical engagements with it. . . . And 'nature' is a meaningless term apart from our will to define it."[15] As we've seen, Marx tried to capture just what was so *queer* (to use a most relevant term) about this relationship between human activity and nature in paradoxical turns of phrase, tracing the mobile logic of a distinction best understood as neither boundary nor line but— what is the shortest distance between two points in practical space—a *convolution*. In productive labor, the human being "opposes himself to Nature as one of her own forces . . . in order to appropriate Nature's productions in a form adapted to his own wants. By thus acting on the external world and changing it, he at the same time changes his own nature."[16] Under the new techniques of bio-power, which perturb any pretense of a logical distinction between the authentic and the contrived, this "production of human nature" is to be taken quite literally.[17]

Such a "nature"—allied with our "bodies" over and against our "minds" and with an impersonal universe outside the subjective self—is of course *in part* a part of culture, a product of culture and its in(ter)ventions: It is that part held to *come before*, to *go beyond*, and to *be other than* culture. The outcome of human intentionalities, labors, representations, and technologies, it also *comes after* culture. As the cultural practice that attempts to name and describe, always partially, those aspects of life that escape volitional, linguistic, and cultural control, this "nature" flourishes in all those conceptions of "man" supposedly dismembered, dispersed, and made impossible under postmodernity. Indeed, in the long contest between man and nature, it is not quite clear who has conquered whom—since "civilization" increasingly conjures "nature," and since the whole image of "conquest" is perpetually gobbled up by Darwinian narratives. The image of man, then, is emptied of humanity and filled with "nature"—a "nature" that sometimes expresses but is also sometimes at war with "culture," a "nature" in and around which contending political claims, identifications, and equivocations circulate.

What could it mean to say that we want to find ourselves anew, there, in a "nature" already marked from the first day by human traces? To seek the truth of gender in a "nature" that is gendered from the start? To delve into the nature of desire among the naturalized desires of a "nature" that is itself the product of so many one-sided longings? To make and remake ourselves in the image of this troubled "nature?"

Reflection on this ambiguous process of invention and self-invention is already under way, not just as a theoretical project among postmodern academics and sci-fi philosophers, but as a phenomenon of mass culture. On certain TV sitcoms, for example—and diametrically opposed to the heteronormative and bioessentialist representations aired in the "serious" media—sex roles and gender norms increasingly appear as malleable, fungible, and in flux. Such images "entertain," in the strictest sense of the word—inter-tenir, to hold between or among us—not just because of efforts over the past thirty years to contest the pervasive heteronormativity of American culture, but also as the result of profound economic changes in American society. As these changes continue to ripple through all of the institutions that touch on gender, sexuality, the family, and states of embodiment in general, they generate anxiety and mirth in equal measure. The way these ambivalences and ambiguities get worked through will affect the shape of both public and personal life in the coming decades, as the following chapters show.

22   Money's Subject

In March 1997, as stories about lesbians who like to have sex with men were making the rounds on the talk-show circuit and while suburban teenagers made weekly pilgrimages to the shopping mall sporting metal-stud piercings and rings in every imaginable body part, a *New York Times* front-page headline blared the serious media's prevailing message of a deep, biological essentialism: "Sexual Identity Not Pliable after All, Report Says."[1]

Such a sweeping conclusion was based on the reevaluation of a single case, first reported by John Money in 1973: that of a biological boy whose penis was accidentally severed when he was an infant. Doctors had created an artificial vagina to replace the destroyed penis, and the child's parents had attempted to rear him as a girl—unsuccessfully, as it turned out.

One might well have expected caution in the interpretation of such a case. The child was fully eight months old when his gender was, as they say, "reassigned." It is logical to imagine that parents and other close relatives could not help but view the child as an emasculated male—a perspective that may have introduced an element of ambiguity into the child's primary socialization. Note also that the child in question had an identical twin brother, that as a girl, s/he was teased by other children because s/he "looked like a boy," that s/he is said to have alternated between periods of acceptance and rejection of the outward signs of femininity (dresses, tresses, dolls), that as a "girl," his/her genitals had been constantly probed, operated upon, and painfully inspected by doctors, and that s/he required estrogen treatments beginning at the age of twelve. Under such circumstances, the failure of a female identity to "take" is unsurprising. At any rate, the subject's experience scarcely warrants the broad conclusions of gender essentialism unleashed in media reportage—although a sober reevaluation of Money's famous case might well give us caution about the limits of technocratic expertise and the dangers of medical hubris.

Instead of voicing caution, however, the avuncular voice of science approvingly chortled over evidence of something deeply embedded in the precultural nature of masculinity. Upon learning of his "true" identity and at the beginning of a long, slow, and only partly successful process of phalloplasty to construct a penis using skin grafts, "he got himself a van with a bar in it," as Dr. Milton Diamond cheerily puts it in the *New York Times*. "He wanted to lasso some ladies."[2] "Wanting to lasso some ladies," of course, is the very definition of natural manhood, or rather, the subject's caricatured performance of heterosexual manhood circa the epoch of *Saturday Night Fever*.

Instead of initiating a conversation directed at developing more nuanced and less invasive medical procedures for boys accidentally unsexed during circumcision, for infants born with ambiguous genitalia, or for anatomically intersexed people, this unhappy medical case is held up as "exhibit A" for the argument that heteronormative nature "always trumps nurture"—or so says Tom Wolfe on the dust jacket of John Colapinto's biographical book on the subject, *As Nature Made Him*.[3] A reviewer in *Elle*, quoted in the publisher's publicity, distills the ubiquitous essentialism that swirled around the book's promotion: "*As Nature Made Him* makes a convincing case that gender has less to do with the signals we send and receive from the world than with ineradicable messages encoded in every cell of our brains and bodies."

Louis Gooren has rightly dubbed similar appeals to the perpetual world of common sense "folklore in a lab coat"—unabashed ideology masquerading as impartial science.[4] We are not far here from "Wilson's" prime-time musings on TV's *Home Improvement*, where the culturally and historically specific outer signs of gender—racing cars, minivans, pork-chop sideburns, lassos—are taken as direct expressions of the inner essence of a timeless and unchanging manhood. But we are also not far from the center of the firestorm—from the real crises that contextualize so much of the present's essentializing, typologizing, and normalizing.

Such claims about the nature of identity flare in an atmosphere of *Stürm und Drang*. The public is entreated to revisit the evolutionary origins of desire—to appreciate the compelling logic of "courtship ritual," to understand the necessity of male/female "pair bonding," to see the beauty of those "eternal laws" of family formation, and to assimilate their implications for essential manhood and womanhood (all of this supposedly deducible from basic genetic principles ineradicably encoded in every cell)—at *precisely* the moment that all of this is in crisis: as gender relations and courtship rituals

Dykes To Watch Out For by Alison Bechdel

Figure 30.   "I.D. Fixe?" Alison Bechdel, from the comic strip *Dykes to Watch Out For*, anthologized in the book *Post-Dykes to Watch Out For*. © 2000 Firebrand Books, Milford, Connecticut. Courtesy LPC Group.

undergo dramatic changes, as it has become the statistical norm for people to pair temporarily, or in sequence, or not at all, as families have come to form in different ways around various gender combinations and with varied reproductive expectations fulfilled by diverse reproductive means.... It is no accident that men, women, and others are given a serious-sounding scientific discourse on eternal man and natural woman, locked in timeless embrace, at the very moment that every sex stereotype has come to seem dated, unreliable, and funny.

THE QUEERING OF PRIME TIME

This was the same decade that witnessed progressively wider circulations of a queer aesthetic in mass culture, a sensibility founded on the premise that there is no "natural" fit between genitalia, affect, and desire. First came Queer Nation's attempted coalition of fags, dykes, switch-hitters, trannies, S&Mers, sissy straight boys, butch wives, and single mothers—all those legions excluded by one means or another from the charmed circle of heteronormalcy. Then, and very quickly, came queer nationalism's academic echo in queer theory, a montage of theories underscoring the flexibility of the body, the performative superficiality of gender, the contrived nature of the heteronorm—in short, the "constructedness" of identity. It was easy enough to see where the intellectual excitement was on campus in the early 1990s: Queer theory succinctly appropriated and pointedly applied the gamut of twentieth-century cultural theory from Saussure and Lacan to Derrida and Foucault as an elegant and sophisticated critique of heteronormativity.

What is sometimes more difficult to see is that ideas about queer performativity simply captured a zeitgeist already well developed across much of America: the idea that sexual identity *is* pliable, the notion that anyone can be anything they want to be. In the dialectical tacking between public representation and public opinion, the academic avant-garde was at best only a step ahead of (and often two steps behind) developments in the wider world of cultural production—in the fashion industry, for example, with its perennial attention to sartorial artifice and the superficiality of identity, in pop music, where Madonna pithily recapped a long tradition of experimentation with body image and sexual personae in the slogan "strike a pose," and in the ever-fragmenting and diversifying realm of commercial entertainment, where the transition from broadcast network to cable TV was facilitating the spread of innovative, niche-market programs like *Xena, Warrior Princess, More Tales of the City,* and *Sex in the City.*

Over the course of the 1990s, images from the scholarly pages of Eve Kosofsky Sedgwick and Marjorie Garber came to feel perfectly at home even on prime-time, network TV: the straight arrow discovered to be gay or bisexual, the tough guy as closet queen, the butch cross-dresser, the straight transvestite, the effeminate heterosexual, the gay man who secretly harbors heterosexual lusts (and the inevitable reverse coming-out), the respectable swinger, the good-natured sexual adventurist, the people you'd least expect who turn out to be endlessly, deliriously, delightfully kinky. . . . Unlike those one-dimensional and self-identical creatures who populate "scientific" reporting in the "serious" media, a growing cast of characters on "frivolous" TV feel lusts and act upon whims and desires in all the ways humanly imaginable. They live the social and institutional changes of their time, as Merleau-Ponty was fond of putting it, "in the flesh." They also experience inner contradictions and conflicts of an order not openly explored in the past.

On top-rated *Seinfeld*, sidekick George Costanza once developed an infatuation with Elaine's affable, athletic boyfriend—a jealous crush so transparent that he was ruthlessly teased by other characters. On more than one occasion, that show's plotline revolved around the homosexual implications of George and Jerry's friendship. The two were once "outed" by a student journalist who overheard their conversation and took it for the comfortable squabbling of a long-term companionship. George, that gelatinous mass of male anxieties, once became convinced he was gay because, while he was being massaged by a male masseur, "it"—George cannot bring himself to name the part—"moved." And during the show's final season, George realized that his current girlfriend bore an unnerving resemblance to Jerry. When he revealed his panic to the latter, the two vowed—rather perfunctorily, I think—to never speak of the matter again.

A spectral homosexuality thus haunted the number-one TV comedy of the 1990s—a vehicle that repeatedly mocked its namesake's masculinity—and homoerotic byplays and queer double entendres were no less frequently foregrounded on other highly rated programs from NBC's Super Thursday lineup. On *Friends*, Ross's ex-wife is a lesbian, while straight roommates Joey and Chandler act very much like a jealous couple—a quarrelsome intimacy underscored for comic effect. And after Chandler moves in with Monica, Ross and Joey discover a forbidden, secret pleasure: the two young men enjoy napping together, one atop the other, on Joey's overstuffed sofa, as innocent as children and as guilty as perverts. I have lost track of how many times the men on *Friends* have found reason to kiss, fondle, or grab each other. The women, too. (I have also lost track of how many times plotlines have challenged Chandler's manhood or questioned his heterosexuality. Chandler's character essentially lives out an attractive and endearing version of George Costanza's male troubles.) On *Just Shoot Me*, Finch's father decides, incorrectly, that his sissy son Finch is gay—but misses all the give-away signs that his other, conventionally masculine, son is gay. Meanwhile, the suggestively titled but ultimately disappointing *Veronica's Closet* mined humor in a supporting character's lack of self-awareness: He is oblivious to the fact that he is gay, although his homosexuality seems self-evident and is perfectly acceptable to everyone around him.

The TV program sometimes promoted as the "anti-*Seinfeld*"—a sitcom set in the Midwest and centered on an ensemble of working-class characters— was, if anything, even queerer at times than NBC's suite of cosmopolitan Manhattan shows. In the season finale of 1997, and after much homosexual innuendo among supporting characters, the entire cast of ABC's *Drew Carey Show* recapped gender-bending movie motifs from *The Rocky Horror Picture Show* to *Priscilla, Queen of the Desert* by facing off on the streets of Cleveland for a full-scale musical drag competition. In subsequent seasons, a male supporting character participated in a pharmaceutical company's cockamamie breast-implant experiment, and Drew's butch, bullying (and heterosexual) big brother came out as a transvestite. On ABC's *Norm*, the lead character briefly became the "kept boy" of an older, wealthy gay man who was infatuated with him. And on even as conventional a vehicle as CBS's *Everybody Loves Raymond*, the homoerotic element of male bonding was mined for sources of humor. When the protagonist tells his wife he "got a funny feeling in the locker room," she tells him—in a deadpan delivery and clearly *not* for the first time—"Ray, you are not gay."

HOORAY FOR HOLLYWOOD

The deepening and expansion of queer themes in public culture has occurred very quickly. In the 1970s and 1980s, gay characters appeared on TV shows simply to assert their existence—usually in one-time-only roles—and to carry simple messages of tolerance. (Gays and lesbians are people, too. They're members of your family. You can't tell who is or isn't gay just by looking at them.) Into the 1990s, gays and lesbians increasingly populated prime-time niches as regular or recurring characters—up to five on brilliant, trend-setting *Roseanne* alone. Acceptance, more than tolerance, was the message. By the close of the 1990s, plotlines were also apt to suggest—of "straight" lead characters—what gay liberation first blurted out in the euphoria of the late 1960s: "We're all just a little bit queer, even if we don't always know it." In the process, the site of humor—the situation of comedy—has migrated a far distance in the sexual imaginary. Where comic writers once exploited gay visibility as a shock effect, today they are more likely to traffic in nervous laughter about the precarious nature of "straight" existence. In mass entertainment, at least, a revolution of sorts has happened: a continuation and deepening of the cultural revolution of the 1960s.

Some will immediately object that I overstate matters. Perhaps I do. Others have developed more nuanced analyses of gay visibility and queer themes on TV and in the movies—I think here of Suzanna Walters's excellent book on the subject.[5] Gay critic Michael Musto crankily underscores the disconnect between gay prime-time minstrelsy and the stalling of gay rights: "Being gay in 2002 means we might not necessarily be able to marry, adopt, or join the armed forces, but honey, we are in every sitcom! We're a regular riot, girlfriend, our flouncy gestures and bons mots providing a cathartic giggle for the entire TV-watching populace. We've got a place at the table, all right, even if it's only a fictional table that's dismantled at 10 P.M.—and by the way, it's usually not even remotely adjacent to the bedroom unless the show is on cable."[6]

One could no doubt point to movies and TV programs that celebrate the eternal verities of gender norms, although I hasten to add that much of what is actually depicted as "timeless" or "universal" on these programs takes on a decidedly local, modern, and up-to-date appearance, as we've seen in the discussion of *Home Improvement.* And although unobjectionable depictions of gay and lesbian supporting characters are becoming more common on successful TV programs *(Friends, Spin City, West Wing, Buffy the Vampire Slayer, Dawson's Creek),* the industry's record on programs with lead gay characters—*Ellen; Will and Grace; Oh, Grow Up; Normal, Ohio; Some of My Best Friends*—has proved uneven. Or, rather, one might say that audience receptions of these vehicles has been varied.[7] Undoubtedly, lesbigay characters are still called upon to serve as objects of a prurient curiosity or to facilitate heterosexual performances of self-assured tolerance. And still, after all these years, news and infotainment programs continue to run stories that weigh the worthiness of gay relationships, assess the fitness of gay parents, or gauge the limits of sexual tolerance.

Still, the situations that give audiences comic *reconnaissance* today seem strikingly different from the comedy of the past, in which lesbian, gay, or bisexual characters were presented for shock value, or were depicted as objects of pity, or were *by definition* funny. The exquisitely nuanced "Lesbian Kiss" episode of *Roseanne* has become the template for a new sort of comic situation. When kissed by a woman (played by Mariel Hemingway) on a "girls' night out" in a lesbian bar, Roseanne feigns nonchalance, but her affected hipness betrays a fundamental nervousness about same-sex desire. As Roseanne protests that she feels "completely secure" in her heterosexuality, the audience sees—and feels—her uneasiness.

In the new comic paradigm, it is not that lesbian, gay, bisexual, transsexual, or otherwise queer existence is funny, but that the presence of LGBTQ figures occasions an existential crisis for straight people, which *is* funny. Lesbigay characters thus dramatize epistemic flux—that is, they serve as an index of confused role expectations, sudden reversals, and the malleability of human dispositions in general. Such comedy offers glimpses into the neurotic, conflicted, and troubled world of heterosexuality—a world where homoerotic feelings and transvestic impulses wrestle with the conventional image of a closed and unfuckable heterosexual manhood, a world in which the liberated woman appropriates "male" habits and dispositions the way she might sport a strap-on phallus. These theatrics open onto real-world conflicts and ambiguities, where identity is in crisis because all the signs of gender have been amputated, scrambled, and recombined. Such were the sparks ignited by the combination of feminism with gay theory, a combustion that produced "queer theory" in the academy. And such are the flashes of a queer sensibility in mass culture, an aesthetic so successful at the end of the 1990s that one is tempted to conclude that it defined the success of one of TV's most successful programs.

Will and Grace succeeds in the ratings where *Ellen* failed in part because the program's scripts, acting, and production values are better and in part because the program's primary relationship, between a gay man and a straight woman, provides a more identifiable situation for mass, straight audiences than did the coming-out travails of a lesbian. But more notably, the program succeeds because it better exploits all the devices of queer humor at its disposal. Scripts knowingly, campily play with stereotypes and expectations. Everyone is shown to have a kinky side. Plotlines explore points of correspondence between gay and straight life without amalgamating either into the other. This is the vital source of the situation's comedy.

An insistent Grace joins a protesting Will in the bathtub, a moment that could have been salacious or offensive, but that instead captures much of the warmth of the show, its attitude toward shared lives, other bodies, and physical intimacy. Will complains to Grace, within Jack's earshot, that Jack is "such a fag"—which gives Jack the occasion to respond, quite eloquently, to gay media critics of the character's flamboyance. On the episode I call "And

Figure 31. Grace Adler (Debra Messing, left) and Will Truman (Eric McCormack, right), NBC's *Will and Grace*. KoMut Entertainment and NBC Studios in conjunction with Three Sisters Entertainment. Courtesy of Photofest, New York, N.Y.

That's Why You'll Never See Two Men Kissing on Network TV" (the episode's recurring phrase), Will does in fact seize the stray opportunity to kiss Jack on prime-time TV. It was thus by means of artful manipulation and devious cunning that an important threshold was crossed. When the usually prim Grace contemplates (and almost consummates) a "threesome" with a straight couple, the resulting performance is very much like watching a young Lucille Ball go kinky, camp homage to a venerable tradition of physical comedy. And in teaching him a dance step, Karen mimics sodomizing Jack, to the mutual confusion of both, a sudden role-reversal that unsettles all the definitions of male/female, gay/straight.

THE MAD SCIENTIST AT THE FESTIVAL OF DISGUISES

The Halloween 1999 episode of ABC's *Two Guys and a Girl* is indicative of its genre: a topsy-turvy world wherein desire is increasingly decoupled from the nature of yore, where gender signs are blended and mixed, and where affect is at least temporarily dislodged from the heterocentric circle. The situation: a mad scientist has switched brains between four of the program's five protagonists. In each of these exchanges, a man and a woman have swapped brains.[8] What follows is a series of affirmations and negations.

The parties to one set of switched brains display a mind-body consonance more stereotypically (though not necessarily "naturally") matched to the characters' temperaments: The brain of an assertive, brusque woman is now lodged inside a man's body, while the brain of a sensitive, reticent man is now in a woman's body. Despite what might seem to be an affirmation of gender essentialism, neither brain is actually happy inhabiting the other sex's body.

The characters in the other switched set experience different conundrums. The young man's brain is delighted to explore the female body from the inside. He cannot keep "his" hands out of *her* pockets, and he thinks, already, of going to a steamroom to be surrounded by naked women. His alter—the woman's brain now stationed inside *his* body—frets that "he" (the male brain) is going to give *her* (body) a "bad reputation," and demands consolation from her confused and reluctant fiancé. *She* (the female brain in the male body) cannot understand how he (who has not been altered) lets a little thing like the sex of the body come between them.

After a series of mishaps, three of the characters' brains are restored to their proper bodies, but the latter woman's brain is ultimately transplanted into the mad scientist's body—a transformation her boyfriend, in the end, accepts. At the denouement of this varied staging of sexual troubles, the woman, whose female brain now resides in the male mad scientist's body, happily drapes her/himself across the boyfriend on a sofa. Someone deadpans: "I guess he really does love her for her brain after all."

If the various permutations on gender and desire explored in this Halloween carnival house might be seen as "affirming" certain essentialist notions of heteronormativity, they might also be seen as undermining or even "negating" them. The relationship between medical science and sexual identity here is especially instructive. The mad doctor's medical manipulations neither confirm nor deny the characters' sexual identities. Rather, they allow for a series of lived experiments whose results are ultimately ambiguous. By means of such merry equivocations, mass entertainment lives the modern identity crisis as a moment of levity. This levity is frequently Rabelaisian in character: It involves "low" humor about the lower-bodily stratum, scatological plays off the "openness" of the body, and jokes that isolate, reify, symbolically amputate, transplant, or metastasize body parts.[9] The sexual identity crisis thus plays out as a social and aesthetic experiment in entertainment culture. The psychic shocks of sexual uncloseting, gender transformation, and sexual chameleonism become so many comic situations.

Laughter, by turns nervous and festive in character, enunciates an overall aesthetics of ambivalence.

LIKE A GAME OF FORT-DA

Meanwhile, in step with changes in the world of entertainment, Madison Avenue was discovering two new principles in addition to its standard "sex sells." Homoeroticism sells. (But you already knew that, even if you pretended not to notice.) And queer sells. (This, I think, came as a surprise to everyone.)

In a decade marked by "lesbian chic," Nike began running ads designed to market its shoes to serious female athletes. These ads conveyed unprecedented images of independent, active, and physically strong women. At the same time, a fruit-juice company was mocking sexual prudery with the advertising come-on: "Flavors Mother Nature Never Intended." Loreal was preparing its camp "Straight Is Boring" ad for eyelash curlers. And by the end of the 1990s, Abercrombie & Fitch stores in suburban shopping malls everywhere had come to resemble nothing so much as temples of Sodom, their walls adorned by Bruce Weber's voyeuristic, adoring, and explicitly homoerotic photographs of scantily clad beautiful young men engaged in various outdoor adventures: wrestling, splashing, reclining together. . . . When A&F issued its quarterly catalog with the warning "Due to mature content, parental consent suggested for readers under 18," sales shot up.

The social changes and psychological complexities involved in the display of such images are, as Herbert Muschamp notes, "there on the surface, for all to see." So, too, their power to move commodities in today's image economy.[10] Dockers, which once pitched its ads at "regular guys," now airs a long-running TV commercial boasting that the company's trousers will draw the amorous attentions of both female *and* male admirers. A recent Minute Maid ad shows Popeye and Bluto frolicking instead of fighting. They get matching tattoos and ride off together on a motorcycle, leaving Olive Oyl behind "in a cloud of sexual ambiguity."[11] A Miller Lite ad, appropriately titled "Switcheroo," begins with two women looking for dates in a pick-up bar. It ends with two attractive men—the objects of the women's interests—holding hands.

As against the disruptions of this ongoing ontological carnival, the conservative and regulatory function lies with solemn, serious culture—with the *un*entertaining media, whose sanctimonious, Sunday-morning homilies project and ratify the image of a closed, self-identical, and unchanging body. With all the more desperation, science writers in serious venues pull down biological "evidence" to sustain heternormalizing stereotypes of stable gender roles and conventional sexual relations—an attempt at outrunning the decidedly queer aesthetics overtaking vast terrains of entertainment and commercial culture.

THE DIVISION OF DESIRE

And so it was that in 1998, Valentine's Day—that quintessentially prefabricated holiday dreamed up by the greeting-card industry in a bygone era of modern love, modern courtship, and stable nuclear families—came in the midst of a minicrisis over whether the president of the United States had had sex with a White House intern and then lied about it under oath. This unseemly drama was attended by a chorus of pundits who lamented the public's indifference to such peccadillos as symptoms of moral decay. Talking heads on *Nightline* debated whether oral sex constitutes "real sex," while conservative Republicans newly converted to feminism attempted to read an Oval Office workplace dalliance as an instance of job discrimination or even child abuse.[12]

It was in such a milieu that on Friday, February 13, ABC's *World News Tonight* took a flight of fancy, devoting a lengthy pre–Valentine's Day segment to one more reiteration of sociobiology's first principles. With an air of calm assurance, Peter Jennings narrated the hardwired genetic script for pick-up rituals in a straight bar: Men are visually oriented, and visually stimulated by signs of fertility (large breasts, wide hips), whereas women are more cautious and calculating—"Does this suitor bring resources?" Men prefer sex with many different partners because a promiscuous strategy broadcasts their genes widely, whereas women's intensive investment in their young naturally predisposes them toward more selective sex with fewer partners. Here, then, the Everlasting Truth of desire—heterosexual complementarity—as prime-time news: man is to woman as plug is to receptacle.

MONEY'S SUBJECTS

Jean Baudrillard, the joker in the deck of cultural theory, inaugurated the final decade of the twentieth century by reflecting on Michael Jackson's "androgynous and Frankensteinian appeal" and by (not altogether happily) noting "a general tendency toward transsexuality." Heralding the advent of a carnivalized image economy, Baudrillard draws the inescapably logical conclusion: In a world of perpetual experimentation with sex and image, and at a moment when everyone plays with identity and difference, "we are all transsexuals."[13] At about the same time, a neoconservative pop music critic lamented the "perverse modernism" that had overtaken rock'n'roll—a "queer" sensibility perhaps best illustrated by Kurt Cobain's resigned couplet: "What more can I say? Everyone is gay."[14] Somewhere, Pearl Jam's Eddie Vedder was wailing out the question of the moment—"Are you woman enough to be my man / bandaged hand in hand"—when the *New York Times* flashed its front-page news: "Sexual Identity Not Pliable after All."

In real life, Money's unhappy subject is not unlike the rest of us: an anonymous individual struggling to find a true identity in an ambiguous body. A nameless soul seeking authenticity in a world of consumer objects, prefabricated meanings, and slippery signs. A self condemned to make himself or herself, as s/he sees fit, in a season of change and turmoil. A lonely, fragmented being in search of connection.

23 History and Historicity
Flow through the Body Politic

Ongoing, uneven transformations in sexual culture since the 1960s have everything to do with reforms wrought by the new social movements, but they also have to do with new forms of production and consumption in contemporary capitalism. "Nature" and its imaginings come into play there, as well. This is also where things get dicey.

THE FORDIST LIBIDINAL ECONOMY

Now is not the first time men and women have found themselves torn, divided, and in the midst of ambiguous metamorphoses—some wrought by their own desires for a better world, others brought on by the institutional avalanches triggered by political-economic upheavals. In his notations on "Fordism," drafted during another period of social turmoil, Italian Marxist Antonio Gramsci devoted considerable attention to the emergence of that constellation of laws, regulations, and conventions that today might best be understood as the historically prevailing variant of "heteronormativity."

The system of Fordism corresponds to an era of large-scale production and marks "the passage from the old economic individualism to the planned economy." Such a transition, Gramsci reasoned, began in the United States during the period historians conventionally understand as "Fordist," but would ultimately touch every institution in a wide belt of societies.

Gramsci correctly identified the salient features of capitalism in the era of mass production: its rationalization of labor and demand, its preference for economies of scale, its centralization of resources, and its tendencies toward regulation, administration, and discipline. But Gramsci's analysis took in more than just economic activity at a given moment of capitalist development. Surveying legal precedents in the Anglophone world, demo-graphic trends in Europe, class conflicts and compromises in the industrial-ized countries, and—of course—news from Henry Ford's America at the end of the 1920s, Gramsci argued that personal life, no less than economic life, was being progressively fed into the calculus of new relations of pro-duction and consumption. In this transition from anarchic individualism to regulated Fordism, the "attempt to create a new sexual ethic suited to the new methods of production and work" would be "extremely complicated," but "necessary." In fact, Gramsci concluded that sex would be a critical link in the emergent institutional nexus, for the "new type of man demanded by the rationalization of production and work cannot be developed until the sexual instinct has been suitably regulated and until it too has been ration-alized."[1] Disciplined by the exigencies of mass production and consump-tion, this new personality type would belong to a total system of associa-tions: in the factory, the conveyor belt; in the world beyond, the mass consumer market; and in the family, a sexual division of labor and a certain mode of life.[2]

Having thus traced the flow of power across a social, cultural, and psy-chological terrain, Gramsci went on to chart points of conflict in the new social formation, social struggles inherent in the newly emerging institu-tional nexus—for example, the challenges of feminism and the panicked resurgence of "masculinism," the deployment of ever more coercive and vi-olent repressions of sexuality versus the florescence of "new forms of en-lightened utopias . . . around the sexual question" and the development of an ongoing "crisis" or "disequilibrium" of "the institutions connected with sexual life"—a crisis differentially refracted through class strata and per-petually at odds with the demands of disciplined work and social authority.[3] It is as though Gramsci peered beyond his present horizon to catch a glimpse of 1969 in 1929.

Or perhaps it was the cultural conflicts and historical coordinates of 1999 and beyond that he foresaw. Writing from a fascist prison cell, Gramsci de-scribed the Fordist social experiment in the following extraordinary passage:

> The sexual question is . . . connected with that of alcohol. Abuse and irregularities of sexual functions is, after that of alcohol, the most dan-gerous enemy of nervous energies. . . . The attempts made by Ford, with the aid of a body of inspectors, to intervene in the private lives of his employees and to control how they spent their wages and how they lived is an indication of these tendencies [toward a regulated, admin-istered social formation]. Though these tendencies are still only "pri-vate" . . . they could become, at a certain point, state ideology, inserting

themselves into traditional puritanism and presenting themselves as a renaissance of the pioneer morality and as the "true" America (etc.). . . . It seems clear that the new industrialism wants monogamy: it wants the man as worker not to squander his nervous energies in the disorderly and stimulating pursuit of occasional sexual satisfaction. . . . The exaltation of passion cannot be reconciled with the timed movements of productive motions connected with the most perfected automatism. This complex of direct and indirect repression and coercion exercised on the masses will undoubtedly produce results and a new form of sexual union will emerge whose fundamental characteristic would apparently have to be monogamy and relative stability.[4]

QUEERING GRAMSCI

It might seem odd—not to say a bit old-fashioned, given the run of recent queer theory—to revisit Gramsci's passages, now over seventy years old, in the context of what this book sets out to accomplish: a critique of the naturalization of heterosexuality in recent science and an understanding of how bioreductivist ideas relate to ongoing changes in sexual culture. Undoubtedly, Gramsci had his blind spots. Although the Mediterranean author was clearly struck by the novelty of the notion that modern marriage might entail monogamy (for men), he does not directly address what varied forms the "exaltation of passion" might actually take on a hot night in Italy, in certain waterfront bars and parks, or along certain backstreets. . . . Gramsci nowhere discusses homosexuality, nor does he examine the heterosexual imperative per se, and it might seem implausible to think that he throws any light on what I have been calling "heteronormativity."

But more so than many recent works in lesbigay studies, Gramsci's proleptic passages on what was then the "new form of sexual union" and its place in a newly emergent institutional nexus ought to alert us to a feature of the heteronormativity not always fully marked or adequately understood: its *historicity*. Here, as throughout this book, the connections I want to establish have to do with the way one thing refers to another—a connectivity to which Gramsci was extraordinarily sensitive.[5]

Claims about sex, whether homosex or heterosex, always refer back to notions of what a man is, what a woman is, what a penis does, what a vagina is for. Discourses and practices of sexuality thus have everything to do with gender. These notions of gender, in turn, occur within institutional settings that govern how affinities are made, how kinship is reckoned, and even what counts as sex. Gender thus has to do with sexuality, but also with kinship and family forms, as we've seen. These institutional arrangements are themselves improvised, if not always in the light of strictly economic activities, then certainly within the horizon of changing political-economic relations. Activity in one realm refers to, supports, and reinforces practices in the other. Sometimes, these ideas and practices are successfully articulated into a self-perpetuating and self-validating mechanism, a system that produces both material objects and the cultural subjects capable of receiving them. And sometimes the one thing imposes upon the other with such force and urgency that waves of changes ripple through the entire social constellation.

Consider what follows, then, to be a thumbnail sketch of Gramscian connectivities—a comparative history of the present in very truncated form. It is also a working hypothesis on the transition from yesterday's Fordism to today's post-Fordism: from a Keynesian, regulated, mass-production economy to a neoliberal, deregulated, flexible system of production and (extrapolating from Gramsci's observations) from one very distinct sort of libidinal economy to another.[6]

FAMILY PORTRAITS:
A SHORT HISTORY OF HUMAN NATURE

By the 1950s, the new human nature Gramsci foresaw had been implanted on a mass basis in most Western societies. Its core—monogamous marriage, the nuclear family—was supported by an elaborate regulatory apparatus. Heteronormative structures of feeling were subsidized by government largesse (tax breaks for marriage, for children, for home ownership), nurtured by Cold War ideology (strong families as the frontline against communism—and in the East, a mirror-image family as the motherland's defense against imperialism), and bolstered by coercive appeals to a mythical "tradition" that never really was (the naturalization of institutional contingencies in romantic film, popular music, and potted histories). The florescence of this "new form of sexual union" historically corresponds to the life cycle of a demographic group. Its trajectory charts the course of the Keynesian economic expansion and follows the curve of the first wave of suburbanization as it affected the lives of white, middle-income people and reshaped the contours of America's race/class formation. Under the proddings of economic carrot and ideological stick, the percentage of adults who were married reached an all-time high. Extended families and compound households were nuclearized. Unprecedented numbers of people were naturalized, normalized, monogamized, heterosexualized.[7]

Figure 32.
Jane Wyatt, Robert
Young, and the cast
of *Father Knows
Best*. Screen Gems/
Columbia Pictures.
Courtesy of Photofest,
New York, N.Y.

At its base, this arrangement proved unstable in many ways. Its sexual division of labor depended upon a "male wage" sufficient to support unremunerated female housework and child care. By extension—and contrary to the logic of prevailing social tendencies in a regulated market economy— "normal" gender relations depended on rigidly structured inequality: on male domination and female subordination. And unlike earlier conjugal forms—especially marriages arranged according to the social and political interests of wider kin groups—modern marriage's sole claim to legitimacy was the reciprocal love of husband and wife, the mutual happiness of the "bonded" pair. In consequence, it was heterosexuals, not homosexuals, who undermined this precarious arrangement, the heteronormative nuclear family, at its very foundations, in sanctioned, not censured desires. No sooner had industrial affluence made the supposedly natural nuclear family broadly feasible—and universally compulsory—for the first time in history than a contradictory thing happened: men and women began exiting unhappy marriages in ever-increasing numbers. And no sooner had heteronormativity triumphed in all its suburban splendor than the regime's disenchanted subjects began waging a grim, twilight struggle against the terms of normal love and moral sex. Playboys, divorcees, beatniks, swingers, flower children, hippies, feminists, and gay revolutionaries followed, in rapid succession—each in the hot pursuit of happiness.

By the 1960s, then, the golden age of the heteronormative nuclear family was over, and all of the institutions connected with sexual life had entered into a period of disequilibrium and crisis. These trends were sometimes experienced as moments of personal breakthrough and liberation and sometimes as moments of acute distress and social breakdown, but they conformed to the deeply romantic ideology of the nuclear family itself, with its accent on amorous love and personal fulfillment. These trends also intensified the basic motivational structure of modern capitalism, with its emphasis on individual autonomy and happiness. If one were to speak of these developments in the language of an old-fashioned Marxism, one might say that the nuclear family carried the "seeds of its own destruction." Or perhaps one might better refer to romantic love as "the progressive element" of the mode of social reproduction—the source of practices and conflicts that tended, over time, to transform the logic of the system. The self-same cult of love and happiness that brought about the end of the stable, pair-based nuclear family—monotonogamy—also contained the kernel of many forms of affinity today: live-in lovers, gay and lesbian families, single parents, open relationships, temporary unions, kin alliances that do not presuppose sexual intercourse, "open-minded, adventurous couples," threesomes . . . modern love in all its varied splendor.[8] The dialectical tension around love and happiness is still with us today, and it is still working its transformative effects.

A SEVENTIES FLASHBACK

I vividly recall my first political struggles, in the late 1970s: Gay activists were organizing "zaps" against Anita Bryant, while advocates of the Equal Rights Amendment sparred with organizations of the nascent religious Right. "Est," it seemed, was on everybody's tongue, and the Village People were taking *Billboard* by storm.

Against this stormy backdrop, where personal liberation, conservative reaction, and commercial cooptation all advanced simultaneously—the one dogging the others, the one briefly outrunning the others before being in turn overtaken—we political activists weighed our strategies. In the gay movement, a still-vigorous socialist Left debated questions of heady import with an emergent libertarian movement. Our arguments over the meaning and direction of history, poised at the crux of actual historical happenings, proceeded with a special animus: they often ended in university quad shouting matches and occasionally in barroom fisticuffs.

Invoking traditions of liberal political theory going back to John Locke and Adam Smith, and drawing on the image of capitalist business practices then being popularized by Milton Friedman, the libertarians argued that freedom in and of the marketplace is the best way to advance freedom for lesbians and gays.[9] The capitalist market, they claimed, has no inherent interest in distinctions based on race, gender, or sexuality. Indeed, since transactors in the marketplace are (theoretically) motivated only by the need to make money, their activity should tend, over time, to undermine social distinctions of any arbitrary sort. (Presumably whatever economic inequalities resulted *from* these transactions would be just and rational, the effects of "natural inequalities" in intelligence or talent combined with differences in industriousness.) The terms of this argument remain basically intact after twenty-plus years. As a gay libertarian succinctly puts it in a recent letter to the editor: "Individual civil and economic liberties prosper best when government stays out of the bedroom and the marketplace."[10]

Socialists, steeped in the traditions of Gramsci, the Frankfurt School, and New Left theories like socialist feminism, countered that actually existing capitalism has seldom worked that way—that business transactions always rest, in practice, atop an institutional nexus. Far from leveling irrational or invidious social distinctions, real-world capitalism has at every turn incorporated racial stratification, ethnic inequalities, a gendered division of labor, and so on, into the basic institutional gridworks of its changing social formations. We also pointed out that the capitalist system has historically resourced narrow, exclusionary structures of family life as its primary site for social reproduction, an arrangement with myriad implications for gays and lesbians.

There was something outrageously idealistic about the abstracted transactors who conducted generic transactions in the libertarians' hypothetical system. But there was something exasperatingly complicated about the sociological and historical arguments mustered by socialists, even leaving aside Freudo-Marxist analyses of the role of surplus sexual repression in labor discipline and the process of capitalist accumulation.[11] I remain convinced to this day that we socialists had—have—the better theory, the better grasp of history. However, real developments *post factum* suggest a double irony: If the weight of historical precedent once suggested that socialists were right, even if we could not compose a pithy draft of our arguments or put together a winning political program, the passage of time now suggests that the libertarians were on to something, after all—that they were substantially right, but for reasons they could not fully understand.

THE WHEEL OF HISTORY

Women still struggle for equality, but today's zone of contestation is very different from that at the beginning of the women's movement, when abortion was illegal, domestic violence went largely undiscussed, and it was taken as an article of faith that women were dispositionally unfit to practice most professions. Gays and lesbians still have too few rights, but our situation is scarcely comparable to that of three decades ago, a time when the police routinely pulled lesbians and gay men out of bars, loaded them into wagons, and took them down to police stations to be photographed and fingerprinted, a time when the official organs of the medical and psychological establishment defined same-sex desire as a disorder and when sodomy laws were actively enforced in many states. Today, the sight of two men or women holding hands would not stop traffic in large parts of any major American city. Today, mayors, city council members, and congressional representatives clamber their way to the front of gay pride parades in far-flung cities.

The social movements of the 1960s and 1970s did not transform everyday life by dint of some Herculean act of will—or even as the cumulative effects of a thousand daily engagements in a thousand different battlefields. Radical changes in gender, sexuality, family life, and public culture were secured by shifts in the deepest substrata of modern society.

It's an old story, one whose logic was captured by John D'Emilio in his classic essay "Capitalism and Gay Identity."[12] Nineteenth-century capitalism encouraged the consolidation of the bourgeois family not by conservative but by revolutionary means: by undermining the social and political power of extended kin groups and by eroding the economic self-sufficiency of domestic units. And although the capitalist culture of the nineteenth century employed law, religion, and medicine to regulate sexuality, industry's insatiable demands for labor also concentrated diverse populations in everbigger cities—in the process creating spaces for the development of cosmopolitan cultures, sexual demimondes, and dissenting subcultures.

Mid-twentieth-century economic rationality oversaw the brief consolidation of a dreadfully conformist social formation, but institutional arrangements under Fordist capitalism also allowed for the relative autonomy of "personal life"—for the pursuit of individual happiness in its own right. And modern economic relations also gave first men and then women a certain independence from "traditional" conceptions of moral personhood—whether those conceptions were fostered by religion, by community, or even by earlier forms of capitalism itself.

Since the 1960s, myriad further adjustments in family life have been stimulated by economic changes—for instance, by the long-term and ongoing decline of the male wage that had secured stable nuclear families and a sexual division of labor in the first place, by the consequent entry of ever-greater numbers of women into the workplace, and by ensuing changes in the nature of appropriate gender roles. And even as the transition from an industrial/producer economy to a service/consumer economy facilitated the mixed successes of feminism, an economy increasingly dependent on advertising culture and "mode of consumption" has also stimulated the proliferation of "lifestyle" forms, subcultures, and subeconomies in great variety.[13] Although such economic transitions alone could not *cause* the emergence of gay, lesbian, and other sexual communities, the service/consumer economy—along with the forms of entrepreneurship it implies—has provided ready supports for modern experiments with identity and affinity.[14] (The seventies slogan "Spend your gay money at gay businesses" begins to capture this dynamic.) That modern techniques of production, advertising, entertainment, and consumption have incorporated these very forms—sometimes in a state of denial, sometimes uneasily, and sometimes with great gusto—also seems undeniable. Recent campaigns by the British tourism industry to capture the "pink dollar" demonstrate both the intensification and the internationalization of a thirty-year trend.

To make a long story short (and to try to capture something of the logic of historical transitions still under way): Mid-twentieth-century relations of mass production and mass consumption logically relied on heteronormative nuclear families, which served as both sites of consumption and sites of social reproduction. In this institutional nexus, the regimentation of gender and the normalization of sexuality produced subject citizens of a certain order: the man who works (and is therefore active, agentic) and the woman who performs domestic chores and emotional work (and is consequently nurturant, affective). It was not human nature that secured the stability of these arrangements—at once cultural and economic—but the male wage, supported by government regulation and Keynesian subvention. Gender nonconformity and sexual rebellion were in fact deeply subversive of these particular arrangements: They threatened the stability of the system, its institutional means of social reproduction.

Now it is not so clear that contemporary capitalism—with its emphasis on flexibility, mobility, and individuation—really requires heteronormativity, a sexual division of labor, stable marriages, monogamy, or nuclear families. Or rather, this is precisely the question in contest today: Whether the state ought to take a more or less laissez-faire approach to family, gender, and sexuality or whether it ought to take a regulatory approach, attempting—against all prevailing trends—to reconstruct the institutional structures of a bygone era.

Not that it's unthinkable that the state might act to reverse those social trends conservatives (and some liberals) view as signs of social degeneration. The Defense of Marriage Act (an attempt to forestall the recognition of gay marriages, even should some states decide to perform them) defines marriage as the union of one man and one woman. DOMA stands as a key exhibit of the regulatory impulse. (Worse yet, it represents an attempt to cheat history by binding future law to current prejudice, in anticipation of what supporters in fact view as probable future actions.) So does the 1996 welfare overhaul, whose Personal Responsibility Act begins "with a hymn to marriage and is based on the theory that poverty and social dysfunction are caused by the untrammeled sexuality of poor women."[15]

Still, the Fordist notion that disciplined labor requires a strict regimentation of gender roles and a puritanical disciplining of sexuality comes off as decidedly quaint in most global and even in very many regional cities. And the very idea of regulating exchanges, images, marriages, and desires seems distinctly out of step with neoliberal doxa in affluent countries today—especially given the role of desire in modern marketing and consumption. In today's libidinal economy, commerce is foremost about harnessing desires and marketing them to disparate populations, thereby soliciting new needs, new wants, new identities, and new experiments in lifestyle.

In this context, the gay subculture that thrives in every major metropolis scarcely seems "subversive" at all. It might better be argued that gay capitalism is actually paradigmatic of the new consumer economy, as Alexandra Chasin's innovative and insightful study suggests.[16] Not even queer politics—which proclaimed its radicalism (and its aversion to the gay subculture's *homo*normativity) with great fanfare at the end of the 1980s—seems very radical anymore, and the passage of time suggests that its "decenterings" pose no real threat to the neoliberal status quo. In fact, the appropriation of queer aesthetics in modern machineries of advertising and entertainment suggests a "fit" more in line with Rosemary Hennessey's observations. The promulgation of "more open, fluid, ambivalent sexual identities" in commercial culture announces "more flexible gender codes and performative sexual identities that are quite compatible with the mobility, adaptability, and ambivalence required of service workers today and with the new more fluid

It's Not a Choice.
It's the Way We're Built.

Subaru All-Wheel Driving System.
In every car we make.

Maximum traction, agility and safety. Experience the performance of the Subaru All-Wheel Driving System in the versatility of the Outback, the ruggedness of the Forester and the get-up-and-go of the Legacy GT Limited. To test drive one of our family of cars, stop by your nearest Subaru dealer, call 1-800-WANT-AWD or visit our Website at www.subaru.com.

Subaru supports the community as the proud founding sponsor of the Rainbow Endowment. The Rainbow benefits health, civil rights and cultural interests. For more information or to apply, call 1-800-99-RAINBOW.

SUBARU
The Beauty of All-Wheel Drive.

Figure 33. Subaru ad, aired in gay publications and neighborhoods, in concert with the 2000 Millennium March on Washington for Equality: A National March for Gay, Lesbian, Bisexual, and Transgender Civil Rights. Reprinted by permission.

forms of the commodity."[17] In the modern consumer economy, the only "bad" object choice is the one that fails to move commodities.

SEXUAL POST-FORDISM

An unreconstructed socialist, I shamelessly sing the praises of contemporary capitalism—in much the same spirit as Marx and Engels, whose *Manifesto* surveyed the dynamic and progressive role of capitalist exploitation in human affairs. "Wherever it has got the upper hand," Marx and Engels enthuse, the bourgeoisie "has put an end to all feudal, patriarchal, idyllic relations." Capitalist profiteering has "torn from the family its sentimental veil," undermined "differences of age and sex," and brought "a cosmopolitan character to production and consumption in every country," paving the way for the emergence of world culture.[18] Capitalism today, like that of the nineteenth century, draws dislocation, emiseration, and pauperism in its wake—but it also provides material conditions for institutional innovations, personal liberations, democratic struggles, and other worthy causes (a balanced and properly dialectical approach perhaps best exemplified in Dennis Altman's book, *Global Sex*).[19]

Despite my heuristic tacking-back to nineteenth-century antecedents, however, I trust that the contingency and uniqueness of the present situation remain plainly foregrounded in my analysis. In my view, it is not sufficient to link contemporary forms of sexual citizenship or modern practices of identity to the long-standing traditions of political liberalism or to general observations about the individual's market situation under historical capitalism. Such generic linkages—the substance of libertarian models from the outset (which are sometimes also evident in recent critical treatments of the question)—fail to establish exactly *how* sexuality became part of the Lockean "procurement of satisfactions" or how emergent identities were incorporated into the all-American "pursuit of happiness." They also fail to account for the volatility of current events, for the unevenness of cultural developments under capitalism, for the close connection between progress and regress in human affairs, and for the relationship between metropolitan freedoms and peripheral superexploitation.

My argument here should not be taken as a species of economic (as opposed to genetic) reductionism. To invoke a durable Marxist distinction, the arrangement I am describing is less an example of "determination" (how a strong system of ideas and practices dominates, controls, or sets limits on a weaker system) and more an instance of "articulation" (how, in the toss and tussle of events, relatively autonomous practices and institutions gradually find a "fit"). I am not suggesting that late-twentieth-century economic transitions have brought about institutional, cultural, and sexual changes in a unidirectional or one-sided way, but rather that sexual revolution and economic transition since the 1960s have been mutually imbricated, the one providing occasion and opportunity for the other to adjust, advancing willy-nilly toward the present state of affairs. That is to say, we arrived at our present conflicts and dilemmas neither solely by means of political struggle nor by means of some abstract or universal tendency in capitalist culture, but by means of a dynamic process of contestation and compromise, played out under very specific conditions.

In tracing the general outlines of these changes, I do not claim that a once-innocent gay movement has lately "sold out" to consumerism or that a market rationality has "invaded" the lesbigay movement from without, and I am skeptical of attempts to "reclaim" a position of authentic, untainted oppositionality. It was already too late for that on the first day, when the Gay Liberation Front called out the media-savvy, marketable catchphrase "Come out, come out, wherever you are." I note only that men's and women's changing desires ever take shape within the horizon of political-economic possibilities, and that the relationship between changes of one sort and changes of another is reciprocal.

As we've seen, when women sought escape from domestic drudgery, they expanded the workforce, a social trend that fueled capitalism's preexisting demand for women's labor and put capital in a new position in relation to paid (and unpaid) work. Subsequent declines in wages further stimulated the entry of more women into the workforce (on disadvantaged terms), with myriad implications for how men and women live today. Similarly, when gay men and lesbians moved to big-city neighborhoods, they sought respite from small-town prejudice and family disapproval. In the process, they gentrified declining neighborhoods, created demand for urban housing, shored up sagging real estate values, established a viable niche market for gay services, and stoked urban growth machines, paving the way for the present "urban renaissance." It goes without saying that the upward pressure on real estate values tends to select over time for increasingly affluent lesbigay communities, that the resulting social and political institutions are "raced" and "classed," and that the forms of life they support express a certain relationship between identity and consumption. It didn't *have* to come out that way, but there's an ex post facto logic to the fit.

Sex radicals and gay leftists have puzzled over the emergence of a visible, vigorous gay conservative movement in recent years, but in view of the obvious linkages—and in view of the ongoing marginalization and displacement of low-income and minority queers from the centers of gay life—what would be more inexplicable would be the failure of such a movement to materialize. I hasten to add, however, that nothing guarantees the continued ascendancy of the gay right. The vast majority of gays still lean to the left. As Richard Goldstein points out, a recent Kaiser Family Foundation survey found that 66 percent of lesbians and gay men identified as "liberal" and only 7 percent called themselves "conservative."[20]

It is not, then, a question of whether sexuality ought to be viewed as either "economic" or "cultural," any more than it is a question of whether lesbigay demands oppose capitalism or subvert its logic.[21] The new social movements never opposed capitalism proper. What they opposed was its Fordist institutional nexus: that arrangement of marriage, family, gender roles, and sexual conformism that was linked to Cold War capitalism in the era of mass production and consumption. To pretend otherwise is to fundamentally distort the material reality of the new social movements. Over time, these oppositional movements developed new institutions, new values, and new social forms. After the fact, their innovations proved profitable resources for a changing capitalist system. Today, these new forms of life provide the elements of a post-Fordist institutional nexus—an apparatus that includes "traditional" heterosexual nuclear families, but also includes flexible relationships, temporary unions, negotiable role expectations, recombinant families, gay families, open relationships, and individuated patterns of consumption.

In short, the social movements (which are no longer "new") are both agents of and beneficiaries of aggregate economic, cultural, social, and political changes associated with the transition from Fordism to post-Fordism. Their position today is analogous to that of labor unions in the era of mass industrial production: oppositional in one sense, they are institutionalized in another. (On either account, they express their political interests either directly or in blunted form through constituency groups in the same Democratic Party.) The relationship between postmodern sexual cultures and neoliberal capitalism is one of open-ended adaptation, opportunistic appropriation, and mutual transformation, as shaped by multiple conflicts and compromises. To crib a line from Baudrillard: What did you expect a successful, if unfinished, sexual revolution to look like? I outline in table 2 the gist of the ongoing transformations that David Harvey has traced in *The Condition of Postmodernity*, amended by a few notations of my own.[22]

When the new social movements politicized personal life, they acted against oppressive social structures, but they also moved with—and advanced—the

TABLE 2. Fordism Contrasted with Post-Fordism

	Mid-Twentieth-Century Fordism	Late-Century Post-Fordism
Production	Mass production of durable goods and homogenous products	Small batch production of disposable goods and individualized products
	Economies of scale, with on-site storage of stocks and inventories	Economies of scope, with outsourcing, subcontracting, and "just-in-time" production
	Concentration	Dispersal
Consumption	Mass consumption of durable goods	Individualized consumption of specialty products and disposables
	Mass markets	Niche markets
Labor	Single task specialization	Multi-tasking, flexibility
	Long-term, full-time employment	Short-term, part-time employment
State	Regulation	Deregulation
	Centralization	Decentralization
	Welfare	Privatization
Culture	Mass culture	Multiculturalism
	Modernism	Postmodernism
	Utopia, unity	Heterotopia, segmentation
	Suburbia	Urban renaissance
Family	Heteronormative nuclear families with marked sexual division of labor, subsidized by government and industry	Variety of domestic/kinship forms, with ongoing contests over norms, marriage, sexual division of labor, and role of state and employers
	Role rigidity, m/f polarity	Role pliability, androgyny
	"Natural" reproduction	"Assisted" reproduction
Politics	Class politics	Identity politics
	Trade unions	Social movements
Entertainment	*I Love Lucy*	*Will and Grace*

wheel of history. Just as nineteenth-century capitalism played a most revolutionary part in transforming family, kinship, and social relations, modern employment and consumer markets have undermined rigid notions of gender, older forms of patriarchal rule, and narrow sexual conventions, while technological developments, especially in biotechnology, have further eroded the props for a strictly heteronormative worldview. In today's transformations as in the former, political struggles, social reform movements, and business practices work in autonomy but achieve their effects in concert.

THE TWILIGHT OF HETERONORMATIVITY?

Now, as in Gramsci's time, shifts in production and consumption imply myriad adjustments in styles of life—and in the politics and disciplining of bodies. Now, as then, masculinity, femininity, and sexual institutions are in a state of disequilibrium, flux, and crisis—and with them, a host of other institutions. We might be tempted to think that manhood, womanhood, and heteronormativity have always been in crisis, the contrived nature of each somehow revealed, its hegemonic status undermined, by the wobbly condition of the others. At any rate, that's what a series of recent scholarly works argue.[23] Perhaps this is just another way of saying that culture fluctuates—that institutions are ephemeral, mapped against the coordinates of specific historical developments. But a more telling point is that heteronormativity is not in crisis the same way today as it was yesterday. It does not quite resolve the same conflicts, do the same work, underwrite the same institutional nexus today as in the past.[24]

Capitalism and its institutional nexus remain a work in progress. The ultimate shape of the emerging social formation is still undecided. It is too early to mark the "end" of heteronormativity, but not—in old-fashioned Marxist style—to mark its reactionary character: It represents the ideology and practices of a waning if not bygone era.

The epoch Gramsci described at its dawn is now at its dusk. Gone is the day when giant institutions—factory, family—centralized all work, monopolized all resources, and met all needs. The economic predominance of masculine, muscular labor and big machinery is also a thing of the past. So, too, the cookie-serving, stay-at-home mother, recalled in sepia and pastels. Who works where and for how long are all changing. The distribution of emotional traits and the satisfaction of emotional needs, too, have broken out of the monopolistic stronghold of the nuclear family. Family forms, systems of kinship, and practices of love are all giving rise to new permutations. You won't catch me mourning the passing of an era. Quite the contrary. But it would be unwise not to take note of contradictions in the emergent social formation.

24   The Politics of Dread and Desire

Deprived of many of its material supports and ideological props, hetero-normativity today is in crisis—and with it, much else besides. I mean to suggest here something of the ambivalence of latter-day American culture, where the industrial world's most deeply religious people engage in some of the planet's most aggressive social reforms and freedoms, where dissenting social movements are intimately coiled with cultural reactions in emergent institutional frameworks, and where the triumph of unfettered capitalism gives everyone the nervous jitters. Describing the emerging political culture is largely a matter of tempering contrary claims—of marking progress and regress with the same breath, of showing how opposed desires resist closure or finality, how they double back on themselves at the last minute, and how the one thing transmutes into the other, party to obfuscatory work and magical transformations. Everything has to be seen and understood in terms of divided motives and nervous ambivalence.[1]

Like the subjects of economic and cultural transformations described by Gramsci at the end of the 1920s, we remain caught somewhere between discipline and intoxication, between a "totalitarian hypocrisy" and a yearning for sexual utopia.[2] And while puritanism ebbs in some quarters, it flows in others.

THE ANTINOMIES OF NEOLIBERAL CULTURE

A popular song based on an advice column matter-of-factly enjoins listeners: "Use your body every way you can."[3] Every indication suggests that sexual options for the well-adjusted heterosexual continue to expand, especially (but not exclusively) if that heterosexual is young and lives in a major urban area. Indeed, sexual experimentation on the free market sometimes seems to challenge the old distinctions between masculinity and femininity, gay and straight. For instance, Debbie Nathan reports on the use of strap-on dildos, marketed by Good Vibrations, a San Francisco–based cooperative that sells erotica and sex toys:

> Originally, strap-ons were considered lesbian-only items. But over her eight years with Good Vibrations, [sexologist Carol] Queen has noticed a new trend. "We see a lot of mixed-gender couples shopping in the strap-on-dildo aisle," she says. "There's a whole male-female phenomenon of doing erotic role reversal with them." Indeed, after Queen produced an anal-sex instructional tape last year for heterosexual couples called "Bend Over Boyfriend," it "immediately became the fastest-selling video we've ever sold." Queen thinks straight folks are "inventing new forms of sodomy."[4]

Attitudes toward sex—even other people's sex—are changing, along with sexual practices. After all, it's hard for reasonable straight people who engage in varied forms of sexual play to condemn gays and lesbians for what *they* do in bed. By every measure, homophobia in the public sphere is diminishing. Seventy-four percent of the respondents in a *Los Angeles Times* poll conducted in June 2000 claimed to be comfortable around gays and lesbians, while 68 percent supported equal rights in the workplace. Polls also indicate that more and more people say they know or are related to someone who's gay. As we've seen, the population of recurring gay and lesbian characters on TV continues to grow, as does the number of openly gay members of Congress, as does the number of cities and corporations providing partner benefits to gay/lesbian couples.

But so does the U.S. prison population—almost as though by means of an inverted ratio whereby liberation in one part of the body politic redoubles the repression at the other end. The decade that had even straight men snapping "You go girl!" and put into general circulation the gender-neutral term "significant other" also distributed new terms and phrases like "tough love," "zero tolerance," and "Three strikes you're out."

> Nearly 1 of every 150 people in this country is in prison or jail . . . a figure no other democracy comes close to matching.
>
> Soon, the total number of people locked up in Federal and state prisons and local jails will likely reach the 2 million mark, almost double the number a decade ago. . . . For an American born this year, the chance of living some part of life in a correction facility is 1 in 20; for black Americans, it is 1 in 4. . . .

At what point does the world's largest penal system hit a plateau—2.5 million inmates, 3 million? Surely, if crime continues to fall, the number of new prisoners must also fall.

Not quite. . . . Some even believe the prison boom could be permanent, at least for another generation.

A big reason is that so many of the new inmates are drug offenders. In the Federal system, nearly 60 percent of all people behind bars are doing time for drug violations; in state prisons and local jails, the figure is 22 percent. These numbers are triple the rate of 15 years ago.

Americans do not use more drugs, on average, than people in other nations; but the United States, virtually alone among Western democracies, has chosen the path of incarceration for drug offenders.[5]

The fear of intoxication that Gramsci associated with Fordist Americanism is perhaps at an all-time high, as is an old-time mania for locking up black, brown, and red-skinned people. The new temperance movements that gained a foothold in Reagan's America—"Just Say No"—have become permanent features of the social environment as we live through perpetual and open-ended crusades against tobacco, marijuana, drugs, and teen sex.[6] In cadences that can only be described as Orwellian, a *New York Times* health report narrates something of the prevailing mood—the anguished desire to keep young adults away from alcohol:

"Parents should emphasize that getting drunk is always to be avoided," said Dr. Henry Wechsler, director of College Alcohol Studies at the Harvard School of Public Health. . . . "Too many kids think that as long as they don't drive, it's O.K. to get drunk." . . . "We're trying to reshape the environment in which students make choices," Dr. [Tim] Marchell [director of substance abuse services at Cornell University] said.[7]

It is almost as though modern culture demands both universal abandon and universal sobriety—abandon to the extent that it subserves modern, libidinized consumer marketing or niche-market lifestyle subeconomies, sobriety to the extent that work in the high-tech world still requires timed movements, automatism, and labor discipline.[8]

THE CRISES OF HETERONORMATIVITY AND CAPITALISM

Americans today are deeply divided over—and deeply anxious about—key issues related to gender and sexuality, bodies and pleasures, kinship, family, and personal life. These divisions are not merely "cultural" misunderstandings, nor is it right to call them strictly "economic" struggles. The conflicts stem from disagreements over and uncertainties about the viability of basic social institutions.

At the same time, Americans are plagued by economic troubles on every front—by everyday economic cataclysms that throw precarious lives out of balance: outsourcing, downsizing, "the race to the bottom," capital flight, plant closures, contagious currency panics, roller-coaster stock markets. . . . One anxiety feeds another, one crisis fuels another.

In consequence, politics today is largely defined by the intertwining of sexual anxieties—or perhaps more properly put, anxieties about the institutions related to sex and reproduction—with economic apprehensions. This "entwinement" is no accidental feature of social life in the United States. As Barbara Ehrenreich notes in her ethnojournalistic account of the low-wage economy today, the majority of American workers—about 60 percent—earn less than the fourteen dollars an hour deemed a no-frills "living wage" by the Economic Policy Institute, an income minimally adequate for the support of one adult and two children. Thirty percent earn less than eight dollars an hour. (The National Coalition for the Homeless estimates that an $8.89 hourly wage will support an adult if s/he rents a one-bedroom apartment.)[9] So how do you make ends meet when your paycheck won't cover rent, groceries, health care, and other basic necessities? The logical options are: conjoin with someone in a stable marriage—"Let us marry our fortunes together"—or alternately, leverage kin ties to stimulate the flow of periodic surpluses from equally strapped parents or siblings.

Either way, family counts. Either way, the intensity of feelings many Americans express over the politics of family and family values—a sentiment not notably vented in any other industrial democracy—has less to do with the idealized, sentimentalized role Christopher Lasch once glossed as "haven in a heartless world" and more to do with the material sustenance provided by family attachments in the absence of supports afforded by social-democratic states in other Northern countries (e.g., universal health care, subsidized child care, and low-income housing).[10] It is in the context of this material function that one should understand the resilience of "family values" in an age of divorce. On the social ground, worries over "bread-and-butter issues," as they used to be called in the era of high Fordism, are by no means disconnected from anxieties about domestic life: marriage, family, child care, divorce. The one provokes, fosters, or stands in for the other. The political just doesn't get any more personal than that. It also doesn't get more ambivalent than that, for this is the point in the body politic where the bravado of sexual liberation chafes against fears of abandonment, loneliness, and loss, where desire transmutes, by way of some parasympathetic reaction, into dread, and where the yearning for freedom suddenly reverses itself, turning into the need for order.

If it is easy enough to see how Left social movements generally orient themselves within a shifting political-economic and institutional framework, it ought to also be clear how reactionary institutional policies and political backlash sometimes find a substantial social base. For millions, the manifold crisis of Fordist institutions plays out as broken marriages, family instability, and personal upheaval, a social reality rendered in compelling terms in Judith Stacey's brilliant, up-close ethnography *Brave New Families*.[11] And for millions, the decline of heteronormative family values and heterosexual privilege is materially linked to downward mobility and personal insecurity, a connection explored in depth in Arlene Stein's vitally relevant ethnography *The Stranger Next Door*, which chronicles conflicts around an anti-gay initiative in a depressed Oregon town.[12] In the resulting perturbations and in the ideological transferences and substitutions they facilitate, class issues resonate within the "culture" wars like the ricochet rhythm of a drum machine at a circuit-party rave.

The result is a two-way traffic in politics. On the one hand, class resentments are readily cathected onto homosexual scapegoats, feminist villains, welfare mothers, the urban poor, and purveyors of sexual permissiveness. When right-wing clergy and conservative politicians tout "the importance of family," they play to sentiment, but they also invoke a complex experience of material dependence deeply inculcated in everyday experience. The politics of hate expresses the reactionary force of this dialectical tension between sexual anxieties and economic troubles, as do demands for state regulation of pornography, abortion, marriage, and sex, or demands for state subsidy for organized religion.

On the other hand, labor unions and advocacy groups representing major sections of the working class have increasingly improvised new forms of class politics. Demands for antidiscrimination workplace protections, for pay equity, for gay domestic partnership benefits, and for gender-neutral family leave policies, for instance, all represent new labor causes that are both indebted to the social movements and fitted to the new institutional realities of the working class. Such demands also point the general way toward a new form of Left class politics—and a way out of the culture wars precipitated and sustained by a nervous, low-wage economy. Simply put: A person's well-being and chance at happiness ought not to be contingent upon his or her conformity to "natural" roles and Fordist institutions. Social rights, the product of social struggles in a democratic con-

Figure 34. Anti-gay protesters from Westboro Baptist Church of Topeka, Kansas, vent their views at Matthew Shepard's funeral in Casper, Wyoming, October 16, 1998. Shepard, a gay university student, was kidnapped, tied to a fence, brutally beaten, and left to die outside Laramie, Wyoming. Photo by Michaell S. Green. AP/World Wide Photos.

text, ought to attach to individuals irrespective of their marital, familial, or sexual status.

Like the 2000 presidential elections, the ongoing culture wars play out most dramatically as a North-South, urban-rural split. But it is not quite right to say that in these conflicts one side squarely faces off against the other. Multitudes of social actors are themselves divided and torn, split by contrary impulses—living embodiments of the manifold contradictions of our era (and here I cite examples from my own circle of friends and family): the independent-minded southern grandmother who believes that women are "just as good as men" and loves her lesbian daughter but also sends donations to the Reverend Dr. James Dobson's Focus on the Family; the middle-class gay man who believes in sexual equality but opposes abortion, dislikes lesbians, and thinks drag queens are "too flamboyant"; the urban black churchgoer whose worldview was shaped in the civil rights struggles and who thinks of herself as progressive but who is also convinced that white feminists and gays are undermining the black family; the socially conservative Latino Pentecostal schoolteacher who opposes "special rights" for gays yet goes out of his way to protect effeminate boys from schoolyard bullying; the white suburban father who attends Promise Keeper rallies and votes Republican on family values but who also socializes with the mixed-race gay neighbors . . . Intellectuals grapple with the same contradictions, even as they attempt to treat the changing politics of gender, sexuality, race, and class critically. George Yúdice evokes the nature of the problem when he frets over the uneasy predicament of progressive straight white men, and Barbara Ehrenreich expresses something of the logic of this ambivalence when she worries that the decline of patriarchy (literally, rule by the father) has left women more, not less, oppressed and vulnerable.[13]

In the current context of volatile contests and divided motives, divorcé Cornel West, coauthor of *The War against Parents*—a text that lends its voice to the Christian choir pining for tougher divorce laws—begins to sound more like Patrick Buchanan, who reciprocates by sounding less like a cryptofascist culture warrior and more like the venerable black socialist.[14] In a March 25, 1998, column, Buchanan writes:

> Reaganism and its twin sister, Thatcherism, create fortunes among the highly educated, but in the middle and working classes, they generate anxiety, insecurity and disparities. . . . Tax cuts, the slashing of safety nets and welfare benefits, and global free trade . . . unleash the powerful engines of capitalism that go on a tear. Factories and businesses open and close with startling speed. . . . As companies merge, downsize, and disappear, the labor force must always be ready to pick up and move on.

The consequences of innovative, restless, global capitalism are "deserted factories," "gutted neighborhoods, ghost towns, ravaged communities and regions that go from boom to bust." Buchanan merely states what the Left has always maintained about "unbridled capitalism"—that it is an "awesome destructive force." He goes on to link personal crises to political-economic relations, evoking causes for the imminent crack-up of the conservative movement: "The cost [of unbridled capitalism] is paid in social upheaval and family breakdown, as even women with toddlers enter the labor force to keep the family's standard of living."[15]

Whereas West would restrain personal freedoms in the name of social democracy, Buchanan would check unbridled capitalism in the name of family values. What undergirds both positions is a theory of human nature and presuppositions about natural law, which in their turn express a desire for order, stability, and certainty. The result: Left meets Right where each doubles back on itself—lost somewhere in the wilds of "nature" just off the information superhighway to the twenty-first century.

25 Sex and Citizenship in the Age of Flexible Accumulation

I'm not taking any bets on which side of capitalism's cultural impulses will ultimately win out in institutional form, its tendency toward innovation and cosmopolitanism or its need for order and discipline. What the invisible hand gives, it can also snatch away. Meanings are slippery, the situation charged and uncertain. New feminisms vie with resurgent masculinisms, each spurred in different ways by shifts in women's and men's respective market situations and life chances. Sexual anxieties transmute into class troubles and racial fears. Longings for liberation abut the need for order, and it is not clear how any of this will be resolved (if any of it is resolvable).

The same recent history that allows—even encourages—men and women to experiment with new designs for living, new forms of sexual relationships, new conceptions of morality, and new ways of reproducing also invites the hatching of various schemes, reactionary and progressive, elitist and populist, to manage and reorganize the primary sites of personal life. That is the toss and tussle of politics today. It is the very stuff of give and take, claim and counterclaim, antagonism and ambivalence in the public sphere. And that is where nature—convoluted, contested nature—comes in. Or rather, it might be said that what's been troubling nature all along is the heave and shove of social institutions within a changing political economy.

SCENES FROM A LIFE

Three decades into the new social movements, televangelists still do battle with homosexuality, adultery, New Age occultism, and other garden variety forms of moral decline—when not themselves encumbered by financial or sexual scandal.

Figure 35. Heather Mechanic of San Diego sings opening hymns at MacAlpine Presbyterian Church in Buffalo, New York, as part of Operation Save America, a project of the anti-abortion group Operation Rescue, on April 18, 1999. Participants in Operation Save America staged events at Buffalo area clinics, schools, and bookstores denouncing abortion, child pornography, and teen sex. Photo by David Duprey. AP/World Wide Photos.

Religious revivalism continues, but so do the new social movements. Public opinion surveys (based on self-reporting) suggest that church attendance is up, but in 1997, one of the largest denominations in America went head-to-head with one of the planet's most powerful entertainment conglomerates and lost in less than a round: The Southern Baptists' much-feared boycott of Disney, producer of recent movies with gay themes and owner of ABC, the network that broadcast *Ellen*, was conspicuously unsuccessful. A state-sponsored "virginity movement," which encourages teens to take vows of abstinence until their wedding night, has attracted the endorsement of politicians and public figures not otherwise noted for chastity. But a close examination of the movement's dynamics suggests that virginity vows are effective at deferring the age of first intercourse only when a relatively small percentage of youth at a given school sign on—that is, when chastity functions essentially as a youth counterculture.[1] Both judicious reflection and polling data suggest that the dip in teen pregnancies registered in recent years probably owes far less to celibacy and continence than to the widespread use of contraception—and the growing popularity of oral sex in the wake of HIV/AIDS.[2]

At the beginning of the 1980s, many had feared (or hoped) that the AIDS epidemic would prod public opinion toward notably more puritanical views. I still remember the cascade of news stories proclaiming the "end of the sexual revolution" in a Reaganized America, even as gay men's organizations across the country were educating constituents about safe sex.[3] In fact, AIDS has inspired a variety of sex panics, but it has not signaled an end to the cultural revolutions that effloresced in the 1960s. The cultural heroine of the 1990s was a single professional, a "fashionable, sexually well-traveled, thirty-something woman," a woman who, as they say, "owns her sexuality" and carries a condom in her purse.[4] Nor has HIV/AIDS led to the wholesale demonization of gay men feared (or wished for) by many at the beginning of the epidemic. AIDS prevention has in fact become the leading edge of progressive sex education efforts in public schools—even public schools in areas like Fairfax County, Virginia, where mainstream suburban Republicans want their children to understand HIV transmission, safe sex practices, and contraception. Still, HIV prevention and safe-sex education efforts in the United States lag far behind successful (i.e., sexually explicit and nonjudgmental) risk-reduction programs in northern Europe. The 1996 welfare reform package, for instance, provided federal funds for abstinence-only indoctrination in the public schools, not for sex, safe-sex, or birth-control education. And when the Clinton administration declined to lift a ban on federal funding for needle exchange programs—despite its own findings that clean needles reduce the spread of HIV without increasing the use of drugs—the president acceded to the argument that "lifting the ban would send the wrong message to the nation's children."[5]

On AIDS policy and in the ongoing campaign against marijuana, as with endless crusades against tobacco and alcohol and in so many of today's political conflicts and logical non sequiturs, "the children"—credulous, vulnerable, innocent, impressionable—loom large, objects simultaneously of government protection, sentimental reverie, indulgent doting, tough love, and public wrath. By one route or another, all of the anxieties in American society ultimately find their expression in discourses on "the children": on the one hand, those "latch-key kids" traumatized by divorce and left unsupervised by working mothers, and on the other, the children of "Take Our Daughters to Work Day," heirs of a feminist revolution in life expectations. In the wake of the child-molestation, child-abuse, and satanic sacrifice panics of the 1980s and 1990s, a new culture of child protection has been crafted, piecemeal, by latter-day social reformers.[6] Entire subfields of junk science have arisen around the contested years of youth, from the psychoanalysis of "repressed memories" to dubious child-rearing practices precariously extrapolated from the results of brain research.[7] Childhood, indeed, has become not just the object of a questionable science, but a special province of politics. Preparations for a second Gulf War, like the inexplicable assault on the Branch Davidian compound, were packaged as "narratives of rescue" and roundly touted in the name of "the children." Public and private misdeeds—philandering, drugging, lying, and so on—are invariably censured not for what the malfeasants might have done or failed to do, but because of the "bad example" such acts set for "the children."

Plainly, all is not well in this brave new world of childhood. If "the nation" is increasingly thought of as "the children," then "the children" themselves necessarily solicit certain acts of paternalism—a paternalism of the state over the parents of the children, a tyranny of childhood over children.[8] Finger wagging by the Daddy state reached a new pitch in the 2000 presidential elections when politicians representing both major political parties donned collars and yarmulkes to scold the entertainment industry for marketing R-rated movies to teen audiences. Meanwhile, in an era of declining crime rates, straight-arrow kids who turn to the age-old practice of infanticide to conceal unwanted and socially stigmatized pregnancies become a special barometer of "the decline of the family," as do boys who get guns and kill their peers—a recurring news story across the Gun Belt. Ironically, in a decade when the rhetoric of child protection cast its shadow across everything from seat-belt laws and curfews to TV programming and Internet regulations, courts in many states diverged from international human rights standards to try juvenile defendants as adults.[9]

It is not just the children, but prechildren, too, who haunt the national consciousness. In the twenty-five years since *Roe v. Wade*, the fetus has been cast in the lead role of a number of ongoing legal dramas, from continuing struggles to define the terms under which abortion might be ethical and legal to the gradual extension of child-endangerment laws to cover the unborn. Although abortion remains legal in the United States, acts of terror by religious extremists have intimidated many doctors into not performing abortions. As a result, the medical procedure is effectively unavailable in many parts of the country.[10] And after so much personal and political turmoil, exactly what might be right or wrong about abortion remains a touchy business. The use of abortion to select the sex of a child— said to be a rare practice in the United States—has sometimes been addressed in public forums, the general consensus being that it would be wrong for parents to "play God." At the same time, prenatal testing for genetic and birth defects has led to the widespread practice of what can only be called selective, eugenic abortion—a trend that has gone largely without comment or reflection. Recent polls indicate that most people cannot quite imagine an America without legal abortion—or even remember just how recently this was the case—but also that even women getting abortions think there are too many women getting abortions.[11] Nervous ambivalence about abortion is likely to be enhanced, not reduced, by advances in reproductive technologies and by the development of new forms of genetic screening.

The antipornography crusades of the 1980s continue, in both neo-Victorian feminist and paleo-Victorian fundamentalist variants. But the modern denizen of pornutopia isn't going to XXX theaters, peep shows, and sex clubs, he's downloading X-rated material from the Internet, joining bisexual chat rooms, and watching she-male videos in the privacy of his own home. Attempts to regulate sexually explicit material on the Internet— almost always in the name of child protection and family values—quickly run afoul not just of the First Amendment but also of a certain undeniable material condition of the cybernetic age: Pornographic sites undoubtedly provide the best-organized, best-illustrated, and most user-friendly material on the net. (It goes to show, as a level-headed colleague once observed, that people put their best effort into the things they really care about.) Industry estimates suggest that more than half of the Internet's bandwidth is taken up with traffic in pornography.[12] Hampering the free flow of sexual representations, which quietly stimulated much of the video revolution of the 1970s and 1980s, would likely do harm to technological development at its most innovative edge.

HOW THE PRESIDENT'S PENIS BECAME NEWSWORTHY

The national political news, which tracks radical contests and institutional compromises from changing perspectives, has come to bear an eerily canny resemblance to those topics treated on afternoon talk shows and in tabloid journals, which also, in their way, track news of a changing social formation: deadbeat dads, partial-birth abortion, the president's penis, Michael Jackson, sex scandals (that fail to scandalize), recreational drug use, Viagra, O. J. Simpson, spiritual emptiness, drunk driving, feel-good religion, DNA, AIDS, SIDS, the disciplining of errant children, hell-fire religion, minors having sex with adults, minors having sex with each other, New Age religion, hate crimes, Satanism, missing children, peer pressure, addiction, medical marijuana, Princess Di, Chandra Levi, kidnapping, kids who smoke, drink, or do drugs, sexual harassment, repressed memories, medical news, diet, tips on how to live longer and healthier lives, fertility treatments, cloning, single motherhood, teenage mothers, parental rights, parental wrongs, bad children, bad babysitters, bad nannies, same-sex parenting, gay weddings, gay ordinations, and—endlessly—the personal lives of royalty, politicians, movie stars, and other celebrities.

So many snapshots of you and me engaged in the various poses of modern life. So many meandering tales spun with a common thread. In this finely woven mesh we find all the subjects of so much pleasure and anxiety, all the objects of personal excess and government regulation: "nature" in its manifold permutations. The crisis of heteronormativity in its myriad implications.

THE FISSIONING PUBLIC SPHERE

Such circumstances complicate efforts to say, once and for all, how public culture works or what it does or how it represents, a set of problems at the core of cultural and media studies.[13] Do the media convey—conserve—cultural rules, or do they destabilize them? Do production conventions and marketing demands "co-opt" or do they "enable" dissenting voices? And do media representations reflect or cause real-world changes?[14]

Lauren Berlant's witty, clever collection of essays, *The Queen of America Goes to Washington City* (whose material I have been rehearsing and updating for the past several paragraphs), engages these questions in terms of the relationship between sexual culture and political economy. Her treatment foregrounds what is perhaps the signature trait of public culture in our era: the erasure of the boundary between "personal life" and "public citizenship," along with an ongoing redefinition of and struggle over what might count as proper objects for public scrutiny and government intervention.

Berlant's analysis turns on two key points. First, Berlant suggests that the "intimacy" issues and "familial politics" theatrically paraded in a downsized and privatized public sphere serve, in large part, as "distractions" from the larger, material issues of the era: economic downsizing, labor outsourcing, global privatization, and the widening gap between rich and poor. Undoubtedly, Berlant makes a strong argument when she claims that a Reaganite "conservative ideology has convinced citizenry that the core context of politics should be the sphere of private life," as when she tracks the development of that ideology through images of childhood vulnerabilities and depictions of imperiled fetuses.[15]

Second, the author emphasizes the heteronormalizing function of mass media representations, their role in stabilizing identity. Berlant wants to show that in the midst of an American "intimacy crisis" (triggered by feminism, gay liberation, and assorted institutional changes) and during a period of economic downsizing and privatization (part of the transition from Keynesian Fordism to post-Keynesian post-Fordism), heteronormalizing images and claims do double time on TV—the better to reinforce the stable, monogamous, nuclear family, with its gender roles and a sexual division of labor, as a locus of order and responsibility in an increasingly disorderly and irresponsible world.

What Berlant critically diagrams, of course, is the conservative variant of the neoliberal dream: the family is to act as a miniature welfare state, modulating consumption, curbing excess desires, improvising child care, and providing social security—in the absence of a Keynesian or social-democratic regulatory state. This is scarcely the message unambiguously projected by the entertainment media, but it *is* the dreamworld conveyed in what I have been culling from the "serious" media—journalism, punditry, science reportage—where the conservative variant of the neoliberal utopia is attributed to the biologically fixed "nature" of desire.

The public sphere is undoubtedly in the midst of ongoing changes. I doubt that Berlant would offer the same arguments in the same terms as those that she proposed in the mid-1990s—but I am also convinced that her analysis missed key trends that were already evident by the time her text appeared. To understand the current situation, its tendencies and contradictions, it might be more useful to invert key terms of Berlant's analysis.

THE SCHIZOGENESIS OF MODERN MEDIA

First, it seems to me that a dispassionate survey of what entertainment does when it entertains, instead of suggesting a uniform or one-dimensional analysis of mass media today, leads more logically toward the sort of analysis Mikhail Bakhtin developed in his book on Rabelais, with its distinction between an open, experimental, and transgressive popular culture best exemplified in the marketplace and a closed, hegemonizing, official culture most pointedly expressed by the church.[16] That is to say, the same "intimacy crisis" that is taken as a grave emergency in the serious cultures of politics, religion, and journalism plays out as a social and aesthetic experiment in entertainment culture.

In more sociological terms, perhaps this is really a question of how different sectors of the media view their bottom lines when they respond to social changes. By definition, the perspective of the journalistic media is sober, serious, and responsible. The charge of journalists, like that of the politicians whose actions they report, is to keep the social order while never questioning the rule of capital. Since it is *precisely* the global movements of capital that are a key source of turbulence and disruption in modern life, the talking classes invariably assume the position and perspective of adults addressing a population of unruly children. That is, by dint of how they define "seriousness," reporters, politicians, and pundits are predisposed to attribute social breakdowns of various sorts to *personal* breakdowns: to private immorality, family problems, or cultural decline. In their depictions of this ongoing "intimacy crisis," they also tend to harbor a nostalgic view of the Fordist institutional nexus—of its well-ordered functionality, if not its overtly sexist, racist, and homophobic forms.

By contrast (and contrary to the unmodulated analysis Berlant conveys), the entertainment media are at least receptive to a different vision within the post-Fordist dispensation, one in which the family, kinship, and domestic arrangements are not so much present as a collective ballast against the disorder of the marketplace but are themselves subject to tests of the marketplace and the needs and desires of changing individuals. The entertainment industry thus makes a very different sense out of the intimacy crisis. Its conjured characters sometimes provoke the crisis through deliberate acts of transgression, its comic situations mine humor from institutional transformations, and its advertising exploits the varied forms of sexual desire. As a result, the entertainment industry—with its pursuit of niche markets, its characterizations of social actors, its need for innovation even at the expense of order, and its necessary peddling of sex—has become, in the eyes of the serious journalists and politicians, the one part of the capitalist economy in need restraint and regulation.

One sector of the media thus refracts the social struggles of the day in one way, the other sector in a different way. The result is a schizoid representational sphere in which images struggle against each other and sometimes threaten to devour each other. Even as a transgressive, queer aesthetic finds a home in commercial and entertainment culture, a conformist and heteronormalizing ethic is deepening its roots in serious culture—in the political discourses of liberals and conservatives alike, where a downsized conception of "responsibility" invariably means individual responsibility according to received sexual norms, gender traditions, and family institutions, and in journalism, with science journalism leading the way. Indeed, as I have been suggesting, science journalism provides the secular template and the supposedly "scientific" rationale for what one variant of neoliberalism wants to claim already.

THE INTIMATE PUBLIC SPHERE
AND THE POLITICAL ECONOMY OF SEX

Second (and again contrary to Berlant's analysis), it seems to me that the journalistic media's focus on "the personal" hides nothing and reveals everything about contemporary material conditions. The decline of the social responsibilities assumed by the Keynesian state—which for much of the twentieth century served to blunt the depredations and ravages of the marketplace—implies an increase in social responsibilities somewhere else.[17] The Left might yet someday elaborate a credible "somewhere else" (it seems to me that we are behind the times and less radical than reality on this front), but on the Right, this "somewhere else" is invariably the heteronormative nuclear family. In the new political situation, talk about "individual responsibility," "spiritual renewal," and "family values" is the more or less civil face of Darwinian social policies. Like recto to verso, reregulation of the libidinal economy is only the other side of deregulation of the wider political economy.

Seen in this light, the downsized and personalized view of citizenship, with its obsessions over sex, is not so much a "distraction" from issues of a political-economic order as it is *another way into* those questions. That is to say, the relationship between the "miniaturized view of citizenship" (with its familial politics, its emphasis on personal morality, its struggles over gender roles, its obsessions over sex) and "economic downsizing" (a perpetual tendency of post-Fordism), is direct, not indirect, much less misdirecting. Politics today is largely a struggle over how domestic institutional structures should mesh with market fluidity and transnational capital flows. In these struggles over how to define and tinker with domestic arrangements, the question of who might legitimately have sex with whom is critical. It touches on all the hot spots of the body politic. It opens onto myriad contestations around race, ethnicity, national origin, and class.

When politicians debate the treatment of single mothers, for example, or the circumstances under which women and their children might qualify for welfare, or the terms and conditions under which marriage might be legally terminated, they are manifestly talking about both sex and domesticity, but they are also deliberating the allocation of resources and the shape of social stratification for present and future generations. They admit as much, even though they typically invert the real terms of their engagement, with much of this distortion happening through the image of imperiled children.

When religious leaders organize politically to decry the decline of the family, they rue the proliferation of nonprocreative sexualities and the availability of legal abortion, but they are also talking about how to best discipline production, organize consumption, and maintain empire. Again, they say as much. It is no mystery what they mean when they say that "strong, God-fearing families were what made this country great," nor is it a secret that they long for the restoration of institutional and economic structures consistent with the America of the 1950s.

When well-heeled populists and think-tank communitarians lament the "culture of disbelief" and fret over the loss of clear-cut "community norms," as when they claim that modern secular culture has eroded the "core moral values" necessary to sustain a good life, they are taking aim at the usual suspects: fags, feminists, village atheists, misfits, modernists, class clowns, scoffers, and bohemians. Parson professors and endowed schoolmarms thus urge the restoration of church, family, and community as panaceas for every conceivable form of postmodern malaise. But whether they know it or not, and no matter how rosy the tint of the glasses through which they contemplate the past, they are also urging the restoration of an institutional authority structure well understood by those of us who grew up in small towns and rural areas, a social order in which the moral hierarchy mimics the class hierarchy, where sexual adventure is tantamount to pissing in the baptistry, and where everyone's treatment by the authorities is at every moment shaped by moral judgments colored by race.

When journalists lay bare the sex lives of public figures, or when they make public the personal lives of private figures, they are offering object lessons on the state of society, surely. But their moral lessons invariably carry cargo of a more tangible sort. That is, they also imply an assessment of the conditions under which a properly disciplined work force might be optimally productive. Economistic tag lines are implicitly present, although they are seldom expressly stated, in such reportage. For instance: "It all goes to show what happens when women work and children lack adult supervision." And "This apparently private behavior [smoking, drinking, doping, sex] will be having public consequences for many years to come." Or, "Here's yet another example of what happens when people become addicted to instant instead of delayed gratification."

When homosexuality is on the agenda (as it invariably seems to be: lesbigay visibility, gays in the military, gay marriage, lesbian teachers, gay ordination, violence against gays and lesbians), what is also on the agenda is how to define the norm, how to mark its acceptable variations and unacceptable deviations, how far to go in sustaining norms—and thus how to understand the limits of those consummate American rights, the pursuit of happiness in civil society and free participation in the democratic process of governance. It is no secret that the conservative agenda is a strategy of containment. Experiments in gay living have been in the vanguard of cultural and institutional changes since the 1960s. Homosexuality is thus curbed in the interests of maintaining the boundaries of that imperiled institution, the heteronormative nuclear family, with its attendant conceptions of proper gender and sexual decency. One need not ask very many questions about the desirability of such families before arriving at a conception in which morality and economics are conjoined in pristinely utilitarian formulae: The family is supposed to oversee, with prudence and good measure, all that is not directly administered by the state. It is there to contain, manage, and domesticate sexuality—to keep well-behaved subjects in line and on the path to a well-adjusted, well-disciplined, and above all else *productive* citizenship.

THE STAKES

At the heart of the matter, when Right and Left contend over how to promote "personal responsibility," how to define marriage, and how to delimit sexual rights, they vie over the very fulcrum of neoliberal capitalism. The ensuing struggles over family values are neither charade nor dumbshow. They are, in essence, conflicts over how to produce men and women whose values allow for the production and accumulation of economic value.

Political struggles have not always revolved around such an axis, and it ought not to be imagined that these will always be the stakes. But it would be a mistake to think that the current zone of contention results from misunderstanding or misrepresentation—or that the contradictions of the age can be overcome by clever strategy or force of will. In the final decades of the twentieth century, cultural and sexual politics displaced class politics neither by force of the new social movements' efficacious conniving nor by dint of conservative conspiracy, but on the impetus of political-economic changes in advanced capitalist societies as these tectonic movements reverberated unevenly in institutional contests and beneath changing geographies of dread and desire.

Ongoing struggles over whether and how to regulate the organization of personal life are thus in no sense "small-scale" conflicts. They ultimately go to the farthest reaches of the world economy. Understood this way, the personal is not merely political, it is political-economic. At bottom, the culture wars are really disagreements over what kind of institutional matrix might provide a fitting context and an appropriate reward structure for social order and prosperity. At issue is whether the post-Fordist social formation should be run along more or less corporatist lines or whether market capitalism implies cultural cosmopolitanism, the innovation and coexistence of varied modes of life. At stake is whether neoliberal capitalism requires at its base a network of total institutions or whether it recommends a laissez-faire approach to personal life. In play are two very different visions of social good: a locked-down, chaste, sober society—random drug tests for everyone—versus a more open society committed to personal freedoms and free expression.

An Open-Ended Conclusion

On Sunday, February 11, 2001, the genomic speculative bubble finally burst—at least for anyone prepared to contemplate the proposition logically. On that date, the *New York Times* front-page headline read: "Genome Analysis Shows Humans Survive on Low Number of Genes."[1] Science reporter Nicholas Wade—who only months before had touted genomania in no uncertain terms—previewed forthcoming reports in the journals *Nature* and *Science*. Researchers at Celera Genomics and at the international public consortium working on the Human Genome Project, he wrote, have found "that there are far fewer human genes than [previously] thought— probably a mere 30,000 or so—only a third more than those in the roundworm" and only a few more than double those of the fruit fly.[2] Previous widely circulated estimates of the number of genes in the human genome had run as high as 142,634.

The logical implications of this finding were widely noted by scientists and journalists over the following week. If it is granted that human beings are considerably more than one-third more complex than roundworms, the additional complexity could not owe to genetic causes. More specifically, if humans not only "survive" but flourish on a mere 30,000 to 50,000 genes, then an entire chain of reductive causal arguments collapses: first, the idea that each gene is responsible for making its own unique protein; second, the derived idea that there is a one-to-one correspondence between "gene" and "trait"; and finally, the notion that the genome could ever give, in micro, a "blueprint" for the organism as a whole.[3]

If 30,000 to 50,000 genes seems too few to explain the *biological* uniqueness of the individual human organism, such a number might seem all the more unconvincing when offered in reductivist accountings for psychological tendencies, social actions, and cultural forms. But not so fast. Genetic reductionism was never really about science, it was about ideology. Since genomania ultimately has no product to push, no scientific application to sell, and admits no tests that might falsify its core premise of a closed and predetermined human nature, the inflated fortunes of genetic reductionism may well prove more resilient than that other ideologically driven windfall of the 1990s, the paper fortunes amassed on the dot-com dominated NASDAQ stock exchange, which have all but disappeared into thin air.

And so only a week after his own headline had puzzled over the oddity that somehow humans manage to scrape by on a small number of genes, Nicholas Wade attempted to rally the sociobiological faithful in the Sunday *Times* Week in Review section.[4] Taking issue with cautionary statements about the follies of genetic reductionism given by the scientists heading both Human Genome Project research teams, Wade writes as though the news of the preceding week had actually *simplified* the task of bioreductive accounting. He begins by wondering what the genome will eventually reveal about "human nature," reciting the untestable maximalist article of faith that "evolution is likely to have molded any and all behaviors that conferred survival value"—not just "fighting and mating," but also "sophisticated human behaviors like art and religion." Rummaging one more time through the handy grab bag of arguments from sociobiology and evolutionary psychology, the editorialist invokes faux universals, speculative evolutionary scenarios, and elusive "genetic algorithms." For the edification of doubters, he even stages an imaginary conversation among a group of fruitflies—who, imbued with the gifts of speech and reason, reject the idea that *their* nature could ever be reduced to simple genetic causes. It goes without saying that in his recitation of reductivist dogma, Wade accuses social constructionists of having an essentially "religious" attitude toward problems of human existence—an attitude that, he allows, no doubt confers some sort of survival advantage. Apparently, we—but not Wade—would be too crushed to go on living if we only knew the truth about human nature. Wade draws his inspiration from Edward O. Wilson when he hypothesizes that "with the human genome in hand, the way is open, at least in principle, to discover how it shapes the architecture of the human mind." He then goes on, in bolder mode, to concur that "all branches of human knowledge . . . , from ethics to economics and aesthetics, will eventually be unified by understanding the genetic rules of the human mind."

Floating above Wade's delirious text is an odd pictorial exhibit: a triptych of three very different self-portraits by the deceased gay artist Robert

Mapplethorpe. "Arguably," suggests the caption—composed by whom is unclear—"the genome made him do it."

A month later, on March 11, 2001, as school districts and states across the country were weighing rules designed to curb predatory school violence, Fox TV's *The Simpsons* opened a new front for pop science interventions into political life. Lisa Simpson, Bart's bright, studious sister, has fallen prey to a school bully. Guided by a hunch and implicitly relying on a mishmash of theories about primate dominance hierarchies and hormonal triggers, she conducts scientific experiments to "prove" that bullying—"older than agriculture"—is triggered by pheromones and is thus an inevitable fact of life. On May 20, 2001, in the Week in Review section of the *New York Times*, Natalie Angier quotes assorted "bully experts" from psychology and primatology to much the same effect.[5]

Later that year, at a moment when everyone was searching for answers to age-old questions about the causes of human suffering, the Reverend Jerry Falwell drew on a hoary theology to assert that feminists, abortionists, lesbians, gays, civil libertarians, and secularists in general bore collective responsibility for the terrorist attacks on New York City and the Pentagon. But if the horrific events of September 11 provided the occasion for familiar sermons on public immorality and God's wrath against an errant nation, they also revealed how dramatically American political culture had changed since the rise of the religious right in the 1970s. The Republican president—whose campaign Falwell had supported from its inception—essentially told the Virginia preacher to knock it off. Falwell retracted his statement and quickly backpedaled his way to an unconditional apology.

Sociobiologists and evolutionary psychologists, too, used the events of September 11 to stage familiar parables about human nature. But which parables? And which human nature?

"Ever since there have been cities," claims John Tierney in the *New York Times*, "there have been merchants paying for walls to protect their families and homes and businesses against invasions by young men without wives." Never mind the obvious fact that the terror cells in question included men in their thirties who had wives, children, and property. For Tierney, the suicide bombers embodied one more instance of aggression by young, unmarried, and presumably testosterone-pumped men in groups.[6]

As against such old-style sociobiological stories, which invariably describe and explain the unpleasant side of human behavior, Natalie Angier was—predictably—accentuating the positive. In her uplifting narrative on the week's events, a broad cast of characters—airplane passengers, policemen, firefighters, rescue workers, and volunteers by the millions—were collectively demonstrating the biological advantages conferred by altruism and heroism. "If not for these twin radiant badges of our humanity," she effused, "there would be no us, and we know it."[7] "No us"—imagine it! Such a thought comes, like a whisper, from the edge of identity, at a moment of crisis.

LOOSE ENDS

No book about contestation, conflict, and crisis should conclude with all the loose ends tied up: a pat statement about the human condition, a mystery solved, the theoretical equivalent of a happy ending. Like reflections on what happened "once upon a time," tying up all the loose ends is the work of fables, biomythologies, and potted histories. Instead, I will mark a few contradictions, note a few ironies, and hazard a few guesses—a skeptical gay socialist's morning prayer.

I have tried to write against and within cross-currents of social happenings, at the point of push and pull, tug and tow, give and take. Sociobiology (this book's object) and queer theory (this book's analytical method), like the social struggles they represent, stand in diametrical opposition to each other: a relationship that resembles one of those "isometrical structures" Claude Lévi-Strauss painstakingly diagrams—or better yet, a configuration that evokes the clash of interests seen by the Bakhtin school of representational theory as inherent in every linguistic sign. But because these theoretical ideas and social movements participate in history—and are open to its effects—the balance of forces and the terms of their engagement are changing, unstable, and volatile. In the social contest for existence, neither denaturalization nor naturalization remains what it once was.

Insofar as they stage essentialist claims about ossified sexual identities and an unchanging human nature, sociobiology, evolutionary psychology, and other forms of naturalization are manifestly risible: their claims fly in the face of both cross-cultural variety and newsworthy happenings. But insofar as queer theory gives the same generic analyses of heteronormativity that described gender, sexuality, and family life of the mid-twentieth century, it, likewise, draws a fixed, fast-frozen picture of sexual culture. In this sense, the prevailing variants of lesbigay studies, queer theory, and related forms of critical cultural studies are unable to account for the history in which they participate and ultimately remain locked in a past-tense politics: addicted to topics like gay invisibility at a time of unprecedented visibility, enamored of questionable concepts like "abjection" in an era of increasing acceptance, entranced by the idea that sexuality per se is radical in an era of soft-porn Calvin Klein ads, and resigned to the ritual denunciation of intolerance and insidious oppressions from the lofty perch of endowed chairs, on op-ed pages in mainstream newspapers, and through mass media outlets.

I have pitched my analysis in strong terms, but I hope my argument will not be misunderstood. It is not my contention that the time has passed for the critique of heteronormativity. I trust that the preceding pages demonstrate the need for continued struggle against heteronormalizing images and claims. Nor do I wish to understate the prevalence of everyday violence and discrimination against gays, lesbians, transgendered persons, androgynous boys, mannish women, and all those other legions perceived as "queer." Institutional heterosexism still affects everyone's life chances—no doubt more so in places like Laramie, Wyoming, or Marriott, Georgia, than in places like Marin County, California, or Brookline, Massachusetts. Still, it seems necessary to mark the distance traveled by queers everywhere. Gay liberation has affected even the lives of teens who live in trailer courts in rural North Carolina.

In gauging this broad sea change in sexual cultures, it is my contention that a substantial amount of the preliminary work of lesbigay politics and lesbigay studies has already been accomplished, and that queer theory will not move forward again until it accounts for the real-world successes and latter-day predicaments of the LGBTQ bloc whose interests it claims to represent, whose multifaceted perspectives it purports to convey and whose existence it supposedly researches. It is also my contention that queer theory will remain ill-equipped to really engage the new social conditions until it attends to political-economic problems and institutional articulations as credibly as it took up questions related to signification, representation, and performance in the nineties.

Although seldom cast in heroic terms, the ferment and creativity of ordinary people at the turn of the millennium puts to shame the social innovations of earlier eras. Today, men, women, and others fashion themselves and improvise their lives both according to the meandering course of their desires and in response to the vagaries of the marketplace. But by virtue of their mixed origins and by dint of postmodernity's unceasing transformations, contemporary social innovations cohabit with reactionary nostalgia and global economic instability. Sexuality in transition and crisis is linked to capitalism in transition and crisis. The one form of creativity fuels the other, the one form of insecurity feeds the other. And if sexual desires have the power to mobilize masses, to perk markets, and to reshape failing institutions, sex panics reverberate unpredictably across a wider social field: in race relations, in anxieties about falling birth rates, in the politics of immigration, and so on.

No one can say how capitalism will organize our lives tomorrow, or what role, if any, heteronormative nuclear families will play in future versions of society. Nor can anyone predict just how social interests will cohere or disintegrate in a volatile political economy or what social movements will arise in response to changing circumstances. New economic crises, whose eruptions are as inevitable as the those of the nineteenth and twentieth centuries, could intensify present-day conflicts, or they could sweep them off the stage of history. The emergence of new social movements—progressive, revolutionary, reactionary, or mixed—could also change the shape of present-day contradictions or produce entirely new ones.

In any event, it seems likely that new technologies will continue to unsettle old ideas about what is natural and what is unnatural. It seems likely that class struggles, slumbering since the demise of Marxism-Leninism, will reawaken and take new forms on a global scale. It also seems likely that the new freedoms precariously established at the end of the twentieth century will continue to deepen and expand: the freedom to invent one's own identity, to choose one's preferred forms of kinship, and to decide for one's self what is right and wrong.[8] And if it seems unlikely that all of this unceasing agitation and industry will ever end in utopia, it might at least be hoped that free men, women, and others would cast off their naturalistic guises— to be as radical as reality itself when confronting their social and moral choices, the unguaranteed means by which they unmake and remake the world. But it seems more likely that most social actors will continue to lean on preconceptions about nature when staging personas and plotting actions.

Either way, everyday life will remain both a social laboratory and a site of intimate contestations. People will continue innovating new identities, elaborating new designs for living, and building hybrid institutions—in the process, discovering new human natures.

Notes

INTRODUCTION

1. See Dean Hamer and Peter Copeland, *The Science of Desire: The Search for the Gay Gene and the Biology of Behavior* (New York: Simon and Schuster, 1994), which includes a reproduction of Hamer et al.'s "A Linkage between DNA Markers on the X Chromosome and Male Sexual Orientation," *Science* 261 (July 16, 1993): 321–27.

2. "Diverse Genes," *Metro Weekly*, May 29, 1997, 23. Simon LeVay conveys geneticist arguments directly from the culture of science to the gay subculture in his column "Queer Science," which was syndicated for a while in Washington, D.C.'s *MW*.

3. Lauren Berlant, *The Queen of America Goes to Washington City: Essays on Sex and Citizenship* (Durham: Duke University Press, 1997), 3–5.

4. Kai Wright, "Transsexuals Can Serve," *Washington Blade*, August 6, 1999. More recently, the by no means gay-friendly state of Texas has allowed a male-to-female transsexual to marry another woman—under a court ruling that defines sex only by chromosomes. *New York Times* (AP), September 18, 2000.

5. Richard Florida's recent book develops an urban-sociological argument about the relationship between gay neighborhoods and urban revitalization and economic health. The emerging "creative class," as Florida understands it—a class that includes entrepreneurs, lawyers, scientists, computer programmers, and others who work in information and technology—is vital to the economic health of the modern metropolis. This new class values cosmopolitanism and tolerance and specifically seeks out cities with a strong Bohemian and gay presence as ideal places to work and live. Richard Florida, *The Rise of the Creative Class: And How It's Transforming Work, Leisure, Community and Everyday Life* (New York: Basic Books, 2002).

6. "In the Time of Tolerance," *Newsweek*, March 30, 1998, 29.

7. A point made by Ronald Steel in "Beyond Monica: The Future of Clinton's Past," *The Nation*, September 7, 1998, 11–18. See also Debbie Nathan, "Sodomy for the Masses," *The Nation*, April 19, 1999, 16–22.

8. Donna Minkowitz, eager to keep an open mind on the Right and its motivations, pushes this point a bit too far when she concludes that the Christian Right actually shares motivations with lesbigay and feminist sex radicals. Minkowitz doesn't quite go so far as to "go native," but her writing on the subject displays one of the oldest pitfalls of ethnographic research: a blurring of one's own notions, goals, and desires with those of one's informants. It is one thing to conclude that the Christian Right is populated by human beings and motivated by desires other than malice. It is another to blur the meanings that define the cultural oppositions of opposing cultures. See "My Encounter with the Right: What I Learned about Marx and Pornography from Focus on the Family," *The Nation*, December 7, 1998, 18–22, and *Ferocious Romance: What My Encounters with the Right Taught Me about Sex, God, and Fury* (New York: Free Press, 1998).

As for my perspective on these matters, I will hide nothing from the diligent reader willing to pursue fine points into the footnotes: As a teenage musician in a Free Will Baptist church, I brought to church music a sound somewhere between Jerry Lee Lewis's honky-tonk and post-Beatles psychedelia. Old women frequently thought themselves filled with the Spirit (and I delighted in hearing them speak in tongues), but I knew this joyful noise actually came from a lower place. Jamming for Jesus, I met a phenomenally successful gospel performer who was likewise drifting into the Devil's music and all that goes with it: He was exploring new musical licks with the help of marijuana, and he was discovering new forms of Christian love with throngs of female and male church groupies. And so, gentle reader, my cynicism about such matters: I understand that Christian rock (and punk, and grunge, and hip-hop) (no doubt Christian club music will be next) is designed to co-opt youth culture on behalf of religious revivalism, that the Promise Keepers is a backward movement tapping rock-concert aesthetics and athletic-event male bonding, and that new forms of Christian therapy tap post-sixties self-actualization movements. But my experience suggests that the spiritual ideal cannot long check the worldly impulse in these hybrid forms.

9. Gustav Niebuhr, "Falwell and Allies to Meet Gay Rights Supporters," *New York Times* (National), September 5, 1999. Rhonda Smith, "Falwell Accepts Dinner Invitation: Infamous Reverend to Meet with 200 Gays," *Washington Blade*, August 27, 1999.

10. Echoing nineteenth-century sex radicals, Marx and Engels put matters bluntly in their Communist coming-out pamphlet—not so as to rebut bourgeois depictions of Communist aims, but to radicalize the question: "[The proletarian's] relation to his wife and children has no longer anything in common with bourgeois family relations." Later, the authors explain: "On what foundation is the present family, the bourgeois family, based? On capital, on private gain. In its completely developed form this family exists only among the bourgeoisie. But this state of things finds its complement in the practical absence of the family among the proletarians, and in public prostitution." The authors leave it to be understood that prevailing forms of family, marriage, and sexual in-

equality would disappear once their material supports were removed—a point Engels expressly developed in a later book. Karl Marx and Frederick Engels, *The Communist Manifesto* (1888, 1948; New York: International Publishers, 1987), 20, 26; see also 26–28. Frederick Engels, *The Origin of the Family, Private Property, and the State* (1884; New York: International Publishers, 1972).

11. Mark Blasius and Shane Phelan, the editors of *We Are Everywhere: A Historical Sourcebook of Gay and Lesbian Politics* (New York: Routledge, 1997), show in their excerpts from the 1876 preface to *Leaves of Grass* that Walt Whitman expressly conceptualized and named something very much like a politics of "Identity" (82–83).

12. Or, as an ad for diamonds declared in 1999: "Men express their feelings in actions, not words. Women don't have a problem with that."

13. I borrow this characterization of nature as a longing for unmediated, organic immediacy with a certain sense of trepidation. Although critical theorists from Georg Lukács to Max Horkheimer and Theodor Adorno correctly identified this desire called "nature," they went right on pining for authentic purity, much to the detriment of the critical element in critical theory. Steven Vogel capably traces the zigzag course of the concept in his book, *Against Nature: The Concept of Nature in Critical Theory* (Albany: State University Press of New York, 1996).

14. Simon LeVay, "A Difference in Hypothalamic Structure between Heterosexual and Homosexual Men," *Science* 253 (1991): 1034–37.

15. I rely on and distill a substantial literature. On the rise of medical models of homosexuality, see Jeffrey Weeks, *Sexuality and Its Discontents: Meanings, Myths and Modern Sexualities* (London: Routledge and Kegan Paul, 1985), esp. 61–95. Vernon A. Rosario's edited collection *Science and Homosexualities* (New York: Routledge, 1997) provides an excellent survey of the history of the topic; see especially Hubert Kennedy, "Karl Heinrich Ulrichs, First Theorist of Homosexuality" (26–45), Harry Oosterhuis, "Richard von Krafft-Ebing's 'Step-Children of Nature'" (67–88), Stephanie Kenen, "Who Counts When You're Counting Homosexuals: Hormones and Homosexuality in Mid-Twentieth-Century America" (197–218), Garland E. Allen, "The Double-Edged Sword of Genetic Determinism: Social and Political Agendas in Genetic Studies of Homosexuality, 1940–1994" (242–70), and Jennifer Terry, "The Seductive Power of Science in the Making of Deviant Communities" (271–95). For analysis of the intertwining histories of science, sexism, racism, and the making of the homosexual body, two papers stand out: Jennifer Terry, "Anxious Slippages between 'Us' and 'Them': A Brief History of the Scientific Search for Homosexual Bodies," in *Deviant Bodies*, ed. Jennifer Terry and Jacqueline Urla (Bloomington: Indiana University Press, 1995), 129–69; and Siobhan Somerville, "Scientific Racism and the Invention of the Homosexual Body," in *The Gender/Sexuality Reader: Culture, History, Political Economy,* ed. Roger N. Lancaster and Micaela di Leonardo (New York: Routledge, 1997), 37–52. Finally, with wit and force of argument, Jonathan Ned Katz shows how the marking of "deviant" bodies gives shape to the norm in *The Invention of Heterosexuality* (New York: Dutton, 1995).

16. Micaela di Leonardo, *Exotics at Home: Anthropologies, Others, American Modernity* (Chicago: University of Chicago Press, 1998), 358.

17. See Marshall Sahlins, *The Use and Abuse of Biology* (Ann Arbor: University of Michigan Press, 1976), xv.

18. Richard Lewontin, Steven Rose, and Leon J. Kamin, *Not in Our Genes: Biology, Ideology, and Human Nature* (New York: Pantheon, 1984), 251.

19. See Stephen Jay Gould's discussion of evolutionary psychology in "Evolution: The Pleasures of Pluralism," *New York Review of Books,* June 26, 1997, 47–52, esp. 50–52.

20. *Points de capiton* ("anchoring points"): I play off one of Jacques Lacan's recurring metaphors for the unstable process of identity formation in a sea of sliding signifiers and uncertain references. "The Agency of the Letter in the Unconscious or Reason since Freud," *Écrits: A Selection,* trans. Alan Sheridan (1966; New York: W. W. Norton, 1977), 154.

21. For ethnographic perspectives on practices of bioscience, the biotech business, and politics, see Paul Rabinow, *Making PCR: A Story of Biotechnology* (Chicago: University of Chicago Press, 1997), and *French DNA: Trouble in Purgatory* (Chicago: University of Chicago Press, 1999).

22. *New York Times,* January 18, 2001.

23. Gina Kolata, "Scientists Report the First Success of Gene Therapy. Three Babies Said to Be Saved. After Years of Dashed Hopes, French Results Promising—Doctors Still Cautious," *New York Times,* front page, April 28, 2000. Gina Kolata, "I'm Not Dead Yet: Genetic Mutation That Lives Up to Its Name Is Found," *New York Times* (National Report), December 15, 2000. Nicholas Wade, "Scientists Report 2 Major Advances in Stem-Cell Work," front page, *New York Times,* April 27, 2001. Will Dunham, "Gene Therapy Used to Restore Sight to Blind Dogs," Reuters, April 28, 2001.

24. Joachim Liepert et al., "Treatment-Induced Cortical Reorganization after Stroke in Humans," *Stroke: Journal of the American Heart Association* 31, no. 6 (June 2000): 1365–69. Emily Yellin, "Stroke Survivors Celebrate Success of Restraint Therapy," *New York Times* (Science Times), June 13, 2000.

25. John Horgan, *The End of Science: Facing the Limits of Knowledge in the Twilight of the Scientific Age* (Reading, Mass.: Addison-Wesley, 1996), and *The Undiscovered Mind: How the Human Mind Defies Replication, Medication, and Explanation* (New York: Free Press, 1999). See also Claudia Dreifus, "A Heretic Takes on the Science of the Mind," *New York Times* (Science Times), September 21, 1999.

26. See Stephen Jay Gould's discussion of this point in *The Mismeasure of Man,* rev. ed. (1981; New York: W. W. Norton, 1996), 32–33: Genetic explanations for *pathological* conditions tell us rather less about healthy variations than scientists sometimes admit.

27. All science has an ideological component, as a vast field of critical science studies has demonstrated. These demonstrations often incur the resentment of scientists, who sometimes imagine their practice to be entirely objective, value-neutral, and apolitical. See, for instance Paul R. Gross and Norman Levitt, *Higher*

Superstition: The Academic Left and Its Quarrels with Science (Baltimore: Johns Hopkins University Press, 1994), and Paul R. Gross, Norman Levitt, and Martin W. Lewis, eds., *The Flight from Science and Reason* (New York: The New York Academy of Sciences, 1996). Still, for reasons that should become clearer in subsequent chapters, I do not believe that all science is equally ideological. I thus employ not a distinction between ideological positions in scientific practice, but a contrast between science and pseudoscience, between the reasonable use of science and the abuse of scientific rhetoric, between "good" and "bad" science. Simply put: Scientists, when they practice science well, pose hypotheses that are subject to either experimental tests or other systematic checks against empirical evidence. When scientists pose untestable hypotheses (as in the origins stories treated in the first part of this book), or when they apply scientistic language to problems not commensurate with the methods of scientific investigation (the meaning of life, the phenomenology of desire), what they practice is not an ideologically inflected science, but ideology outright.

28. For a fuller discussion, to which I am indebted, see Kennedy, "Karl Heinrich Ulrichs," especially 29–30, 36, 39, and Oosterhuis, "Krafft-Ebing's 'Step-Children of Nature,'" 71–72, in *Science and Homosexualities*.

29. Thomas Laqueur, *Making Sex: The Body and Gender from the Greeks to Freud* (Cambridge, Mass.: Harvard University Press, 1990), 152. Also Kennedy, "Karl Heinrich Ulrichs," in *Science and Homosexuality*, 33.

30. Quoted by Kai Wright, "Where Gender, Mental Health Collide: Youth Pushed into 'Therapy' for Gender Identity Disorder," *Washington Blade*, October 2, 1998.

31. I have deliberately used the archaic terminologies of a defunct medical science, the better to underscore their hidden presence in today's discourses.

32. See Daniel Goleman, "The 'Wrong' Sex: A New Definition of Childhood Pain," *New York Times* (Science News), March 22, 1994.

33. Anne Fausto-Sterling, *Myths of Gender: Biological Theories about Men and Women*, rev. ed. (1985; New York: Basic Books, 1992), 223–59. See Simon LeVay, *The Sexual Brain* (Cambridge, Mass.: MIT Press, 1993). LeVay's reasoning about the "sex" of the brain, here as elsewhere, is embedded in familiar stereotypes and just-so stories about the behavior of men and women—the sort of evolutionary folklore to be considered at length in this book.

34. See William Byne, "Science and Belief: Psychological Research on Sexual Orientation," in *Sex, Cells, and Same-Sex Desire: The Biology of Sexual Preference*, ed. John P. De Cecco and David Allen Parker (New York: Harrington Press Park, 1995), 303–44.

35. The ad concludes: "If you really love someone, you'll tell them the truth." *New York Times*, July 13, 1998, A11. Or, Truth = Heterosexuality = Nature, by divine design.

36. The cultlike practices of the "ex-gay" Exodus movement emphasize group confessional therapeutics and same-sex bonding as key ingredients of "healthy" gender identity and Christian fellowship. They also stress a heavily amatorized notion of Christ's love, a series of displacements and sublimations

that allows one to live a homosexual lifestyle without living the homosexual lifestyle.

37. *Good Morning America,* August 5, 1998, ABC Transcript #3270.

38. Urvashi Vaid, personal communication, e-mail, August 6, 1998.

39. Either proposition—"I am powerless before my desires" or "I exercise volitional control over my desires"—distorts the experience of desire, as I will argue subsequently. See more sustained phenomenological arguments in my essay, "Coming Out Stories: Recent Videos on Gay and Lesbian Themes," *American Anthropologist* 98, no. 3 (September 1996): 604–16, and in "Transgenderism in Latin America: Some Critical Introductory Remarks on Identities and Practices," *Sexualities* 1, no. 3 (1998): 263–76. See also Vera Whisman's innovative reflections on interviews with lesbigay informants in *Queer by Choice: Lesbians, Gay Men, and the Politics of Identity* (New York: Routledge, 1996).

40. Lisa Duggan, "Queering the State," in *Sex Wars: Sexual Dissent and Political Culture,* by Lisa Duggan and Nan Hunter (New York: Routledge, 1995), 179–93, 183.

41. Alfred C. Kinsey et al., *Sexual Behavior in the Human Male* (Philadelphia: W. B. Saunders, 1948) and *Sexual Behavior in the Human Female* (Philadelphia: W. B. Saunders, 1953). Laud Humphreys, *Tearoom Trade: Impersonal Sex in Public Places,* enlarged ed. (1970; Hawthorne, N.Y.: Aldine de Gruyter, 1975). See also Bill Leap, ed., *Public Sex/Gay Space* (New York: Columbia University Press, 1999).

42. Maybe we should be running ads on the sides of buses, beginning with a quote from John Lennon—"Whatever gets you through the night is all right"—followed by a line from the Kinks—"Girls will be boys and boys will be girls"—and a string of gender-bending, sexually experimental lyrics from classic rock to hip-hop and from Bowie-glam to Seattle grunge, the better to underscore just how *queer* the basic building blocks of popular entertainment culture have actually become.

43. Uneasiness, in general, with what Jeffrey Weeks has called "the politics of choice." See Jeffrey Weeks, *Invented Moralities: Sexual Values in an Age of Uncertainty* (New York: Columbia University Press, 1995).

44. "Going Down Screaming," *New York Times Magazine,* October 11, 1998, 49.

45. Carey Goldberg, "Vermont Gives Final Approval to Same-Sex Unions," *New York Times,* April 26, 2000.

46. Mark Lilla, "Still Living with '68," *New York Times Magazine,* August 16, 1998, 34–37.

47. Michael Taussig, *Mimesis and Alterity: A Particular History of the Senses* (New York: Routledge, 1993), xix. Camille Paglia, *Sexual Personae: Art and Decadence from Nefertiti to Emily Dickinson* (New Haven: Yale University Press, 1990), 2.

48. The classics of cultural feminism, with its essentialist approach to gender roles and temperament, include Adrienne Rich, *Of Woman Born: Motherhood as Experience and Institution* (New York: W. W. Norton, 1976); Mary Daly,

Gyn/Ecology: The Metaethics of Radical Feminism (Boston: Beacon Press, 1978); and Susan Griffin, *Pornography and Silence: Culture's Revenge against Nature* (New York: Harper and Row, 1981).

49. John O. McGinnis, "The Origins of Conservatism," *National Review* 49, no. 22 (December 1997): 31–35. See the critical response by Stephen Jay Gould, "Let's Leave Darwin out of It," *New York Times* (Op Ed), May 29, 1998.

50. Jeffrey Weeks, *Sexuality and Its Discontents: Meanings, Myths, and Modern Sexualities* (London: Routledge Kegan Paul, 1985), 62.

51. Max Horkheimer and Theodor W. Adorno, *Dialectic of Enlightenment*, trans. John Cumming (1944; New York: Continuum, 1995).

52. Karl Marx, *The Eighteenth Brumaire of Louis Bonaparte*, excerpted in *Karl Marx: Selected Writings*, ed. David McLellan (1852; Oxford: Oxford University Press, 1977), 300.

53. David Harvey, *The Condition of Postmodernity* (Cambridge, Mass.: Basil Blackwell, 1990), 171.

54. The general connection between sexual flux, ideas about homosexuals, and new political possibilities was perhaps first evoked by Edward Carpenter a century ago. See the excerpts from Carpenter's *Intermediate Sex* in Blasius and Phelan, *We Are Everywhere*, 114–31, esp. 117.

55. Writing for the *Village Voice* (August 8–14, 2001), Chris Nutter describes the coming into existence of "poststraight" men this way: "Glossy magazines have noticed that straight men are looking more gay, but the influence is more than a matter of working out, waxing, and wearing Prada. It involves a profound change in consciousness, reflected in everything from greeting gay buddies with a kiss to treating women the way other women—and most gay men—do."

56. In an article that turns on existentialist arguments that human *activity* defines what it is to be human, Barash concludes: "In the end, genes just aren't that important." "DNA and Destiny," *New York Times*, November 16, 1998.

57. Natalie Angier, *Woman: An Intimate Geography* (New York: Anchor Books, 1999).

58. See Barbara Ehrenreich, *Blood Rites: Origins and History of the Passions of War* (New York: Metropolitan Books, 1997).

59. I say *supposedly* "queering" science because the idea that sexuality and sexual identity are innate, immutable traits seems a most *unqueer* concept. "Queer," presumably, represents a nonessentialist form of sexual politics and culture.

60. "Were You Born That Way?" Photographs by Karen Kuehn, text by George Howe Colt, reporting by Anne Hollister, *Life*, April 1998.

61. In some cases, the essay cites studies whose results could not be replicated or whose findings had been quietly withdrawn at the time of the essay's publication.

62. Edward O. Wilson, "The Biological Basis of Morality," *Atlantic Monthly* 281, no. 4 (1998).

63. I have wrestled with tone at almost every turn in this book, beginning with the question of where and how frequently to enclose words like "nature,"

"culture," and "science" in eyebrow-raising scare quotes. I can scarcely claim to have always selected wisely, nor am I prepared to say that no preconception of nature rests beneath my debunkings of claims about an abstract and objectified nature. "What I claim," as Roland Barthes said in 1957, "is to live to the full the contradictions of my time, which may well make sarcasm the condition of truth." *Mythologies,* trans. Annette Lavers (1957; New York: Hill and Wang, 1972), 12.

64. It will be clear that I am attempting to articulate, in clear language, the basic moves of what has come to be known as "queer theory." Some readers will also notice here a correspondence to the scope of arguments in my ethnography of everyday life in Nicaragua, in which I show how discourses and representations of same-sex desire shape the experience of gender and sexuality in general—and how it is difficult to extricate the "political economy of the body" from political economy as it is usually understood. See my *Life Is Hard: Machismo, Danger, and the Intimacy of Power in Nicaragua* (Berkeley: University of California Press, 1992).

CHAPTER 1. IN THE BEGINNING, NATURE

1. I steal this narrative device from Alexander Cockburn's "A Short, Meat-Oriented History of the World. From Eden to the Mattole," a brilliant survey of some of the ways human beings have related to animals and to nature, published in *New Left Review* 215 (1996): 16–42.

2. Maurice Merleau-Ponty, quoting Lucien Herr, commenting on Hegel. From *In Praise of Philosophy and Other Essays* (1953; Evanston: Northwestern University Press, 1963), 133.

3. Roy Porter, *The Greatest Benefit to Mankind: A Medical History of Humanity* (New York: W. W. Norton, 1997), 7.

4. Carol P. MacCormack and Marilyn Strathern, eds., *Nature, Culture, and Gender* (Cambridge: Cambridge University Press, 1980). See especially MacCormack's overview essay, "Nature, Culture and Gender: A Critique," 5–11.

5. Thomas Laqueur, *Making Sex: The Body and Gender from the Greeks to Freud* (Cambridge, Mass.: Harvard University Press, 1990). See also Laqueur's "Orgasm, Generation, and the Politics of Reproductive Biology," reprinted in *The Gender/Sexuality Reader: Culture, History, Political Economy,* ed. Roger N. Lancaster and Micaela di Leonardo (New York: Routledge, 1996), 219–43.

6. Martha McCaughey, "Perverting Evolutionary Narratives of Heterosexual Masculinity: Or, Getting Rid of the Heterosexual Bug," *GLQ* 3, nos. 2–3 (1996): 261–87.

7. Michael Warner, introduction, to *Fear of a Queer Planet: Queer Politics and Social Theory,* ed. Michael Warner (Minneapolis: University of Minnesota Press, 1993), xxi.

8. Donna J. Haraway, *Primate Visions: Gender, Race, and Nature in the World of Modern Science* (New York: Routledge, 1989), 4, 8.

9. Edward Said, *Beginnings: Intention and Method* (Baltimore: Johns Hopkins University Press, 1975), 5.

10. This is the gist of Chandler Burr's argument in *A Separate Creation: The Search for the Biological Origins of Sexual Orientation* (New York: Hyperion, 1996).

CHAPTER 2. THE NORMAL BODY

1. Frederick Turner, "Biology and Beauty," in *Incorporations*, ed. Jonathan Crary and Sanford Kwinter (New York: Zone Books, 1992), 406–21, 413.

2. Note that I draw this example neither from the lowest vulgate of science journalism nor from the rantings of conservative sociobiology, but from a vaguely progressive volume of academic essays: a compilation whose selections supposedly sample up-to-date ways of thinking about human embodiment.

3. Turner, "Biology and Beauty," 414.

4. Clyde Kluckhohn, *Mirror for Man: The Relation of Anthropology to Modern Life* (New York: Whittlesey House, 1949), esp. 9, 20.

5. "The charmed circle of nature" combines historian David Hollinger's apt phrase "the circle of the we" with anthropologist Gayle Rubin's contrast between the "charmed circle" of normal sexuality and the "outer limits" of deviant sex. See David Hollinger, "How Wide Is the Circle of the 'We'? American Intellectuals and the Problem of the Ethnos since World War II," *American Historical Review* 89, no. 2 (April 1993): 317–33. See Micaela di Leonardo's sustained critical discussion of Hollinger's essay in *Exotics at Home: Anthropologies, Others, American Modernity* (Chicago: University of Chicago Press, 1998), 140–43. See also Gayle Rubin, "Thinking Sex: Notes for a Radical Theory of the Politics of Sexuality," in *Pleasure and Danger: Exploring Female Sexuality*, ed. Carole S. Vance (1984; London: Pandora, 1989), esp. 280–83.

6. Physical anthropologists have for a long time reasoned that early human societies and their hominid antecedents were nomadic foragers without permanent dwellings.

7. Sigmund Freud, *Civilization and Its Discontents*, trans. James Strachey (New York: Norton, 1961), 34. John Updike, "Can Genitals Be Beautiful?" *Egon Schiele: The Leopold Collection, Vienna*, exhibit at the Museum of Modern Art, October 12, 1997–January 4, 1998. *The New York Review of Books*, December 4, 1997, 10–12.

8. The technical term *"venus observa"* has been around considerably longer than the folkloric term, "missionary position." In Christian doctrine dating back to Saint Augustine and Saint Paul, other sexual positions were deemed "sinful" or "against nature." *Merriam-Webster's Collegiate Dictionary* dates the first recorded use of the term "missionary position" to 1948. Robert Priest concurs, tracing the term to Alfred Kinsey's 1948 invocation of (but not quotation from) Bronislaw Malinowski's 1929 ethnography of the Trobriand Islands. See Robert J. Priest, "Missionary Positions: Christian, Modernist, Postmodernist," *Current Anthropology* 42, no. 1 (February 2001): 29–46, with comments and reply on 46–68; Alfred C. Kinsey et al., *Sexual Behavior in the Human Male* (Philadelphia: W. B. Saunders, 1948), 373; and Bronislaw Malinowski, *The Sex-*

ual Life of Savages in North-Western Melanesia (New York: Harcourt, Brace and World, 1929). Priest, an anthropologist and seminary professor, constructs an odd argument around this bit of sexual folklore—a recent spin on the "I'm rubber, you're glue" retort (duly noted by a number of *Current Anthropology* commentators): When anthropologists, modernists, or postmodernists criticize conservative Christians, evangelicals, and colonial missionaries for being sexually intolerant or culturally ethnocentric, they are themselves being intolerant or ethnocentric. (By means of the same rhetorical trampoline, evangelical fundamentalists—many of whom used their considerable clout to oppose civil rights in the 1960s—now shamelessly depict themselves as an oppressed minority group in the industrial world's most religious country. Their real beef is that a multiethnic and largely secular public culture refuses to privilege their bigotry and intolerance.)

Not everyone who commented on Priest's paper is convinced that Kinsey in fact coined the term "missionary position." Commentator Kenelm Burridge notes having heard the term right after World War II among "old India hands" in English pubs and surmises that the term "must have been part of Indian Army lore for generations" (47). On the sexual impositions of Christian missionaries on peoples of the Pacific, see Robert C. Suggs, *Marquesan Sexual Behavior* (New York: Harcourt, Brace and World, 1966) and Donald S. Marshall and Robert C. Suggs, eds., *Human Sexual Behavior: Variations in the Ethnographic Spectrum* (New York: Basic Books, 1971).

9. According to Barthes, this is "the very principle of myth: It transforms history into nature." Thus, "everything happens as if the picture *naturally* conjured up the concept, as if the signifier *gave a foundation* to the signified." Myth is "speech justified *in excess*." Roland Barthes, *Mythologies*, trans. Annette Lavers (1957; New York: Hill and Wang, 1972), 129–30.

10. Turner, "Biology and Beauty," 415.

11. Ibid., 416.

12. A brief discussion of Galton's role in the history of the concept of normalcy appears in chapter five.

13. Natalie Angier, "A Couple of Chimps Sitting around Talking," review of *Grooming, Gossip, and the Evolution of Language*, by Robin Dunbar, *New York Times Book Review*, March 9, 1997, 10.

14. Robert Wright, *The Moral Animal: The New Science of Evolutionary Psychology* (New York: Pantheon Books, 1994), 67. Wright, in turn, is recycling David Barash's musings in *The Whisperings Within: Evolution and the Origin of Human Nature* (Harmondsworth: Penguin, 1979), 39.

15. Marshall Sahlins, "The Original Affluent Society," in *Stone Age Economics* (Chicago: Aldine-Atherton, 1972), 1–39. On the relative *and* absolute decline in standards of living signaled by the advent of agriculture and settled civilization, see also Alan H. Goodman and George J. Armelagos, "Disease and Death at Dr. Dickson's Mounds," *Natural History* 9, no. 85: 12–18, and Jared Diamond, "The Worst Mistake in the History of the Human Race," *Discover Magazine*, May 1987, 64–66.

16. See James Woodburn's classic essay "Egalitarian Societies," *Man*, n.s., 17, no. 3 (September 1982): 431–51. See also Richard Lee, "Politics, Sexual and Nonsexual in an Egalitarian Society: The !Kung San," in *Social Inequality: Comparative and Developmental Approaches*, ed. Gerald D. Berreman, with the assistance of Kathleen M. Zaretsky (New York: Academic Press, 1981), 83–102.

17. Dr. Spock, too, says as much about early childhood diet and the development of adult cravings—a vernacular finding that contends with its opposite folk explanation, the idea that certain cravings for fats and sweets are hardwired into human biology. Benjamin Spock and Michael Rothanberg, *Dr. Spock's Baby and Child Care*, 6th ed., fully revised and updated for the 1990s (New York: Dutton, 1992), 185–87.

18. I use the term "method" advisedly here, since scientific method is *supposed* to check stories or hypotheses against empirical and preferably experimental evidence.

19. Wright, *The Moral Animal*, 38.

20. Yohannes Haile-Selassie, "Late Miocene Hominids from the Middle Awash, Ethiopia," and Henry Gee, "Palaeontology: Return to the Planet of the Apes," *Nature* 412 (July 12, 2001): 178–81 and 131–32. Michael D. Lemonick and Andrea Dorfman, "How Apes Became Human," *Time*, July 23, 2001, 54–61.

21. See Sally Slocum's classic of evolutionary reinterpretation, "Woman the Gatherer: Male Bias in Anthropology," in *Toward an Anthropology of Women*, ed. Rayna R. Reiter (New York: Monthly, 1975), 36–50. The essays in Joan M. Gero and Margaret W. Conkey's pioneering edited volume, *Engendering Archaeology: Women and Prehistory* (Oxford: Blackwell, 1991), express the nature of the problem, as does Margaret Conkey and Sarah Williams's essay "Original Narratives: The Political Economy of Gender in Archaeology," in *Gender at the Crossroads of Knowledge: Feminist Anthropology in the Postmodern Era*, ed. Micaela di Leonardo (Berkeley: University of California Press, 1991), 102–39. Traditional archaeology was conceived from an unselfconsciously *male* point of view. Feminist attempts at "engendering" archaeology grapple with both the androcentric biases of the discipline and the frequently ambiguous nature of the data. It is not always clear what tools, relics, and representations were made and/or used by women, which were made/used by men—much less what any of these practices might indicate about the structure of gender relations in antiquity. See subsequent books and collections in the field: Elisabeth A. Bacus, Alex W. Barker, and Jeffrey D. Bonevich, eds., *A Gendered Past: A Critical Bibliography of Gender in Archaeology*, Technical Report no. 25 (Ann Arbor: University of Michigan Museum of Anthropology, 1993); Rita P. Wright, ed., *Gender and Archaeology* (Philadelphia: University of Pennsylvania Press, 1996); Kelley Ann Hays-Gilpin and David S. Whitley, eds., *Reader in Gender Archaeology* (New York: Routledge, 1998); Roberta Gilchrist, *Gender and Archaeology: Contesting the Past* (London: Routledge, 1999); and Marie Louise Stig Sorensen, *Gender Archaeology* (Cambridge: Polity, 2000). The question of sexuality has more recently been problematized.

See Robert A. Schmidt and Barbara L. Voss, eds., *Archaeologies of Sexuality* (New York: Routledge, 2000).

22. Sahlins, *Stone Age Economics;* Woodburn, "Egalitarian Societies"; Lee, "Politics, Sexual and Nonsexual in an Egalitarian Society." See also Richard Lee, *The Dobe Ju/'hoansi,* 2d ed. (Fort Worth, Tex.: Harcourt Brace College Publishers, 1993); Eleanor Leacock and Richard Lee, eds., *Politics and History in Band Societies* (Cambridge: Cambridge University Press, 1982); and Marshall Sahlins, *Tribesmen* (Englewood-Cliffs, N.J.: Prentice-Hall, 1968).

23. Lionel Tiger and Robin Fox, *The Imperial Animal* (1971; New Brunswick, N.J.: Transaction, 1998). See esp. "Political Nature," "Bond Issue: Women and Children First," and "The Benign Oppression."

24. Here and elsewhere, I draw on Ian Hacking's discussion of the slippery relation between the "is" and the "ought" in scientific models. See *The Taming of Chance* (Cambridge: Cambridge University Press, 1990), esp. 160–69.

25. Edward O. Wilson, *On Human Nature* (Cambridge, Mass.: Harvard University Press, 1978), 119, 171.

26. Helen Fisher, *The First Sex: The Natural Talents of Women and How They Are Changing the World* (New York: Random House, 1999).

CHAPTER 3. THE HUMAN DESIGN

1. This combines Raymond Williams's felicitous phrase "structures of feeling" with the title of David P. Barash's sociobiological text, *The Whisperings Within* (New York: Harper and Row, 1979).

2. Michael Segell, "The Second Coming of the Alpha Male," *Esquire* 126, no. 4 (October 1996): 74–81.

3. At the close of the 1980s, it was still a good joke that sociobiology lacked a necessary sense of irony about its own flat-footed narratives. Donna J. Haraway, *Primate Visions: Gender, Race, and Nature in the World of Modern Science* (New York: Routledge, 1989), 366. Popular culture has since supplied that missing ingredient in abundance. If artifact, detachment, and irony are all conditions of postmodern culture, it is not always clear whether our "nature" is being serious with us or whether s/he is being kitschy.

4. Segell, "Alpha Male," 76.

5. Philomena Mariani, "Law and Order Science," in *Constructing Masculinity,* ed. Maurice Berger, Brian Wallis, and Simon Watson (New York: Routledge, 1995), 135–56.

6. Cited in ibid., 137–38.

7. Terry Burnham, an economics professor at Harvard's Kennedy School of Government, and Jay Phelan, a professor of biology at UCLA, have published a self-help book under the title *Mean Genes: From Sex to Money to Food: Taming Our Primal Instincts* (Boulder, Colo.: Perseus, 2000).

8. The use of the term "tribe" here and throughout reflects the shallow ethnological background of Fishbein's claims. Most ethnographers ceased using the term "tribe" decades ago. See R. Brian Ferguson and Neil L. Whitehead's dis-

cussion of "tribes" and "tribalization" as an effect of state expansion in "The Violent Edge of Empire," as well as Neil Whitehead's essay, "Tribes Make States and States Make Tribes: Warfare and the Creation of Colonial Tribes and States in Northeastern South America," in *War in the Tribal Zone: Expanding States and Indigenous Warfare,* ed. R. Brian Ferguson and Neil L. Whitehead (Santa Fe, N.M.: School of American Research, 1992). The term "tribe" historically denoted an extended kin group of some size and complexity. See Marshall Sahlins, *Tribesmen* (Englewood Cliffs, N.J.: Prentice-Hall, 1968). Whatever forms of kinship were practiced by hominids and early humans, it seems improbable that they were anything as complex as an extended lineage system involving thousands of members—a "tribe." Anthropologists use the word "band" to describe the small-scale, technologically simple, low-density foraging societies observed in some parts of the world today and theorized as a model for archaic forms of subsistence and social organization. See Kent V. Flannery's distinction between "band" and "tribe" in "The Cultural Evolution of Civilization," *Annual Review of Ecology and Semantics* 3 (1972): 401. But for the kind of claim Fishbein is advancing, the word "band" simply would not do. He needs the word "tribe," for it carries all the resonances of a xenophobic "tribalism," associations difficult to convey with the word "band."

9. Richard Morin, "The Deep Roots of Racism," *Washington Post,* February 9, 1997, C5. Harold D. Fishbein, *Peer Prejudice and Discrimination: Evolutionary, Cultural and Developmental Dynamics* (Boulder, Colo.: Westview, 1996).

10. Andrew Sullivan, "What's So Bad about Hate," *New York Times Magazine,* September 26, 1999.

CHAPTER 4. OUR ANIMALS, OUR SELVES

1. Some readers will note that I am leaning on an old semiological distinction between rules for sequencing, ordering, or combining signs into meaningful utterances ("the syntagm") and rules for selecting signs from among other signs in the vast storehouse of language (the "system," "association," or "paradigm"). See Roland Barthes, *Elements of Semiology,* trans. Annette Lavers and Colin Smith (1964; New York: Hill and Wang, 1967), 58–88.

2. See Claude Lévi-Strauss, "Sketches for an American Bestiary (1964–5)," in *Anthropology and Myth: Lectures 1951–1982,* trans. Roy Willis (Oxford: Basil Blackwell, 1987), 86–88.

3. Bonobos "are as close to us as chimpanzees, the species on which much ancestral human behavior has been modeled." But the behavioral repertoire of the bonobo carries rather different implications from that of the chimpanzee: equality and promiscuity, rather than male domination and sexual control. Frans de Waal maintains that bonobos keep the peace and maintain equality in no small part through constant, promiscuous lovemaking. "Bonobos engage in sex in virtually every partner combination: male-male, male-female, female-female, male-juvenile, female-juvenile, and so on. The frequency of sexual contact is

also higher than among most other primates." Thus, "if bonobo behavior provides any hints, very few human sexual practices can be dismissed as 'unnatural.'" Frans de Waal, *Bonobo: The Forgotten Ape*, photographs by Franz Lanting (Berkeley: University of California Press, 1997), 1, 4, 5.

4. One might say that the first 90 percent of Edward O. Wilson's *Sociobiology: The New Synthesis* works this way, in that the material for neo-Darwinian explanations for animal behaviors serves to found the argument, developed in the final pages, that complex human activities are under genetic control and governed by the same genetic logic. Others tell fables of the bestiary in far less disciplined ways.

5. Jonathan Marks, chapter 5, "Behavioral Genetics," in *What It Means to Be 98 Percent Chimpanzee* (Berkeley: University of California Press, 2002).

6. Stephen Jay Gould, *Ever since Darwin: Reflections in Natural History* (New York: W. W. Norton, 1977), 258–59.

7. Claude Lévi-Strauss, *Totemism* (1962; Boston: Beacon, 1963), 81.

8. Stephen Jay Gould, *The Mismeasure of Man*, rev. ed. (1981; New York: W. W. Norton, 1996), 33, 395–98.

9. Donna J. Haraway, *Primate Visions: Gender, Race, and Nature in the World of Modern Science* (New York: Routledge, 1989), esp. 186–230.

10. Richard B. Lee and Irven DeVore, "Problems in the Study of Hunters and Gatherers," and Sherwood L. Washburn and C. S. Lancaster, "The Evolution of Hunting," in *Man the Hunter*, ed. Richard B. Lee and Irven DeVore (Chicago: Aldine, 1968), 3–12 and 293–303, respectively. Irven DeVore and Sherwood L. Washburn, "Social Behavior of Baboons and Early Man," in *Social Life of Early Man*, ed. Sherwood L. Washburn (Chicago: Aldine, 1961), 91–105.

11. Susan Sperling, "Baboons with Briefcases versus Langurs with Lipstick: Feminism and Functionalism in Primate Studies," in *The Gender/Sexuality Reader: Culture, History, Political Economy*, ed. Roger N. Lancaster and Micaela di Leonardo (New York: Routledge, 1997), 249–64, 249.

12. Susan Sperling, personal communication.

13. See, for instance, Lila Leibowitz's discussion of how various primate species behave in various environments, "Perspectives on the Evolution of Sex Differences," in *Toward an Anthropology of Women*, ed. Rayna R. Reiter (New York: Monthly, 1975), 20–35.

14. Stanford primatologist Robert Sapolsky summarizes the state of baboon studies in a fresh popular magazine article. It turns out that baboon alpha males "don't achieve dominance by winning fights, but by avoiding them." Dominance is not associated with "aggressiveness" or testosterone but with "impulse control" and "political behavior" (the cultivation of alliances with other males). "Friendly," empathic male baboons—"the best friends"—receive more social and perhaps sexual attention from females than do dominant males. Genetic studies of baboon troops have shown that "being a so-called alpha male" doesn't "necessarily translate into increased reproductive success." "Gorilla Tactics: Beasts Can Teach Us How to Bed Beauties," *Men's Health* 17, no. 2 (March 2002): 68–70. See also Sapolsky's summary article "Sex and the Single Mon-

key: Among Some Primates, Nice Guys—Not Just Studs—Get the Girl," *The Sciences* 38, no. 4 (July–August 1998): 10–12, and his book, *A Primate's Memoir* (New York: Scribner, 2001), on his life as a primatologist.

Anne Fausto-Sterling surveys studies on dominance, hormones, and reproductive success in *Myths of Gender* (141–47). A number of scholars have disaggregated the descriptive terms and behaviors usually lumped together under the term "dominance" (leadership, aggression, hierarchy) or have suggested that behaviors often associated with "dominance" are artifacts of the laboratory environment, captivity, or conditions of scarcity. See Thelma Rowell, "The Concept of Social Dominance," *Behavioral Biology* 11, no. 2 (June 1974): 131–54; Gina Bari Kolata, "Primate Behavior: Sex and the Dominant Male," ("Research News"), *Science* 191 (January 9, 1976): 55–56; A. F. Dixson, "Androgens and Aggressive Behavior in Primates: A Review," *Aggressive Behavior* 6, no. 1 (1980): 37–67; Linda Marie Fedigan, "Dominance and Reproductive Success in Primates," *Yearbook of Physical Anthropology* 26 (1983): 91–129; Irwin S. Bernstein, "The Evolution of Nonhuman Primate Social Behavior," *Genetica* 73, nos. 1–2 (August 31, 1987): 99–116.

Recent research has investigated the roles of "female dominance," "female choice," "male friendliness," "sperm competition," and other factors in reproductive success. See Anne Pusey, "The Influence of Dominance Rank on the Reproductive Success of Female Chimpanzees," *Science* 277 (August 8, 1997): 828–31; Julie L. Constable et al., "Noninvasive Paternity Assignment in Gombe Chimpanzees," *Molecular Ecology* 10 (2001): 1279–1300; Ute Radespiel et al., "Sexual Selection, Multiple Mating and Paternity in Grey Mouse Lemurs, *Microcebus murinus*," *Animal Behaviour* 63 (2002): 259–68.

15. See, for instance, the first edition of Richard Lee's widely used textbook, *The Dobe !Kung* (New York: Holt, Rinehart and Winston, 1984), esp. 33, 99, which derives from and extracts from Lee's earlier work, including *The !Kung San: Men, Women, and Work in a Foraging Society* (Cambridge: Cambridge University Press, 1979).

16. Richard Lee, "Politics, Sexual and Nonsexual in an Egalitarian Society: The !Kung San," in *Social Inequality: Comparative and Developmental Approaches*, ed. Gerald D. Berreman, with the assistance of Kathleen M. Zaretsky (New York: Academic Press, 1981), 83–102.

17. Marjorie Shostak, *Nisa: The Life and Words of a !Kung Woman* (Cambridge, Mass.: Harvard University Press, 1981).

18. Compare Richard Lee's descriptions of !Kung all-male big-game hunting in the desert with Colin M. Turnbull's description of BaMbuti coed rainforest hunting in *The Forest People* (New York: Simon and Schuster, 1961). I play here off the titles of a vastly influential book and an important feminist counterargument, Richard B. Lee and Irven DeVore, eds., *Man the Hunter* (Chicago: Aldine, 1968) and Sally Slocum, "Woman the Gatherer: Male Bias in Anthropology," in *Toward an Anthropology of Women*, 36–50.

19. Richard Lee, *The Dobe Ju/'hoansi* (1984; Fort Worth, Tex.: Harcourt Brace College Publishers, 1993), 83.

20. See Nancy Makepeace Tanner's brilliant empirical and theoretical survey *On Becoming Human* (Cambridge: Cambridge University Press, 1981), esp. 23–28, 139–47, and 191–223. See also Ruth Bleier, chapter 5, "Theories of Human Origins and Cultural Evolution: Man the Hunter," in *Science and Gender: A Critique of Biology and Its Theories on Women* (Elmsford, N.Y.: Pergamon, 1984), esp. 121–23. See Lila Leibowitz's critique of bioreductivism in evolutionary accountings of sex differences, "Origins of the Sexual Division of Labor," in *Woman's Nature: Rationalizations of Inequality,* ed. Marian Lowe and Ruth Hubbard (Elmsford, N.Y.: Pergamon, 1983), 123–47. Especially interesting is Leibowitz's conclusion that patterns of production appearing in the prehuman hominid line actually contributed to the *reduction* of sexual dimorphism in human beings. Of closely related interest, see Linda Marie Fedigan's discussion of hunting, gathering, foraging, and gender, "The Changing Role of Women in Models of Human Evolution," *Annual Review of Anthropology* 15 (1986): 25–66. Donna Haraway provides a critical overview of these and other approaches to subsistence and human evolution in *Primate Visions,* 186–230, 316–48.

21. Napoleon Chagnon, *Yanomamo, The Fierce People* (New York: Holt, Rinehart and Winston, 1968). Patrick Tierney, *Darkness in El Dorado: How Scientists and Journalists Devastated the Amazon* (New York: W. W. Norton, 2000). See also Frank A. Salamone, *The Yanomami and Their Interpreters: Fierce People or Fierce Interpreters?* (Lanham, Md.: University Press of America, 1997).

22. Edwin N. Wilmsen, *Land Filled with Flies: A Political Economy of the Kalahari* (Chicago: University of Chicago Press, 1989). Robert J. Gordon, *The Bushman Myth: The Making of a Namibian Underclass* (Boulder, Colo.: Westview, 1992). See Micaela di Leonardo's discussion in *Exotics at Home: Anthropologies, Others, American Modernity* (Chicago: University of Chicago Press, 1998), 286–91.

23. Gerald D. Berreman, "The Incredible 'Tasaday': Deconstructing the Myth of a 'Stone-Age' People," *Cultural Survival Quarterly* 15, no. 1 (1991): 3–44. See also Jean-Paul Dumont, "The Tasaday, Which and Whose? Toward the Political Economy of an Ethnographic Sign," *Cultural Anthropology* 3, no. 3 (1988): 261–75.

24. In spite of—or perhaps because of—the simplicity of their technology, most foragers live and eat well, especially by comparison to the poor majorities of most peasant societies. Foragers also meet their daily needs with far fewer work hours than your average suburban commuter, leaving more time for leisure, inactivity, or—on the downside of such efficiency—boredom. By most ethnographic accounts, modern foragers value both social equality and individual independence, which they attribute to geographical mobility and the flexibility of local groups, in contrast to the law-governed life of settled farmers around them. It might seem tempting to discern in modern foragers a counter-Hobbesian picture of man and woman in a state of nature.

25. See Eric Wolf's monumental *Europe and the People without History* (Berkeley: University of California Press, 1982), a text that shows how all modern cultures live the history of colonialism and global capitalism.

26. Richard Lee, "Art, Science, or Politics? The Crisis in Hunter-Gatherer Studies," *American Anthropologist* 94, no. 1 (1992): 40–43.

27. Haraway, *Primate Visions*, 3. Paul Rabinow, *Reflections on Fieldwork in Morocco* (Berkeley: University of California Press, 1977), 150. Bruno Latour, *Pandora's Hope: Essays on the Reality of Science Studies* (Cambridge, Mass.: Harvard University Press, 1999), 127.

28. Haraway, *Primate Visions*, 9–10.

CHAPTER 5. THE SCIENCE QUESTION

1. Alan Sokal, "A Physicist Experiments with Cultural Studies," *Lingua Franca*, May–June 1996, 62.

2. On January 2, 1996, Natalie Angier inaugurated the silly season in genomania, reporting the results of two separate studies in the *New York Times* under the front-page headline: "Variant Gene Tied to a Love of New Thrills." Almost immediately, Gina Kolata queried: "Is a Gene Making You Read This?" *New York Times*, January 7, 1996. Soon, drug abuse, criminality, risky sex, President Clinton's shenanigans—you name it—were being attributed to the "thrill-seeking gene" in newspaper articles and journal stories. *Time* ran a cover story on the thrill-seeking gene ("Why We Take Risks"), announcing an evolutionary-psychological perspective on the subject ("For Our Ancestors, Taking Risks Was a Good Bet," September 6, 1999)—three years *after* it had become evident that the two studies on which the whole concept was based could not be replicated. See Natalie Angier, "Maybe It's Not a Gene behind a Person's Thrill-Seeking Ways," *New York Times*, November 1, 1996.

3. Judith Butler, *Bodies That Matter: On the Discursive Limits of "Sex"* (New York: Routledge, 1993), 1.

4. Better put, from the same text: "Culture is not merely nature expressed in another form. Rather the reverse: the action of nature unfolds in the terms of culture; that is, in a form no longer its own but embodied as meaning." Marshall Sahlins, *Culture and Practical Reason* (Chicago: University of Chicago Press, 1976), 209.

5. See Bruno Latour's chapter, "The Historicity of Things," in *Pandora's Hope: Essays on the Reality of Science Studies* (Cambridge, Mass.: Harvard University Press, 1999), 145–73.

6. Ian Hacking, *The Taming of Chance* (Cambridge: Cambridge University Press, 1990), 162–63.

7. I am much indebted to Gina Maranto, whose coverage of a Cornell University workshop ("Making People: The Normal and Abnormal in Constructions of Personhood," *New York Times* [Health Tuesday], May 26, 1998) stimulated a series of conversations and further readings, including Maranto's own superb *Quest for Perfection: The Drive to Breed Better Human Beings* (New York: Scribner, 1996) and Georges Canguilhem's *Ideology and Rationality in the History of the Life Sciences* (Cambridge, Mass.: MIT Press, 1988), especially its brilliant chapter "The Question of Normality in the History of Biological Thought."

8. Hacking, *The Taming of Chance*, 163–64.

9. Ibid., 81–86 and 165–69.

10. Georges Canguilhem, *The Normal and the Pathological* (1966; New York: Zone Books, 1989), 47–64.

11. Canguilhem, *The Normal and the Pathological*, 154–60; Hacking, *The Taming of Chance*, 105–14.

12. Francis Galton, *Hereditary Genius: An Inquiry into Its Laws and Consequences* (1892; New York: Horizon Press, 1952); Hacking, *The Taming of Chance*, 180–88.

13. Roger Smith, *The Norton History of the Human Sciences* (New York: W. W. Norton, 1997), 580–83.

14. See Earnest Albert Hooton's reassuring and avuncular account of the Grant study conducted by the Department of Hygiene at Harvard: *Young Man, You Are Normal: Findings from a Study of Students* (New York: G. Putnam's Sons, 1945). The book's title alone immediately locates it within the dilemmas posed by anthropometrics: a stingy tendency to define almost everything as deviant, counterweighted by a willingness to include virtually everyone within a generous norm.

15. I have tried to capture in this sentence something of the gist of Michel Foucault's approach to institutional and scientific histories, from *Madness and Civilization: A History of Insanity in the Age of Reason*, trans. Richard Howard (1961; New York: Pantheon, 1965), which treats how the elaboration of medical ideas changed both the social position of the mad and the shape of Western cultures, to *The Birth of the Clinic: An Archaeology of Medical Perception*, trans. A. M. Sheridan Smith (1963; New York: Pantheon, 1973), which traces the elaboration of a distinctly medical body through institutional practices, to *The History of Sexuality, Volume I: An Introduction*, trans. Robert Hurley (1976; New York: Vintage, 1980), which argues that modern sexual identities emerged in and through the discursive institutional practices of psychiatry, sexology, and medicine. On "normalization," see also Michael Warner's famous "Introduction," in *Fear of a Queer Planet: Queer Politics and Social Theory* (Minneapolis: University of Minnesota Press, 1993), vii–xxxi, and his subsequent book, *The Trouble with Normal: Sex, Politics, and the Ethics of Queer Life* (New York: The Free Press, 1999).

16. Readers will note my debt to Pierre Bourdieu, whose theory of practice brings together concepts usually blocked off under rubrics like "ideology" and "hegemony." *The Logic of Practice* (1980; Stanford, Calif.: Stanford University Press, 1990). They will undoubtedly also see the influence of Judith Butler, whose application of semiotic and performance theory to models of gender/sexuality and embodiment were, throughout the 1990s, more precise and more disciplined than similar attempts by many others (including myself. See my arguments, dating from earlier published papers, in *Life Is Hard: Machismo, Danger, and the Intimacy of Power in Nicaragua* [Berkeley: University of California Press, 1992], xviii–xix, 19–20, 222–30, 235–37, 265–71, and 274–78). However, the careful reader might also detect my discomfort with certain poststructuralist for-

mulations, due, in part, to what I see as their lack of historicity. When Judith Butler claims that "the materiality of sex is constructed through a ritualized repetition of norms," it seems to me that she invokes as part of the nature of discourse the very thing that ought to be historically problematized: "normativity." No less problematic is the formulaic proposition that "the fixity of the body, its contours, its movements" are best understood as "the effect of power, as power's most productive effect." Naming every corporeal contour and carnal practice a "power" relation does nothing to clarify either power relations or practices of embodiment. The generic claim that "ideal norms" are somehow "sedimented" as corporeal "materiality" through discursive "reiteration" is, at best, a misleading shorthand for how people construct their sense of the world and body: It perpetually sets precept in front of percept and schema before practice. Butler rightly attempts a non-Cartesian, materialist approach to embodiment, but her perpetual turns on the term "posit" might well have sent critical readers back to the first several chapters of Merleau-Ponty's *Phenomenology of Perception,* trans. Colin Smith (London: Routledge and Kegan Paul, 1962), wherein the author dissects what he calls "intellectualist" fallacies about the body and physical experiences. See Judith Butler, *Bodies That Matter,* x, xi, 2, 10, 13, 15, 250 n. 5, 251 n. 12. It is not in saying or representing alone, but in *doing* that we might speak of the world or the body as "made." Abstracted from both history *and* flesh, poststructuralist propositions of this sort begin with gestures toward a theory of practice, but end as airy idealism masquerading as a variant of materialism.

17. Feminist science studies has been particularly sensitive to the interplay of culture, presupposition, and objectification in scientific inquiry. See Carolyn Merchant, *The Death of Nature: Women, Ecology, and the Scientific Revolution* (1980; New York: Harper and Row, 1989); Marian Lowe and Ruth Hubbard, eds., *Woman's Nature: Rationalizations of Inequality* (New York: Pergamon, 1983); Cynthia Eagle Russett, *Sexual Science: The Victorian Construction of Womanhood* (Cambridge, Mass.: Harvard University Press, 1989); Ruth Hubbard, *The Politics of Women's Biology* (New Brunswick, N.J.: Rutgers University Press, 1990); Mary Jacobus, Evelyn Fox Keller, and Sally Shuttleworth, eds., *Body/Politics: Women and the Discourses of Science* (New York: Routledge, 1990); Sandra Harding, "Why 'Physics' Is a Bad Model for Physics," in *Whose Science? Whose Knowledge: Thinking from Women's Lives* (Ithaca, N.Y.: Cornell University Press, 1991); Lynda Birke and Ruth Hubbard, eds., *Reinventing Biology: Respect for Life and the Creation of Knowledge* (Bloomington: Indiana University Press, 1995); and Nancy Hartsock, *The Feminist Standpoint and Other Essays* (Boulder, Colo.: Westview, 1998).

18. Donna J. Haraway, *Primate Visions: Gender, Race, and Nature in the World of Modern Science* (New York: Routledge, 1989), 12.

19. Stephen Jay Gould, *The Mismeasure of Man,* rev. ed. (1981; New York: W. W. Norton, 1996), 54.

20. Margaret Wertheim, "The Odd Couple," *The Sciences* (March–April 1999): 42.

21. Donna J. Haraway, "Situated Knowledges: The Science Question in Feminism and the Privilege of Partial Perspective," in *Simians, Cyborgs, and Women: The Reinvention of Nature* (New York: Routledge, 1991), 183–201.

22. See Anne Fausto-Sterling, *Sexing the Body: Gender Politics and the Construction of Sexuality*, chapter 5: "Sexing the Brain: How Biologists Make a Difference" (New York: Basic Books, 2000), 115–45.

23. Quoted by John Noble Wilford, "New Answers to an Old Question: Who Got Here First?" *New York Times* (Science Times), November 9, 1999.

24. Stephen Jay Gould, "Cardboard Darwinism," *New York Review of Books*, September 25, 1986, 50–51. See also Susan Sperling, "Baboons with Briefcases and Langurs with Lipstick: Feminism and Functionalism in Primate Studies," in *The Gender/Sexuality Reader: Culture, History, Political Economy*, ed. Roger N. Lancaster and Micaela di Leonardo (New York: Routledge, 1997), 250.

25. In *The Origin of Species*, Darwin himself states that natural selection is not the sole mechanism of evolution, a point often lost on subsequent Darwinists. Not every trait is "adaptive," nor is every trait "selected." "I am convinced that natural selection has been the main but not the exclusive means of modification." Charles Darwin, *The Origin of Species by Means of Natural Selection, or, The Preservation of Favored Races in the Struggle for Life, and The Descent of Man and Selection in Relation to Sex* (1859, 1871; New York: The Modern Library, 1977), 14, 367.

26. On the production of scientific knowledge in and through discursive networks, see Bruno Latour and Stephen Woolgar, *Laboratory Life: The Social Construction of Scientific Facts* (London: Sage, 1979), esp. 154–67. See also Bruno Latour's *We Have Never Been Modern*, trans. Catherine Porter (1991; Cambridge, Mass.: Harvard University Press, 1993) and "Science's Blood Flow," in *Pandora's Hope: Essays on the Reality of Science Studies* (Cambridge, Mass.: Harvard University Press, 1999), 80–112.

27. Londa Schiebinger, *Nature's Body: Gender in the Making of Modern Science* (Boston: Beacon, 1993), 41, 40–74.

28. See Ashley Montagu's old classic, *Man's Most Dangerous Myth: The Fallacy of Race*, 5th ed., rev. (1942; New York: Oxford University Press, 1974). Stephen Jay Gould, "Why We Should Not Name Human Races: A Biological View," in *Ever since Darwin: Reflections in Natural History* (New York: W. W. Norton, 1977), 231–36. See also various issues of the American Anthropological Association's *Anthropology Newsletter*, whose annual theme for 1997–98 was the "race" concept.

29. On physiognomic fables at the crossroads of sex and race, see Stephen Jay Gould, "The Hottentot Venus," *Natural History* 91 (1982): 20–27; Sander L. Gilman, "Black Bodies, White Bodies: Toward an Iconography of Female Sexuality in Late Nineteenth-Century Art, Medicine, and Literature," in *"Race," Writing, and Difference*, ed. Henry Louis Gates Jr. (Chicago: University of Chicago Press, 1985), 223–61; Anne Fausto-Sterling, "Gender, Race, and Nation: The Comparative Anatomy of 'Hottentot' Women in Western Europe:

1815–1817," in *Deviant Bodies: Critical Perspectives on Difference in Science and Popular Culture*, ed. Jennifer Terry and Jacqueline Urla (Bloomington: Indiana University Press, 1995), 19–48.

30. The first quotation is from Friedrich Engels, *Socialism: Utopian and Scientific*, in *The Marx-Engels Reader*, ed. Robert C. Tucker (New York: W. W. Norton, 1978), 697. The longer passage is Karl Marx, in a letter to Engels, quoted in Alfred Schmidt's *The Concept of Nature in Marx* (1962; London: NLB, 1971), 46.

31. A dialectic first sketched by Georg Lukács in *History and Class Consciousness: Studies in Marxist Dialectics*, trans. Rodney Livingstone (1922; Cambridge, Mass.: MIT Press, 1971) and given eloquent explanation by Marshall Sahlins in *Culture and Practical Reason* (Chicago: University of Chicago Press, 1976).

32. See Carol MacCormack, "Nature, Culture and Gender: A Critique," in *Nature, Culture and Gender*, ed. Carol MacCormack and Marilyn Strathern (Cambridge: Cambridge University Press, 1980), 6.

CHAPTER 6. SEXUAL SELECTION

1. Charles Darwin, *The Descent of Man and Selection in Relation to Sex*, approvingly quoted in Robert Wright, *The Moral Animal: The New Science of Evolutionary Psychology* (New York: Pantheon, 1994), 33. See *The Origin of Species by Means of Natural Selection, or, The Preservation of Favored Races in the Struggle for Life, and The Descent of Man and Selection in Relation to Sex* (1859, 1871; New York: The Modern Library, 1977), 915, 579. Subsequent citations from this paired edition are identified either as *The Origins of Species by Means of Natural Selection* or *The Descent of Man and Selection in Relation to Sex*.

2. Darwin, *The Descent of Man and Selection in Relation to Sex*, 873.

3. Ibid.

4. Such conceits still resonate with much of what goes without saying about heterosexual courtship in the Anglophone world, although today, in a world reshaped by women's liberation and male sensitivity, they might seem increasingly antiquated and quaint, like period-piece costumes donned for special occasions.

5. "The ideal of female 'passionlessness' allowed women to claim the moral high ground. At the same time, the new feminine ideal also firmly ensconced women on a pedestal. From this exalted position, her unique role in the new republic would be to tame men's passions and maintain the purity of the home, not to participate in the public world of work and politics." Carol Groneman, *Nymphomania: A History* (New York: W. W. Norton, 2000), xix.

6. Elizabeth Abbott, *A History of Celibacy* (New York: Scribner, 2000), 34. Groneman's and Abbott's books were intelligently previewed by Dinitia Smith on the Arts and Ideas page of the *New York Times*, May 13, 2000, but without any overt reference to the fables of evolutionary psychology so often developed in the Science Times section.

7. Londa Schiebinger, *Nature's Body: Gender in the Making of Modern Science* (Boston: Beacon, 1993), 1. See Aristotle, *Historia Animalium*, 3 vols., with an English translation by A. L. Peck (Cambridge, Mass.: Harvard University Press, 1970), vol. 2, books 4–6, on the sexual assertiveness of pidgeon hens (233), and mares and cows (299–301, 319–20). Aelian gives countless instances of amorous and aggressive female animals.

8. In *Historia Animalium*, Aristotle flatly asserts that "female animals are most fierce just after parturition" and matter-of-factly states: "It is no easy business to catch a bear when pregnant" (297, 339). Aelian's tales about the maternal fierceness of dolphins, she-bears, mares, female elephants, and other female creatures are more touching. See *On the Characteristics of Animals*, with an English translation by A. F. Scholfield (Cambridge, Mass.: Harvard University Press, 1958), vol. 1, books 1–5 (35), and vol. 2, books 6–11 (21, 67, 123, 229). Pliny's *Natural History*, vol. 3, books 8–11 (Cambridge, Mass.: Harvard University Press), relates a beautiful tale of the female tiger's ferocious heroism in protecting her litter against trappers and hunters (51).

9. Patrick Bateson and Paul Martin give a surprisingly complacent invocation of Paley in *Design for a Life: How Behavior and Personality Develop* (New York: Simon and Schuster, 2000), 13–14.

10. See, for instance, Sarah Blaffer Hrdy's discussion of the difference between Darwinian and Spencerian conceptions of "fitness" in *Mother Nature: A History of Mothers, Infants, and Natural Selection* (New York: Pantheon Books, 1999), 13–14.

11. Stephen Jay Gould, "Darwinian Fundamentalism," *The New York Times Review of Books*, June 12, 1997, 34. Gould's *The Structure of Evolutionary Theory* (Cambridge, Mass.: Harvard University Press, 2002) sums up the author's attempts at thinking and applying evolutionary theory outside the constraints of the received ultra-Darwinism, according to which every trait represents a positive adaptation to the pressures of natural and sexual selection.

12. Darwin, *The Origin of Species by Means of Natural Selection*, 68–69.

13. Geoffrey Miller, interviewed by Natalie Angier, "A Conversation with Geoffrey Miller: Author Offers Theory on Gray Matter of Love," *New York Times* (Science Times), May 30, 2000. See also Miller's book of evolutionary-psychological whimsy, *The Mating Mind: How Sexual Choice Shaped the Evolution of Human Nature* (New York: Doubleday, 2000).

14. See Darwin's summary, *The Descent of Man and Selection in Relation to Sex*, 916.

15. See Anne Fausto-Sterling, *Myths of Gender: Biological Theories about Women and Men*, rev. ed. (1985; New York: Basic Books, 1992), 179.

16. Darwin, *The Descent of Man and Selection in Relation to Sex*, 623, 616, and 915–16.

17. See Bruce Bagemihl's corrective text, *Biological Exuberance: Animal Homosexuality and Natural Diversity* (New York: St. Martin's Press, 1999).

18. What might count as sex and how animal behaviors and human acts

might be meaningfully compared are complex questions incommensurate with easy answers, as Edward Stein shows in *The Mismeasure of Desire: The Science, Theory, and Ethics of Sexual Orientation* (Oxford: Oxford University Press, 1999), 164–98 and 190–228. In a different vein, Yudhijit Bhattacharjee describes new zoological research into the long-overlooked realm of female markings, ornamentation, and flashy behavior. He summarizes: The standard "account of mate choice, in which males do all the dancing-to-impress and females sit on the judging panel, is being increasingly viewed by scientists as too simplistic." "In the Animal Kingdom, a New Look at Female Beauty," *New York Times* (Science Times), June 25, 2002.

19. Darwin, *The Descent of Man and Selection in Relation to Sex*, 901.

20. Although evolutionary biologists were from the very beginning keen on the idea of "male competition," they proved reluctant to embrace the notion of "female choice." See Blaffer Hrdy, *Mother Nature*, 36–37, and Fausto-Sterling, *Myths of Gender*, 180–81.

21. Rosemary Jann, "Darwin and the Anthropologists: Sexual Selection and Its Discontents," *Victorian Studies* (winter 1994): 287–306.

22. Darwin, *The Descent of Man and Selection in Relation to Sex*, 919–20.

23. Ibid., 919.

24. See Gayatri Spivak, "Can the Subaltern Speak?" in *Marxism and the Interpretation of Culture*, ed. Cary Nelson and Lawrence Grossberg (London: Macmillan, 1988), 271–313, and Elizabeth Povinelli, "Sex Acts and Sovereignty," in *The Gender/Sexuality Reader: Culture, History, Political Economy*, ed. Roger N. Lancaster and Micaela di Leonardo (New York: Routledge, 1997), 513–28.

25. Fausto-Sterling, *Myths of Gender*, 179. Janet Sayers, *Biological Politics: Feminist and Anti-Feminist Perspectives* (London: Tavistock, 1982). Vernon Rosario, "Homosexual Bio-Histories: Genetic Nostalgias and the Quest for Paternity," 12, Harry Oosterhuis, "Richard von Krafft-Ebing's 'Step-Children of Nature': Psychiatry and the Making of Modern Homosexual Identity," 71, and James D. Steakley, "Per scientiam ad justitiam: Magnus Hirschfeld and the Sexual Politics of Innate Homosexuality," 137, 143, 146, all in *Science and Homosexualities*, ed. Vernon A. Rosario (New York: Routledge, 1997). Ann Laura Stoler, "Carnal Knowledge and Imperial Power: Gender, Race, and Morality in Colonial Asia," 22–23, and Siobhan Somerville, "Scientific Racism and the Invention of the Homosexual Body," in *The Gender/Sexuality Reader*, 42. Jennifer Terry, *An American Obsession: Science, Medicine, and Homosexuality in Modern Society* (Chicago: University of Chicago Press, 1999), 36–37, 47, 54–55, and 100–103.

26. Then, as today, in science as with a new generation of "wild" men, dreams of primitive authenticity serve as a screen for the projection of hopes and fears about ongoing institutional changes. See Gail Bederman, *Manliness and Civilization: A Cultural History of Gender and Race in the United States, 1880–1917* (Chicago: University of Chicago Press, 1995). See also the chapters in the following section, "Venus and Mars at the Fin de Siècle."

CHAPTER 7. THE SELFISH GENE

1. Marie-Jean-Antoine Nicolas de Caritat Condorcet, "On Giving Women the Right of Citizenship," in *Condorcet: Foundations of Social Choice and Political Theory*, ed. and trans. Iain McLeane and Fiona Hewlette (Hants, England: Edward Elgar Publishing, 1994), 338–39. See Londa Schiebinger, *Nature's Body: Gender in the Making of Modern Science* (Boston: Beacon, 1993), 143–44.

2. Jean-Jacques Rousseau, *The Social Contract and the Discourses*, trans. G. D. H. Cole, rev. J. H. Brumfitt and John C. Hall (1754, 1913; New York: Alfred A. Knopf, 1993), "A Discourse on the Origin of Inequality," 49–50.

3. Lionel Tiger and Robin Fox, *The Imperial Animal* (1971; New Brunswick, N.J.: Transaction, 1998), 18. Emphasis in the original.

4. All too handily, the high-tech analogy situates sociobiology in its proper social milieu. But the notion of a cybernetic design is no loose metaphor. It is central not just to how sociobiologists express their ideas, but also to what they mean to say. This computer-code analogy takes care of several of sociobiology's main arguments: first, the idea that genes "act" as a "program" that "controls" the development of the organism; second, the idea of a "coherent design" through which every aspect of the organism is biologically adaptive; and third, the idea that higher human practices usually associated with learned culture (family values, a tendency to hate outsiders) might be transmitted mechanically from generation to generation as part of a basic "program" of human nature.

5. Edward O. Wilson, *Sociobiology: The New Synthesis* (Cambridge, Mass.: Harvard University Press, 1975), 3.

6. Richard Dawkins's vastly influential book *The Selfish Gene* (1976; Oxford: Oxford University Press, 1989) spins its tale of origins and design in a quirky, insider's language redolent of that sci-fi subgenre, the space opera. His chapter "The Replicators" begins: "In the beginning was simplicity." It ends:

> do not look for them floating loose in the sea; they gave up that
> cavalier freedom long ago. Now they swarm in huge colonies, safe
> inside gigantic lumbering robots, sealed off from the outside world,
> communicating with it by tortuous indirect routes, manipulating
> it by remote control. They are in you and in me; they created us,
> body and mind; and their preservation is our ultimate rationale for
> existence. They have come a long way, those replicators. Now they
> go by the name of genes, and we are their survival machines. (12,
> 19–20)

7. Dean Hamer and Peter Copeland, *The Science of Desire: The Search for the Gay Gene and the Biology of Behavior* (New York: Simon and Schuster, 1994), 180.

8. I have registered *"[sic]"* only once in the above passage, which is altogether typical of its genre, but I might just as easily have sprinkled the passage

with dozens of *"[sic]"*s. The same logical error occurs throughout: the attribution of human agencies and characteristics to genes, nature, and evolution.

9. Richard Hofstadter, *Social Darwinism in American Thought* (1944; Boston: Beacon, 1955), 8.

10. Herbert Spencer, quoted by Hofstadter, in *Social Darwinism in American Thought*, 37. Herbert Spencer, *First Principles* (1862; New York: De Witt Revolving Fund, 1958), 511.

11. Wilson, *Sociobiology*, 242. Robert Wright, *The Moral Animal* (New York: Pantheon, 1994), 391. See also Erica Goode's preview of Randy Thornhill and Craig T. Palmer's *A Natural History of Rape: Biological Bases of Sexual Coercion* (Cambridge, Mass.: MIT Press, 2000), "What Provokes a Rapist to Rape? Scientists Debate Notion of an Evolutionary Drive," *New York Times* (Arts and Ideas), January 15, 2000.

12. Wilson, *Sociobiology*, 572, and *On Human Nature* (Cambridge, Mass.: Harvard University Press, 1978), 99–120. Tiger and Fox, *The Imperial Animal*, 204–31. Robin Fox, "Aggression: Then and Now," in *Man and Beast Revisited*, ed. Michael H. Robinson and Lionel Tiger (Washington, D.C.: Smithsonian Institution Press, 1991).

13. Wilson, *Sociobiology*, 564–65.

14. Hofstadter, *Social Darwinism in American Thought*, 35. Greg Grandlin gives a good snapshot of the incubation of sociobiological concepts among conservatives on university campuses in the 1960s in his review of Patrick Tierney's *Darkness in El Dorado: How Scientists and Journalists Devastated the Amazon* (New York: Norton, 2000), *The Nation*, December 11, 2000, 12–17.

15. On the relevance of eugenic theory to biological models of homosexuality, see Jennifer Terry, *An American Obsession: Science, Medicine, and Homosexuality in Modern Society* (Chicago: University of Chicago Press, 1999), esp. 100–103.

16. I will elaborate on this point in "Varieties of Human Nature."

17. Stephen Jay Gould, "Biological Potentiality versus Biological Determinism," in *Ever since Darwin: Reflections in Natural History* (New York: W. W. Norton, 1977), 251–59.

18. The term "biogram" (roughly, "biological program") was freely invoked by 1970s sociobiologists—as in "the search for the human biogram"—to lend an aura of scientific inquiry to essentialist notions of human nature. See Tiger and Fox, *The Imperial Animal*, 1–23.

19. Ruth Hubbard and Elija Wald, *Exploding the Gene Myth: How Genetic Information Is Produced and Manipulated by Scientists, Physicians, Employers, Insurance Companies, Educators, and Law Enforcers* (Boston: Beacon, 1993), 11, 44, 50, 53, 55, 64; see also 2 and 36. See also Abby Lippman on "geneticization": "Prenatal Genetic Testing and Screening: Constructing and Reinforcing Inequities," *American Journal of Law and Medicine* 17 (1991): 15–50, esp. 19.

20. Richard Lewontin, *The Triple Helix: Gene, Organism, and Environment* (Cambridge, Mass.: Harvard University Press, 2000), 17.

21. Richard Lewontin, *Biology as Ideology: The Doctrine of DNA* (New York: HarperPerennial, 1991), 27, 26.

22. See also Richard Levins and Richard Lewontin, *The Dialectical Biologist* (Cambridge, Mass.: Harvard University Press, 1985).

23. Nicholas Wade, "Genetic Code of Human Life Is Cracked by Scientists," and Natalie Angier, "A Pearl and a Hodgepodge: Human DNA," *New York Times*, June 27, 2000, front page, above the fold. Wade's somewhat more nuanced article ("Now, the Hard Part: Putting the Genome to Work") appeared farther back in the Science Times supplement, along with Erica Goode's cautionary review of the abysmal record of genetic science in understanding chronic illnesses ("Most Ills Are a Matter of More Than One Gene"). The *Washington Post* gave a more modest headline less prominent position, at the top of the front page, but spanning only three of five columns: "Teams Finish Mapping Human DNA." The subhead noted the tentative and limited nature of the day's announcement while still paying homage to the determinist logic of genomania: "Clinton, Scientists Celebrate 'Working Draft' of Human Genetic Blueprint."

24. Nicholas Wade, "The Four-Letter Alphabet That Spells Life: The Genome Project Brings Humans Back to the Garden of Eden, or at Least to the Tree of Knowledge," *New York Times* (Week in Review,) July 2, 2000.

25. Margaret Lock, "Genetic Bodies and the Decline of Culture: From Inborn Errors to Gene Regulation" (a paper presented at "Anthropology United: Challenging Bio-Social Reductivism in the Academy, Popular Media, and Public Policy," an invited session of the American Anthropological Association Executive Program Committee, at the ninety-ninth annual meeting of the American Anthropological Association, San Francisco, November 15–19, 2000).

CHAPTER 8. GENOMANIA AND HETEROSEXUAL FETISHISM

1. The critique of sociobiology is now as old as sociobiology itself—older, perhaps, because sociobiology reiterates much of Herbert Spencer's mechanistic, Social Darwinian worldview, so that much of its critique retraces the earlier arguments of Boasian anthropology, Marxian cultural criticism, and sophisticated biology. In addition to Ruth Hubbard and Elija Wald, *Exploding the Gene Myth: How Genetic Information Is Produced and Manipulated by Scientists, Physicians, Employers, Insurance Companies, Educators, and Law Enforcers* (Boston: Beacon, 1993), and Stephen Jay Gould, *Ever since Darwin: Reflections on Natural History* (New York: W. W. Norton, 1977), see R. C. Lewontin, Steven Rose, and Leon J. Kamin, *Not in Our Genes: Biology, Ideology, and Human Nature* (New York Pantheon, 1984), Ruth Bleier, *Science and Gender: A Critique of Biology and Its Theories on Women* (New York: Pergamon, 1984), and of course, Marshall Sahlins, *Use and Abuse of Biology: An Anthropological Critique of Sociobiology* (Ann Arbor: University of Michigan Press, 1977).

2. Karl Marx, "*Capital*, Volume One," in *The Marx-Engels Reader*, 2d. ed., ed. Robert Tucker (New York: W. W. Norton, 1978), 320–21, 328.

3. Georg Lukács develops Marx's analysis of fetishism and reification in *History and Class Consciousness: Studies in Marxist Dialectics* (1922; Cambridge, Mass.: MIT Press, 1971), a text that greatly influenced Horkheimer, Adorno, and other members of the Frankfurt School. For more recent Freudian and Marxist approaches to fetishism, see the essays in Emily Apter and William Pietz, eds., *Fetishism as Cultural Discourse* (Ithaca, N.Y.: Cornell University Press, 1993), esp. William Pietz's "Fetishism and Materialism: The Limits of Theory in Marx," 119–51.

4. Lewontin, Rose, and Kamin, *Not in Our Genes*, 235.

5. Michael Rothschild, *Bionomics: The Inevitability of Capitalism* (New York: Armonk, 1992). See Slavoj Žižek's remarks in "Multiculturalism, or, The Cultural Logic of Multinational Capitalism," *New Left Review* 225 (September–October 1997): 28–51, 36. See also Paulina Borsook, *Cyberselfish: A Critical Romp through the Terribly Libertarian Culture of High Tech* (New York: Public Affairs, 2000), with its incisive insider's critique of "bionomics."

6. Bleier, *Science and Gender*, 1–5, 13, 18. For other feminist critiques of sociobiology's "selfish gene" from the same period, see also Ruth Hubbard and Marian Lowe, eds., *Genes and Gender* (Staten Island, N.Y.: Gordian Press, 1979) and Ruth Hubbard, Mary Sue Henifer, and Barbara Fried, eds., *Women Look at Biology Looking at Women: A Collection of Feminist Critiques* (Cambridge, Mass.: Schenkman, 1979).

7. I believe the first time I ever heard the phrase "heterosexual fetishism" used, it was being invoked by Michael Higgins, sometime around 1989. See Michael James Higgins and Tanya L. Coen, *Streets, Bedrooms, and Patios: The Ordinariness of Diversity in Urban Oaxaca. Ethnographic Portraits of the Urban Poor, Transvestites, Discapacitados, and Other Popular Cultures* (Austin: University of Texas Press, 2000), 292–93, 293 n. 1.

8. Gayle Rubin, "The Traffic in Women: Notes on the Political Economy of Sex," in *Toward an Anthropology of Women*, ed. Rayna Reiter (New York: Monthly, 1975), 179–80. Rubin does not actually use the term "heteronormativity" in her classic essay. But she might have. At any rate, it is her depiction of the coworkings of gender and sexuality that I appropriate as a proper understanding of heteronormativity—a social system understood not as a universal form of kinship (this universalist casting was one of the main flaws in Rubin's early draft of the thought), but as a very specific model of North American ideas and institutions, as these were practiced for much of the twentieth century.

9. Dennis Altman, *Homosexual: Oppression and Liberation* (1971; New York: New York University Press, 1993), 78–79, 90, 92. See also Steven Seidman, "Identity and Politics in a 'Postmodern' Gay Culture: Some Historical and Conceptual Notes," in *Fear of a Queer Planet: Queer Politics and Social Theory*, ed. Michael Warner (Minneapolis: University of Minnesota Press, 1993), 114.

10. Judith Butler, *Gender Trouble: Feminism and the Subversion of Identity* (New York: Routledge, 1990), 151 n. 6.

11. Monique Wittig, *The Straight Mind* (Boston: Beacon, 1992), 43. See also Michael Warner, introduction to *Fear of a Queer Planet*, xxi.

12. Warner, *Fear of a Queer Planet*, xxiii.

13. I draw freely here not only on Rubin's work, but also on Eve Kosofsky Sedgwick's, whose "Introduction: Axiomatic," in *Epistemology of the Closet*, remains axiomatic (Berkeley: University of California Press, 1990), 1–63.

14. In *Homosexual: Oppression and Liberation*, Dennis Altman puts it this way: "The Greeks extolled both bisexuality and the supremacy of men" (91). This point has been taken up and elaborated by a substantial current of subsequent gay historiography. See David M. Halperin, *One Hundred Years of Homosexuality and Other Essays on Greek Love* (New York: Routledge, 1990).

15. On Chaddock's coining of the term "homosexual" and on the distinctly modern phenomenon of heteronormativity, see Halperin, *One Hundred Years of Homosexuality*, 15–40 and 41–53, esp. 15.

16. Jeffrey Weeks, *Coming Out: Homosexual Politics in Britain from the Nineteenth Century to the Present* (London: Quartet Books, 1979), and *Sex, Politics, and Society: The Regulation of Sexuality since 1800*, 2d. ed. (1981; London: Longman Group, 1989). Michel Foucault, *The History of Sexuality, Volume I: An Introduction*, trans. Robert Hurley (1976; New York: Vintage, 1990).

17. Some readers will point out that there can never really be a "first" nature. According to the arguments I myself have forwarded, *every* knowledge of nature draws on preexisting ideas, categories, and distinctions, such that even the "first" nature is *also* culturally elaborated, socially constructed—in a word: "produced." I hope the reader has been appropriately forewarned about the lure of origins stories, which give the effect of transporting their audience to the remote past, but which actually transport readers to a facsimile of the present. But I also think that notwithstanding its problematic structure, the opposition between a "first" and a "second" nature marks a useful contrast here: It is, of course, the difference between what one might claim based on rigorous, empirically tested scientific study and what one might claim based on received wisdom or prejudicial common sense.

18. Lionel Tiger and Robin Fox, *The Imperial Animal* (1971; New Brunswick, N.J.: Transaction, 1998), 18. Emphasis in the original.

19. Quoted in Anne Fausto-Sterling, *Myths of Gender: Biological Theories about Women and Men*, rev. ed. (1985; New York: Basic Books, 1992), 156.

20. Robert Trivers, "Parental Investment and Sexual Selection," in *Sexual Selection and the Descent of Man, 1871–1971*, ed. Bernard Grant Campbell (Chicago: Aldine Publishing, 1972), 136–79.

21. Edward O. Wilson, *On Human Nature* (Cambridge, Mass.: Harvard University Press, 1978), 125.

22. Fausto-Sterling, *Myths of Gender*, 183. Ruth Hubbard presents similar arguments in "Have Only Men Evolved?" in *Discovering Reality: Feminist Perspectives on Epistemology, Metaphysics, Methodology, and Philosophy of Science*, ed. Sandra Harding and Merrill B. Hintikka (Dordrecht: Reidel, 1983), esp.

60–61. See also Donna J. Haraway, "Investment Strategies for the Evolving Portfolio of Primate Females," in *Body/Politics,* ed. Mary Jacobus, Evelyn Fox Keller, and Sally Shuttleworth (New York: Routledge, 1990), 155–56.

23. Emily Martin, "The Egg and the Sperm: How Science Has Constructed a Romance Based on Stereotypical Male-Female Roles," in *Gender and Scientific Authority,* ed. Barbara Laslett, Sally Gregory Kohlstedt, Helen Longino, and Evelynn Hammonds (Chicago: University of Chicago Press, 1996), 323–39.

24. Edward O. Wilson, *Sociobiology: The New Synthesis* (Cambridge, Mass.: Harvard University Press, 1975), 320.

25. Ibid., 553. See David M. Schneider, "The Nature of Kinship," *Man* 64 (November–December 1964): 180–81, and *American Kinship: A Cultural Account* (Englewood Cliffs, N.J.: Prentice-Hall, 1968), as well as his *A Critique of the Study of Kinship* (Ann Arbor: University of Michigan Press, 1984).

26. Dawkins, *The Selfish Gene,* 146. See also Ruth Bleier's discussion in *Science and Gender,* 20.

27. Randy Thornhill, "Rape in Panorpa Scorpionflies and a General Rape Hypothesis," *Animal Behavior* 28 (1980): 57. David Barash, *The Whisperings Within: Evolution and the Origin of Human Nature* (New York: Harper and Row, 1979), 54–55, 30–31. Anne Fausto-Sterling, in *Myths of Gender,* 156–67, 190–95, and Ruth Bleier in *Science and Gender,* 31–33, provide excellent surveys of sociobiological pronouncements on rape, and I draw heavily on their work. Yet another quotable quote from Fausto-Sterling's cabinet of horrors: "We suggest that *all* men are potential rapists. . . . We expect the probability of a particular individual raping will be a function of the average genetic cost/benefit ratio associated with the particular condition he faces. . . . Ultimately, men rape because it increases their biological fitness." W. M. Shields and L. M. Shields, "Forcible Rape: An Evolutionary Perspective," *Ethology and Sociobiology* 4 (1983): 119, 120, 122.

28. Robert L. Trivers, "Parental Investment and Sexual Selection," in *Sexual Selection and the Descent of Man,* 149–50. Pierre L. Van Den Berghe and David P. Barash, "Inclusive Fitness and Human Family Structure," *American Anthropologist* 79 (1977): 809–23. See also Wilson, *Sociobiology,* 324–27.

29. Robert Wright, *The Moral Animal: The New Science of Evolutionary Psychology* (New York: Pantheon, 1994), 54.

30. "Survival of the Rapist," review of Randy Thornhill and Craig T. Palmer's *A Natural History of Rape* (Cambridge, Mass.: MIT Press, 2000), *New York Times Book Review,* April 2, 2000.

31. See Anne McClintock's discussion of fetishism in *Imperial Leather: Race, Gender, and Sexuality in the Colonial Contest* (New York: Routledge, 1995), 184.

32. David F. Greenberg, *The Construction of Homosexuality* (Chicago: University of Chicago Press, 1988), 10, 10 n. 21.

33. On heteronormativity and the suppression of intersexuality in modern medicine, see Anne Fausto-Sterling's *Sexing the Body: Gender Politics and the Construction of Sexuality* (New York: Basic Books, 2000), esp. 45–114.

34. John Eliot, "Lively Sex Life of Peppermint Shrimps," *National Geographic* 197, no. 3 (March 2000).

35. Bruce Bagemihl, *Biological Exuberance: Animal Homosexuality and Natural Diversity* (New York: St. Martin's Press, 1999), 9.

CHAPTER 9. BIOLOGICAL BEAUTY

1. Roland Barthes, preface to the 1970 edition, *Mythologies* (1957; New York: Hill and Wang, 1972), 9.

2. Geoffrey Cowley, "The Biology of Beauty: What Science Has Discovered about Sex Appeal," *Newsweek*, June 3, 1996, 60–69.

3. Ibid., 62.

4. See two reports along these lines in the *New York Times:* Natalie Angier, "Why Birds and Bees, Too, Like Good Looks" (Science Times), February 8, 1994, C1, and Jane E. Brody, "Notions of Beauty Transcend Culture, New Study Suggests" (National), March 21, 1994, A14.

5. Natalie Angier convincingly argues that women's cinched waists, celebrated in an ample heterosexual lore, actually weigh *against* the usual logic of bioreductive fables about sex. Her reasoning about hip-waist proportions is straightforward. In evolutionary psychology, the coy, choosy female attempts to capture the support of a male breadwinner for herself and her offspring, whereas the eager, aggressive male attempts to impregnate as many women as possible, avoiding or postponing a long-term commitment with any one woman. Under these terms, any semblance of an hourglass shape puts the woman at a considerable reproductive disadvantage, since it implies that her pregnancy "shows"—the sign (if we take the perspective of evolutionary psychology) that the man's work is done and it's time to move on. (Other primate females have waists roughly the same size as their hips; their pregnancies are not terribly obvious.) "Perhaps," Angier deadpans, "a woman's body isn't designed to attract the long-term investment of a mate." *Woman: An Intimate Geography* (New York: Random House, 1999), 370–72.

6. Micaela di Leonardo, *Exotics at Home: Anthropologies, Others, American Modernity* (Chicago: University of Chicago Press, 1998), 354–55.

7. Compare Sigmund Freud's *Three Essays on the Theory of Sexuality,* trans. James Strachey (New York: Basic Books, 1962) with Teresa de Lauretis's interpretation, "Freud, Sexuality, and Perversion" in *The Practice of Love: Lesbian Sexuality and Perverse Desire* (Bloomington: Indiana University Press, 1994). See also Ann McClintock's history of fetishism in *Imperial Leather: Race, Gender and Sexuality in the Colonial Contest* (New York: Routledge, 1995).

8. Raymond Williams, *Problems in Materialism and Culture* (1980; London: Verso, 1997), 69.

9. See George L. Hersey's discussion of body canons in *The Evolution of Allure: Sexual Selection from the Medici Venus to the Incredible Hulk* (Cambridge, Mass.: MIT Press, 1996), 41–59. See also Philippe Comar, *Images of the Body,*

trans. Dorie B. Baker and David J. Baker (1993; New York: Harry N. Abrams, 1999), 13–37, 130–31.

10. Hersey, *Evolution of Allure*, 42–43, 59.

11. Ibid., 43–44.

12. This mode of argument begins with Franz Boas's critique of the comparative method of ethnology and his elaboration on what would become known as "cultural relativism." Franz Boas, "The Limitations of the Comparative Method of Anthropology" (1896) and "The Methods of Ethnology" (1920), in *High Points in Anthropology*, ed. Paul J. Bohannan and Mark Glazer (New York: Knopf, 1973), 84–92 and 92–99, respectively. See especially A. L. Kroeber's critical discussion of "universal patterns" in his textbook *Anthropology* (1923; New York: Harcourt, Brace and Company, 1948), 311–12. In a related vein, see Kroeber's essays, "The Concept of Culture in Science" and "Values as a Subject of Natural Science Inquiry," in his *Nature of Culture* (Chicago: University of Chicago Press, 1952), 118–35 and 136–38. Kroeber sets forth general terms for anthropology's relationship to natural science. I draw here on the lithe cadences Clifford Geertz used to express a Boasian anthropology in *The Interpretation of Culture* (New York: Basic Books, 1973), 39–40.

13. Bronislaw Malinowski, *The Sexual Life of Savages in North-Western Melanesia: An Ethnographic Account of Courtship, Marriage and Family Life among the Natives of the Trobriand Islands, British New Guinea* (New York: Harcourt, Brace and World, 1929), 286.

14. Ibid., 296–305.

15. Ibid., 308.

16. Ibid., 307.

17. Oscar Wilde, "The Decay of Lying," in *Intentions: The Complete Works of Oscar Wilde*, vol. 5 (1891; Garden City: Doubleday, Page, 1923), 47.

18. I have written at length about carnival masks and disguises in "Guto's Performance: Notes on the Transvestism of Everyday Life," in *Sex and Sexuality in Latin America*, ed. Daniel Balderston and Donna J. Guy (New York: New York University Press, 1997), 9–32.

19. Franz Boas, "The Limitations of the Comparative Method of Anthropology," in Bohannan and Glazer, eds., *High Points in Anthropology*, 87–88.

20. I recast here as a discussion of "beauty" what Marshall Sahlins (periscoping a Boasian form of argument) writes in a Sartrean language about "warfare." *The Use and Abuse of Biology* (Ann Arbor: University of Michigan Press, 1976), 8.

21. See parallel arguments from Stephen Jay Gould, *Ever since Darwin: Reflections on Natural History* (New York: W. W. Norton, 1977), 252–53.

22. David M. Buss, *The Evolution of Desire: Strategies of Human Mating* (New York: Basic Books, 1994).

23. Martha McCaughey, "Perverting Evolutionary Narratives of Heterosexual Masculinity, or, Getting Rid of the Heterosexual Bug," *GLQ* 3 (1996): 261–87.

24. Stephen Pinker, *How the Mind Works* (New York: W. W. Norton, 1997), 471–72.

25. Pinker alleges that *Playgirl* is not designed for women, but for gay men— a claim most gay men would find implausible, even astonishing. The physical definition, comportment, and situational posing of *Playgirl* models do not conform to the aesthetic patterns found in any nude journal produced by and for gay men. Gay aesthetics invariably idealize far more (how shall I put this?) "masculine" images than what one encounters in *Playgirl*.

26. Margaret Conkey, with the collaboration of Sarah Williams, "Original Narratives: The Political Economy of Gender in Archaeology," in *Gender at the Crossroads of Knowledge: Feminist Anthropology in the Postmodern Era*, ed. Micaela di Leonardo (Berkeley: University of California Press, 1991), 117–21. See also Margaret W. Conkey, "Mobilizing Ideologies: Paleolithic 'Art,' Gender Trouble, and Thinking about Alternatives," in *Women in Human Evolution*, ed. Lori D. Hager (New York: Routledge, 1997), 172–207.

27. McDermott is cited by Shahrukh Husain, *The Goddess* (Boston: Little, Brown, 1997), 11.

28. Susan Weed, *The Menopausal Years: The Wise Woman Way* (Woodstock, N.Y.: Ash Tree Publishing, 1992), 126. I am grateful to Christa Craven, a graduate student of anthropology at American University, for a delightful conversation on the "Venus" figures and for putting me onto the trail of these alternative readings of what they might represent.

29. See Cynthia Eller's discussion of the ambiguities associated with so-called Venus figures—and with objects said to represent vulvas, breasts, or other aspects of the female form in prehistoric art and architecture—in *The Myth of Matriarchal Prehistory: Why an Invented Past Won't Give Women a Future* (Boston: Beacon, 2000), 116–56.

30. John R. Clarke, *Looking at Lovemaking: Constructions of Sexuality in Roman Art, 100 B.C.–A.D. 250* (Berkeley: University of California Press, 1998). Lynn Hunt, ed., *The Invention of Pornography: Obscenity and the Origins of Modernity, 1500–1800* (New York: Zone Books, 1996). Carole Vance, "Negotiating Sex and Gender in the Attorney General's Commission on Pornography," in *The Gender/Sexuality Reader: Culture, History, Political Economy*, ed. Roger N. Lancaster and Micaela di Leonardo (New York: Routledge, 1997), 440–52.

31. Susan Bordo, *The Male Body: A New Look at Men in Public and in Private* (New York: Farrar, Straus and Giroux, 1999), esp. 170–71, 177–78.

32. Samuel R. Delany, *Times Square Red, Times Square Blue* (New York: New York University Press, 1999), 78.

33. Ibid., 78–79.

34. Cowley, "The Biology of Beauty," 68–69.

35. Ibid., 66.

36. Ralph Waldo Emerson, "Nature," in *The Portable Emerson*, ed. Carl Bode and Malcolm Cowley (New York: Penguin, 1946), 25, 29.

37. Crowley, "The Biology of Beauty," 62.

38. Natalie Angier, "Nothing Becomes a Man More Than a Woman's Face," *New York Times* (Science Times), September 1, 1998. Magnus Enquist and Stefano Ghirlanda, "The Secrets of Faces" (826–27), and D. I. Perrett et al., "Effects of Sexual Dimorphism on Facial Attractiveness" (884–91), *Nature* 394 (August 27, 1998).

39. John O'Neil, "For Women, Ideal Man Has 2 Faces, Study Finds," *New York Times,* June 24, 1999. I. S. Penton-Voak, D. I. Perrett, et al., "Menstrual Cycle Alters Face Preference," *Nature* 399 (June 24, 1999): 741–42.

40. Susan Bordo, personal communication.

CHAPTER 10. HOMO FABER, FAMILY MAN

1. Sherry B. Ortner, "Is Female to Male as Nature Is to Culture?" in *Woman, Culture, and Society,* ed. Michelle Zimbalist Rosaldo and Louise Lamphere (Stanford, Calif.: Stanford University Press, 1974), 67–87.

2. Carol Gilligan, *In a Different Voice: Psychological Theory and Women's Development* (Cambridge, Mass.: Harvard University Press, 1982). Sara Ruddick, *Maternal Thinking: Towards a Politics of Peace* (New York: Ballantine Books, 1989). Lillian Glass, *He Says, She Says: Closing the Communication Gap between the Sexes* (New York: Putnam, 1992).

3. Robert Bly, *Iron John: A Book about Men* (Reading, Mass.: Addison-Wesley, 1990).

4. John Gray, *Men Are from Mars, Women Are from Venus: A Practical Guide for Improving Communication and Getting What You Want in Your Relationships* (New York: HarperCollins, 1992); Warren Farrell, *Why Men Are the Way They Are* (New York: McGraw-Hill, 1986).

5. All review quotes are from Rob Becker's *Defending the Caveman* Web site (http://www.cavemania.com) as it appeared on April 13, 1997, and November 21, 1999.

6. See Louis Althusser's famous essay, "Ideology and Ideological State Apparatuses (Notes toward an Investigation)," in *Lenin and Philosophy and Other Essays,* trans. Ben Brewster (1969; New York: Monthly, 1971), esp. the subsection "Ideology Interpellates Individuals as Subjects," 170–77. I also draw here on the discussion of myth by Roland Barthes in *Mythologies* (1957; New York: Hill and Wang, 1972): "Myths are nothing but this ceaseless, untiring solicitation, this insidious and inflexible demand that all men recognize themselves in this image, eternal yet bearing a date, which was built of them one day as if for all time" (155).

7. Randy Thornhill and Craig T. Palmer, *A Natural History of Rape: Biological Bases of Sexual Coercion* (Cambridge, Mass.: MIT Press, 2000).

8. Richard Herrnstein and Charles A. Murray, *The Bell Curve: Intelligence and Class Structure in American Life* (New York: The Free Press, 1994).

9. Compare Wilson's grisly disquisition on genocide in *Sociobiology: The New Synthesis* (Cambridge, Mass.: The Belknap Press of Harvard University Press, 1975), 572–74, with the pictures of human nature he paints in more re-

cent works, e.g., *Biophilia* (Cambridge, Mass.: Harvard University Press, 1984), with its key premise of a benevolent and nature-loving human nature, or *Consilience: The Unity of Knowledge* (New York: Alfred A. Knopf, 1998), esp. chapter 7, "From Genes to Culture," and chapter 8, "The Fitness of Human Nature." One could easily imagine the emergence of a socialistic sociobiology—part Wilson, part Chomsky—staged in terms of claims about how human beings are "hardwired" to cooperate. See Peter Singer, *A Darwinian Left: Politics, Evolution and Cooperation* (New Haven, Conn.: Yale University Press, 2000).

10. After surveying the supposed options for eliminating sexual inequality by standardizing the socialization of males and females, Edward O. Wilson sets the tone for much of the subsequent models: "I am suggesting that the contradictions [between men and women, between sexual equality and sexual stratification] are rooted in the surviving relics of our prior genetic history, and that one of the most inconvenient and senseless, but nevertheless unavoidable, of these residues is the modest predisposition toward sex role differences." *On Human Nature* (Cambridge, Mass.: Harvard University Press, 1978), 135. Or as Anne Fausto-Sterling summarizes this argument: "In Wilson's view, the best bet is to live with our differences." *Myths of Gender: Biological Theories about Men and Women*, rev. ed. (1985; New York: Basic Books, 1992), 202. Such emphasis on "difference" appeals to the sensibilities of cultural feminism, in which women's "difference" deserves recognition and celebration, while still allowing conservatives to understand "difference"—in mathematical skills, thinking style, emotional responses, and so on—as the basis for inequality.

11. Alice Echols describes the emergence of cultural feminism in the early 1970s as a critique of Left politics that neglected women's issues in *Daring to Be Bad: Radical Feminism in America, 1967–1975* (Minneapolis: University of Minnesota Press, 1989), esp. chapter 6, "The Ascendance of Cultural Feminism," 243–95.

12. Micaela di Leonardo and Roger N. Lancaster, "Gender, Sexuality, Political Economy," *New Politics*, n.s., 21, no. 6 (summer 1996): 32–33, reprinted in *The Socialist Feminist Project: A Reader*, ed. Nancy Holmstrom (New York: Monthly, forthcoming).

13. Wendy Kaminer, *I'm Dysfunctional, You're Dysfunctional: The Recovery Movement and Other Self-Help Fashions* (Reading, Mass.: Addison-Wesley, 1992); *Sleeping with Extra-Terrestrials: The Rise of Irrationalism and the Perils of Piety* (New York: Random House, 1999); *Free for All: Defending Liberty in America Today* (Boston: Beacon, 2000).

14. Early works of cultural feminism include Adrienne Rich, *Of Woman Born: Motherhood as Experience and Institution* (New York: W. W. Norton, 1976); Mary Daly, *Gyn/Ecology: The Metaethics of Radical Feminism* (Boston: Beacon, 1978); Phyllis Chesler, *Women and Madness* (New York: Avon, 1972); and Susan Griffin, *Pornography and Silence: Culture's Revenge against Nature* (New York: Harper and Row, 1981). It is sometimes claimed that cultural feminism no longer exists, but this is not quite accurate. Today, cultural femi-

nism continues to generate popular texts on subjects such as woman-centered spirituality, ecofeminism, and women's self-help psychology. See, for example, Anne Wilson Schaef, *Women's Reality: An Emerging Female System in a White Male Society* (1981; Minneapolis: Winston Books, 1985); Kim Chernin, *Reinventing Eve: The Search for a Feminine Culture* (New York: TimesBooks, 1987); Clarissa Pinkola Estes, *Women Who Run with the Wolves* (New York: Ballantine Books, 1992); and Charlene Spretnak, *The Resurgence of the Real: Body, Nature and Place in a Hypermodern World* (Reading, Mass.: Addison-Wesley, 1997). Cultural feminism has also left its imprimatur on the broad current of scholarship that Katha Pollitt dubs "difference feminism," that is, feminism predicated on the idea that women speak, think, act, and relate to others differently than men do, and that this difference ought to be the basis for women's recognition and empowerment. See Carol Gilligan's *In a Different Voice,* Sara Ruddick's *Maternal Thinking,* and Lillian Glass's *He Says, She Says.*

15. Alice Rossi, "Maternalism, Sexuality and the New Feminism," in *Contemporary Sexual Behavior: Critical Issues in the 1970s,* ed. Joseph Zubin and John Money (Baltimore: Johns Hopkins University Press, 1973), 145–73, and "A Biosocial Perspective on Parenting," *Daedalus* (spring 1977): 1–31. For an early, thoughtful critique of Rossi's glorification of the mother-child bond, see Wini Breines, Margaret Cerullo, and Judith Stacey, "Social Biology, Family Studies, and Antifeminist Backlash," *Feminist Studies* 4, no. 1 (February 1978): 43–67, esp. 45–51. Although Rossi's arguments about gender and biology are undoubtedly more sophisticated than what one encounters in either the simplified models of cultural feminism or in the sociobiological vulgate, I have always thought that they were of a different order entirely from Sarah Blaffer Hrdy's presentation of sociobiological arguments in *The Woman That Never Evolved* (Cambridge, Mass.: Harvard University Press, 1981). As Donna Haraway points out, Hrdy emphasizes male-female sexual competition (over cooperation) to position primate and human women as somewhat aggressive agents endowed with sexual drives. See Donna J. Haraway, *Primate Visions: Gender, Race, and Nature in the World of Modern Science* (New York: Routledge, 1989), 357–67. In other words, Hrdy's naturally sexed and innately passionate females are more allied with the politics of radical feminism than with those of cultural feminism. Similar arguments could be made of Hrdy's *Mother Nature: A History of Mothers, Infants, and Natural Selection* (New York: Pantheon Books, 1999), which further develops a sociobiological model of female agency and even female alliances. Still, the main thrust of these works was and remains starkly heteronormative, and their social residue—the naturalization of gender relations and sexuality, however these might be conceived politically—has contributed to the more familiar staging of arguments about female nurture and its role in male-female complementarity. See Helen Fisher, *The First Sex: The Natural Talents of Women and How They Are Changing the World* (New York: Random House, 1999).

16. Bruce Mau, *Life Style,* ed. Kyo Maclear with Bart Testa (London: Phaidon, 2000), 39–87.

17. Michel Foucault, *The History of Sexuality, Volume I: An Introduction,* trans. Robert Hurley (1976; New York: Vintage, 1980), 92–102, esp. 93, 100–101.

18. On "strategic essentialism," see Gayatri Spivak, *In Other Worlds: Essays in Cultural Politics* (New York: Methuen, 1987), 205. See also Diana Fuss, *Essentially Speaking: Feminism, Nature and Difference* (New York: Routledge, 1989), 119.

19. Has the patriarchy no shame? Yesterday's effeminacy is today's masculinity!

20. Andrew Levison, "Who Lost the Working Class," *The Nation,* May 14, 2001, 31.

CHAPTER 11. T-POWER

1. Harrison G. Pope Jr., Katharine A. Phillips, and Roberto Olivardia, *The Adonis Complex: The Secret Crisis of Male Body Obsession* (New York: The Free Press, 2000); Lionel Tiger, *The Decline of Males* (New York: Golden Books, 1999); Christina Hoff Sommers, *War against Boys: How Misguided Feminism Is Harming Our Young Men* (New York: Simon and Schuster, 2000).

2. Andrew Sullivan, "Dumb and Dumber," *The New Republic,* June 26, 2000, 6.

3. Tiger, *The Decline of Males.* Note that here, as elsewhere, Sullivan mimics Tiger's arguments when he codes feminism as "powerful" or "influential" and men as imperiled. When women get equal pay for equal work or constitute half of Congress, I'll cede the point.

4. There will always be a soft spot in my heart for the former Navy SEAL, who once said of gays in the military, "You don't need to be 'straight.' . . . You just need to shoot straight." A fiscal conservative, Ventura nevertheless supported legal recognition of same-sex partnerships with a plainspoken appeal to reason and decency: "I have two [gay] friends who have been together 41 years, and if one of them becomes sick, the other one is not even allowed to be at the [hospital] bedside. I don't believe government should be so hostile, so mean-spirited." Ventura once chided a group of reporters who were attempting to make scandal over an aide's arrest for soliciting sex in a gay cruising area with the straightforward question: "Who among us hasn't had sex somewhere he wasn't supposed to?"

5. David McBride, in conversation.

6. Andrew Sullivan, "The He Hormone," *New York Times Magazine,* April 2, 2000.

7. Emily Eakin, "In This Woman's World, What's a Guy to Do? Sweat! As Women Gain Independence and Power, Men of Steel Feel like 97-Pound Weaklings," *New York Times* (Arts and Ideas), October 7, 2000.

8. See especially arguments by Lionel Tiger in *The Decline of Males.*

9. Lionel Tiger, quoted in Eakin, "In This Woman's World, What's a Guy to Do?" See also Tiger, *The Decline of Males,* 29–60.

10. Marjorie Shostak, *Nisa: The Life and Words of a !Kung Woman* (Cam-

bridge, Mass.: Harvard University Press, 1981), 19, 66–67. Richard B. Lee, *The !Kung San: Men, Women and Work in a Foraging Society* (Cambridge: Cambridge University Press, 1979), 451–52.

11. Richard B. Lee, "Technology and the Organization of Production," in *The !Kung San: Men, Women and Work in a Foraging Society,* 119.

12. Anne Fausto-Sterling, *Myths of Gender: Biological Theories about Women and Men,* rev. ed. (1985; New York: Basic Books, 1992), 126–32.

13. Gail Vines, *Raging Hormones: Do They Rule Our Lives?* (Berkeley: University of California Press, 1994), 79.

14. Natalie Angier, *Woman: An Intimate Geography* (New York: Anchor, 1999), 271–74.

15. Moreover, there is no correlation between testosterone and male dominance or rank in primate species. See Robert Sapolsky, *The Trouble with Testosterone and Other Essays on the Biology of the Human Predicament* (New York: Scribner, 1997), 147–59.

16. Robert M. Rose, "Androgen Excretion in Stress," in *The Psychology and Physiology of Stress, with Reference to Special Studies of the Viet Name War,* ed. Peter G. Bourne (New York: Academic, 1969), 117–47.

17. C. Wang et al., "Testosterone Replacement Therapy Improves Mood in Hypogonadal Men," *Journal of Clinical Endocrinology and Metabolism* 81, no. 10 (October 1996): 3578–83. Natalie Angier, "Does Testosterone Equal Aggression? Maybe Not," *New York Times,* June 20, 1995.

18. Tiger, *The Decline of Males,* chapter 6, "Mother Courage in an Abrasive World," 156–78.

19. Gail Bederman, *Manliness and Civilization: A Cultural History of Gender and Race in the United States, 1880–1917* (Chicago: University of Chicago Press, 1995).

20. For a brilliant rendering of the implications of the chiasmus, see Drew Leder, *The Absent Body* (Chicago: University of Chicago Press, 1990), esp. 62–68.

21. I distill, perhaps too quickly, a queer reading of Lacan, whose shell game with the signifying phallus and desire as lack says much the same thing. See *Écrits: A Selection,* trans. Alan Sheridan (1966; New York: W. W. Norton, 1977), especially "The Signification of the Phallus," 281–91.

22. I crib the gist of key works from the Bakhtin school, whose theory of the subject rested on the intersubjective and dialogical circulation of signs. V. N. Vološinov, *Marxism and the Philosophy of Language,* trans. Ladislav Matejka and I. R. Titunik (1929; Cambridge, Mass.: Harvard University Press, 1986). Mikhail Bakhtin, *The Dialogic Imagination,* ed. Michael Holquist, trans. Caryl Emerson and Michael Holquist (Austin: University of Texas Press, 1981).

CHAPTER 12. NATURE'S MARRIAGE LAWS

1. Jane Brody, "Genetic Ties May Be Factor in Violence in Stepfamilies," *New York Times* (Science Times), February 10, 1998.

2. See Judith Stacey's insightful and important work on neocommunitari-

anism and the pro-marriage movement, *In the Name of the Family* (Boston: Beacon, 1996); also her "The Neo-Family Values Campaign," in *The Gender/ Sexuality Reader: Culture, History, Political Economy,* ed. Roger N. Lancaster and Micaela di Leonardo (New York: Routledge, 1997), 453–70; and "Family Values Forever," *The Nation,* July 9, 2001, 26–30.

3. Brody, "Genetic Ties May Be Factor in Violence in Stepfamilies."

4. Anne Innis Dagg, "Infanticide by Male Lions Hypothesis: A Fallacy Influencing Research into Human Behavior," *American Anthropologist* 100, no. 4 (December 1998): 940–95. See also Joan Silk and Craig Stanford, "Infanticide Article Disputed," *Anthropology Newsletter,* September 1999, 27, and Dagg's reply, "Sexual Selection Is Debatable," *Anthropology Newsletter,* December 1999, 20.

5. Natalie Angier, *New York Times* (Women's Health), June 21, 1998.

6. Linda J. Waite and Maggie Gallagher, *The Case for Marriage: Why Married People Are Happier, Healthier, and Better Off Financially* (New York: Doubleday, 2000), 185. See also Linda J. Waite et al., eds., *The Ties That Bind: Perspectives on Marriage and Cohabitation* (New York: Aldine de Gruyter, 2000).

7. Waite and Gallagher, *The Case for Marriage,* 203.

8. Waite and Gallagher, *The Case for Marriage,* 25–27, 55–57.

9. The author is invoking a now-ubiquitous argument connecting the corpus callosum to "female intuition," a general line of argument about sex differences whose circulation Anne Fausto-Sterling traces to a 1992 *Time* illustration. See her *Sexing the Body: Gender Politics and the Construction of Sexuality* (New York: Basic Books, 2000), 116.

10. Ibid., chapter 5: "Sexing the Brain: How Biologists Make a Difference," 115–45, quote at 145.

11. Ian Fisher, "Health Care for Lesbians Gets a Sharper Focus," *New York Times,* June 21, 1998.

12. Oscar Wilde, "The Decay of Lying," in *Intentions: The Complete Works of Oscar Wilde, Volume V* (1891; Garden City: Doubleday, Page, 1923), 26.

13. On the lesbian seagulls reported along the coast of Southern California, see Edward Stein, *The Mismeasure of Desire: The Science, Theory, and Ethics of Sexual Orientation* (Oxford: Oxford University Press, 1999), 170–71.

14. Natalie Angier, "Birds' Design for Living Offers Clues to Polygamy," *New York Times* (Science Times), March 3, 1998.

CHAPTER 13. MAROONED ON *SURVIVOR* ISLAND

1. John Tierney, " 'Survivor' Finale: Testosterone vs. Indirect Aggression: Negotiating the Sexual Politics of a Hit TV Show," *New York Times* (Region), August 22, 2000. See also James McBride Dabbs and Mary Godwin Dabbs, *Heroes, Rogues, and Lovers: Testosterone and Behavior* (New York: McGraw-Hill, 2000).

2. Caryn Jones, "Machiavelli, on a Desert Isle, Meets TV's Reality. Unreal," *New York Times,* front page, August 24, 2000. Mary Catherine Bateson, "It's Just a Game, Really," *New York Times* (Op Ed), August 27, 2000.

3. Randolph M. Nesse, "Is Depression an Adaptation?" *Archives of General Psychiatry* 57, no. 1 (January 2000): 14–20. Erica Goode, "Viewing Depression as Tool for Survival," *New York Times* (Science Times), February 1, 2000.

4. "Depression usually makes it impossible for patients to spend constructive time on anything.... Research evidence shows that depression almost never permits a person to take stock and figure out what to do next because depressive thinking is most always colored by a global hopelessness about the future in general, and one's fate in specific." Dr. Michael M. Gindi, letter to the editor, "Depression Debunking," *New York Times* (Science Times), February 8, 2000.

5. Stephen Jay Gould, "Biological Potentiality versus Biological Determinism," in *Ever since Darwin: Reflections in Natural History* (New York: W. W. Norton, 1977), 251–59.

6. This analogy has a long and venerable tradition. See, for example, Marshall Sahlins's invocation of material "levels" in *The Use and Abuse of Biology: An Anthropological Critique of Sociobiology* (Ann Arbor: University of Michigan, 1977), 63–64.

7. For brevity's sake, my example skips over intermediate sciences such as chemistry and transitional ones such as biochemistry, as well as certain problematic transitional areas where the definition of "life" might be less than clear.

8. Emile Durkheim, "What Is a Social Fact?" in *The Rules of Sociological Method*, 8th ed., trans. Sarah A. Solovay and John H. Mueller, ed. George E. G. Catlin (1895; New York: The Free Press, 1966), 1–13, esp. 3, 5–6. I repeat here a rendition of Durkheim's argument on language.

9. Ferdinand de Saussure, *Course in General Linguistics*, ed. Charles Bally and Albert Sechehaye in collaboration with Albert Riedlinger, trans. Wade Baskin (1915; New York: McGraw-Hill, 1966), 67–78. The import of this concept is explicated in detail by Jonathan Culler in *Ferdinand de Saussure*, rev. ed. (1976; Ithaca, N.Y.: Cornell University Press, 1986), 28–33.

10. Alan Sokal, "Transgressing the Boundaries: An Afterword," in Alan Sokal and Jean Bricmont, *Fashionable Nonsense: Postmodern Intellectuals' Abuse of Science* (New York: Picador USA, 1998), 270 n. 4.

11. Karl Marx, "Theses on Feuerbach," in *Karl Marx: Selected Writings*, ed. David McLellan (Oxford: Oxford University Press, 1977), 157.

12. Karl Marx, "The German Ideology," in *Karl Marx: Selected Writings*, ed. David McLellan (Oxford: Oxford University Press, 1977), 161.

13. Karl Marx, *Economic and Philosophic Manuscripts of 1844* (New York: International, 1964), 137–38. I invoke here Merleau-Ponty's brilliant rendition of Marxist philosophy in *Sense and Non-Sense*, trans. Hubert L. Dreyfus and Patricia Allen Dreyfus (1948; Evanston, Ill.: Northwestern University Press, 1964), 128.

14. Marx, *Economic and Philosophical Manuscripts of 1844*, 191, 137, 143, 141.

15. Marx problematizes the very oppositions on which Durkheim leans: nature/culture, social facts/natural facts, individual/society. Marx, "The German Ideology," 175. See also Merleau-Ponty, *Sense and Non-Sense*, 126.

/ Notes to Pages 175–178

16. Merleau-Ponty, *Sense and Non-Sense,* 129.

17. Marx, "The German Ideology," 174.

18. Marx, "Theses on Feuerbach," 156. See also Merleau-Ponty, *Sense and Non-Sense,* 129.

19. Donald L. Donham, *History, Power, Ideology: Central Issues in Marxism and Anthropology* (Cambridge: Cambridge University Press, 1990), 57.

20. Terry Eagleton, "Self-Realization, Ethics, and Socialism," *New Left Review* (September–October 1999): 154.

21. See Marx, *Economic and Philosophic Manuscripts of 1844,* 113, 139–41. Merleau-Ponty, *Sense and Non-Sense,* 139.

22. Marx, *Economic and Philosophic Manuscripts of 1844,* 145.

CHAPTER 14. SELECTIVE AFFINITIES

1. I hasten to add that it is not just in evolutionary psychology or in socially conservative circles that one finds such appeals to commonality. The rhetoric of a "universal family" exercises a powerful draw in liberal and progressive circles, where notions like "international brotherhood," "feminist sisterhood," "the family of man," and so on, evoke cross-cultural and transhistorical universals in the service of the ends of humanitarian suasion or to promote activist solidarity.

2. Emile Durkheim, "What Is a Social Fact?" in *Rules of the Sociological Method,* 8th ed., trans. Sarah A. Solovay and John H. Mueller, ed. George E. G. Catlin (1895; New York: The Free Press, 1966), 1–13.

3. For the moment, I leave aside the argument that the "consistency" of social forms could flow from the exigencies of social organization rather than from biogenetic causes. The following chapter will take up this question in some detail.

4. Marilyn Strathern develops some of the implications of this variation in her classic essay "Culture in a Netbag: The Manufacture of a Subdiscipline in Anthropology," *Man,* n.s., 16 (1981): 665–88, which compares gendered labor and the circulation of values in three Melanesian societies.

5. Lila Leibowitz, "Perspectives on the Evolution of Sex Differences," in *Toward an Anthropology of Women,* ed. Rayna R. Reiter (New York: Monthly, 1975), 20.

6. In the early days of second-wave feminist scholarship, Michelle Rosaldo tried to hang a universalist theory of women's subordination on just such a hook, but subsequently backtracked. See her "Woman, Culture, and Society: A Theoretical Overview," in *Woman, Culture, and Society,* ed. Michelle Zimbalist Rosaldo and Louise Lamphere (Stanford, Calif.: Stanford University Press, 1974), 17–42.

7. Judith K. Brown, "Iroquois Women: An Ethnohistoric Note," in *Toward an Anthropology of Women,* 235–51.

8. For a recent cross-cultural collection, see Linda J. Seligmann, ed., *Women*

Traders in Cross-Cultural Perspective: Mediating Identities, Marketing Wares (Stanford, Calif.: Stanford University Press, 2001). See also Florence E. Babb, *Between Field and Cooking Pot: The Political Economy of Marketwomen in Peru*, rev. ed. (Austin: University of Texas Press Sourcebooks on Anthropology, 1998), and Gracia Clark, *Onions Are My Husband: Survival and Accumulation by West African Market Women* (Chicago: University of Chicago Press, 1994).

9. Much of sociobiology's emphasis on male hunting derives from a current of mid-twentieth-century physical anthropology. See Sherwood L. Washburn and C. S. Lancaster, "The Evolution of Hunting," in *Man the Hunter*, ed. Richard B. Lee and Irven DeVore (Chicago: Aldine, 1968), 293–303, and Irven DeVore and Sherwood L. Washburn, "Social Behavior of Baboons and Early Man," in *Social Life of Early Man*, ed. Sherwood L. Washburn (Chicago: Aldine, 1961), 91–105. See also Robert Wright's arguments in *The Moral Animal: Evolutionary Psychology and Everyday Life* (New York: Pantheon, 1994), esp. 38, 60.

10. See Matt Ridley's speculations on the supposed implications of male hunting and female gathering on marital pair bonding in the evolution of human nature. *The Origins of Virtue: Human Instincts and the Evolution of Cooperation* (New York: Penguin, 1996), 92–93.

11. Richard B. Lee, *The Dobe Ju/'hoansi*, 2d. ed. (Fort Worth, Tex.: Harcourt Brace College Publishers, 1993) and *The !Kung San: Men, Women, and Work in a Foraging Society* (Cambridge: Cambridge University Press, 1979). Eleanor Leacock and Richard Lee, eds., *Politics and History in Band Societies* (Cambridge: Cambridge University Press, 1982). Marjorie Shostak, *Nisa: The Life and Words of a !Kung Woman* (New York: Vintage Books, 1981).

12. Jane C. Goodale, *Tiwi Wives: A Study of the Women of Melville Island, North Australia* (Seattle: University of Washington Press, 1971).

13. Agnes Estioko-Griffin and P. Bion Griffin, "Woman the Gatherer: The Agta," in *Gender in Cross-Cultural Perspective*, 2d ed., ed. Caroline B. Brettell and Carolyn F. Sargent (1993; Upper Saddle River, N.J.: Prentice-Hall, 1997), 219–29.

14. Micaela di Leonardo, "Introduction: Gender, Culture, and Political Economy: Feminist Anthropology in Historical Perspective," in *Gender at the Crossroads of Knowledge: Feminist Anthropology in the Postmodern Era*, ed. Micaela di Leonardo (Berkeley: University of California Press, 1991), 10–17. See also Nadine R. Peacock, "Rethinking the Sexual Division of Labor: Reproduction and Women's Work among the Efe," in *Gender at the Crossroads of Knowledge*, 339–60; Ellen Ross and Rayna Rapp, "Sex and Society: A Research Note from Social History and Anthropology," in *The Gender/Sexuality Reader: Culture, History, Political Economy*, ed. Roger N. Lancaster and Micaela di Leonardo (New York: Routledge, 1997), 153–68; and Sylvia Junko Yanagisako and Jane Fishburne Collier, "Toward a Unified Analysis of Gender and Kinship," in *Gender and Kinship: Essays toward a Unified Analysis*, ed. Jane Fishburne Collier and Sylvia Junko Yanagisako (Stanford, Calif.: Stanford University Press, 1987), 14–50.

15. Robert Murphy and Yolanda Murphy, *Women of the Forest* (New York: Columbia University Press, 1985).

16. Kathleen Gough, "Variation in Residence," in *Matrilineal Kinship*, ed. David M. Schneider and Kathleen Gough (Berkeley: University of California Press, 1961), 547–48.

17. Carol Stack, *All Our Kin: Strategies for Survival in a Black Community* (1974; New York: Basic Books, 1997).

18. Jack Goody, "Women, Class and Family," *New Left Review* 219 (September–October 1996): 123. See also Jack Goody, *Domestic Groups* (Reading, Mass.: Addison Wesley, 1972).

19. Roger N. Lancaster, *Life Is Hard: Machismo, Danger, and the Intimacy of Power in Nicaragua* (Berkeley: University of California Press, 1992). See Richard Adams's comparisons of family forms in different highland and lowland areas of Central America, "Cultural Components of Central America," *American Anthropologist* 58 (1956): 881–907, and *Cultural Surveys of Panama–Nicaragua–Guatemala–El Salvador–Honduras*, Scientific Publications no. 33 (Washington, D.C.: Pan-American Sanitary Bureau, Regional Office of the World Health Organization, 1957).

20. John Boswell, *Same-Sex Unions in Pre-Modern Europe* (New York: Villard Books, 1994), xxi–xxii, 9.

21. Carol R. Ember and Melvin Ember, *Cultural Anthropology*, 8th ed. (Upper Saddle River, N.J.: Prentice-Hall, 1996), chapter 11, "Marital Residence and Kinship," 197–219. Burton Pasternak, Carol R. Ember, and Melvin Ember, *Sex, Gender, and Kinship: A Cross-Cultural Perspective* (Upper Saddle River, N.J.: Prentice-Hall, 1997). Kathleen Gough, "The Nayars and the Definition of Marriage," in *Readings in Kinship and Social Structure*, ed. Nelson Grayburn (New York: Harper and Row, 1971), 365–77.

22. Lawrence Stone, "The Rise of the Nuclear Family in Early Modern England: The Patriarchal Stage," in *The Family in History*, ed. Charles E. Rosenberg (Philadelphia: University of Pennsylvania Press, 1975), 14–15.

23. Ellen Ross and Rayna Rapp, "Sex and Society," in *The Gender/Sexuality Reader*, 164.

24. Elizabeth Povinelli, "Sex Acts and Sovereignty: Race and Sexuality in the Construction of the Australian Nation," in *The Gender/Sexuality Reader*, esp. 521–23. Lee, *The Dobe Ju/'hoansi*, 71–78. Stephen O. Murray, " 'Sentimental Effusions' of Genital Contact in Amazonia," in *Latin American Male Homosexualities* (Albuquerque: University of New Mexico Press, 1995), 264–73. Sidney Mintz and Eric Wolf, "An Analysis of Ritual Co-Parenthood (Compadrazgo)," *Southwest Journal of Anthropology* 6 (1950): 341–68. E. E. Evans-Pritchard, *Kinship and Marriage among the Nuer* (Oxford: Oxford University Press, 1951). Melville J. Herskowitz, "A Note on 'Woman-Marriage' in Dahomey," *Africa* 10 (1937): 335–41. Stephen O. Murray and Will Roscoe, eds., *Boy-Wives and Female Husbands: Studies in African Homosexualities* (New York: St. Martin's Press, 1998). Will Roscoe, *Changing Ones: Third and Fourth Genders in Native North America* (New York: St. Martin's Press, 1998), and *The Zuni Man-Woman*

(Albuquerque: University of New Mexico Press, 1991). John Boswell, *Same-Sex Unions in Pre-Modern Europe*. Micaela di Leonardo, "Warrior Virgins and Boston Marriages: Spinsterhood in History and Culture," *Feminist Issues* (fall 1985): 47–68. Carroll Smith-Rosenberg, "The Female World of Love and Ritual: Relations between Women in Nineteenth-Century America," in *A Heritage of Her Own*, ed. Nancy F. Cott and Elizabeth H. Pleck (New York: Simon and Schuster, 1979), 311–42. Lillian Faderman, *Surpassing the Love of Men: Romantic Friendship and Love between Women from the Renaissance to the Present* (New York: William Morrow, 1981).

25. William N. Stephens, *The Family in Cross-Cultural Perspective* (1963; Lanham, Md.: University Press of America, 1982), 3–4.

26. Jane Collier, Michelle Z. Rosaldo, and Sylvia Yanagisako, "Is There a Family? New Anthropological Views," in *The Gender/Sexuality Reader*, 76.

27. David D. Gilmore, *Manhood in the Making: Cultural Concepts of Masculinity* (New Haven, Conn.: Yale University Press, 1990).

28. Susan Faludi, *Stiffed: The Betrayal of the American Man* (New York: William Morrow, 1999). Margaret Mead, *Sex and Temperament in Three Primitive Societies* (1935; New York: Morrow Quill Paperbacks, 1963), 245–64. I see nothing implausible about the transition Faludi traces but do find odd her unconcealed nostalgia for the Golden Age of Manhood's industrial-era masculinity. Why not long for the high-Victorian daddy, who was independent, autonomous, and entrepreneurial? Or the agrarian paterfamilias, who took a close interest in what went on in his household? On these distinctions, see Gail Bederman's thumbnail sketch on changes in what it means to be a man: *Manliness and Civilization: A Cultural History of Gender and Race in the United States, 1880–1917* (Chicago: University of Chicago Press, 1995), 10–20.

29. Robert Briffault, *The Mothers*, abridged ed. (1927; New York: Universal Library, 1963). John Bowlby, *Attachment* (New York: Basic Books, 1969). Alice Rossi, "A Biosocial Perspective on Parenting," *Daedalus* 106, no. 2 (spring 1977): 1–32. See Nancy Scheper-Hughes's perspicuous discussion of "maternal thinking," as opposed to "maternal instinct," in *Death without Weeping: The Violence of Everyday Life in Brazil* (Berkeley: University of California Press, 1992), 340–412.

30. Margaret Mead, *Coming of Age in Samoa: A Psychological Study of Primitive Youth for Western Civilization* (New York: Morrow, 1928), and *Growing Up in New Guinea: A Comparative Study of Primitive Education* (New York: Morrow, 1930); Abram Kardiner, *The Individual and His Society* (New York: Columbia University Press, 1939); John Whiting and Irvin Child, *Child Training and Personality* (New Haven, Conn.: Yale University Press, 1953); Beatrice Whiting, ed., *Six Cultures: Studies of Child Rearing* (New York: Wiley, 1963). See also Margaret Mead and Martha Wolfenstein, eds., *Childhood in Contemporary Cultures* (Chicago: University of Chicago Press, 1955). Not all perspectives on culture, personality, and socialization were constructionist. Some anthropologists believed they could identify biologically grounded cultural universals. See various pieces in George D. Spindler, ed., *The Making of*

Psychological Anthropology (Berkeley: University of California Press, 1978), and Robert A. LeVine, ed., *Culture and Personality: Contemporary Readings* (Chicago: Aldine, 1974).

31. Gregory Bateson and Margaret Mead, *Bathing Babies in Three Cultures,* videorecording (1952; New York: Institute for Intercultural Studies, 1999).

32. Nancy Scheper-Hughes, *Saints, Scholars, and Schizophrenics: Mental Illness in Rural Ireland* (Berkeley: University of California Press, 1979), 147.

33. See Nancy Scheper-Hughes's discussion of "good enough" mothering in *Death without Weeping: The Violence of Everyday Life in Brazil* (Berkeley: University of California Press, 1992), 359–61. For a review of psychoanalytic constructions of the "good-enough" mother, see Janice Doane and Devon Hodges, *From Klein to Kristeva: Psychoanalytic Feminism and the Search for the "Good Enough" Mother* (Ann Arbor: University of Michigan Press, 1992).

34. Francis Fukuyama, *The Great Disruption: Human Nature and the Reconstitution of Social Order* (New York: The Free Press, 1999), 99. Lionel Tiger and Robin Fox, *The Imperial Animal* (1971; New Brunswick, N.J.: Transaction, 1998), 67.

35. Sarah Blaffer Hrdy, *Mother Nature: A History of Mothers, Infants, and Natural Selection* (New York: Pantheon Books), 357.

36. Ibid., 351.

37. Scheper-Hughes, *Death without Weeping,* 316–26. See Donna J. Haraway's reprise of Scheper-Hughes's arguments in *Modest_Witness@Second_Millennium. FemaleMan©Meets_OncoMouse™* (New York: Routledge, 1997), 202–12.

38. Barry Hewlett, *Intimate Fathers: The Nature and Context of Aka Pygmy Paternal-Infant Care* (Ann Arbor: University of Michigan Press, 1991).

39. John Boswell, *The Kindness of Strangers: The Abandonment of Children in Western Europe from Late Antiquity to the Renaissance* (Chicago: University of Chicago Press, 1988), 15–16. The passage includes footnotes on sources for these figures.

40. Boswell, *The Kindness of Strangers,* 16. See also David L. Ransel, *Mothers of Misery: Child Abandonment in Russia* (Princeton, N.J.: Princeton University Press, 1988).

41. Lancaster, *Life Is Hard: Machismo, Danger, and the Intimacy of Power in Nicaragua.*

42. Lori Heise, "Violence, Sexuality, and Women's Lives," in *The Gender/Sexuality Reader,* 421. Lori Heise et al., "Ending Violence against Women," *Population Reports,* series L, no. 11 (December 1999): 17.

43. Penelope Harvey and Peter Gow, eds., *Sex and Violence: Issues in Representation and Experience* (London: Routledge, 1994).

44. For a recent spin on this old argument, see Stephanie Gutmann, *The Kinder, Gentler Military: Can America's Gender-Neutral Fighting Force Still Win Wars?* (New York: Scribner, 2000).

45. See Lucinda J. Peach's succinct and compelling arguments, "Gender and War: Are Women Tough Enough for Military Combat?" in *Gender in Cross-*

Cultural Perspective, 20–29. For a robustly feminist perspective on gender and militarism, see Cynthia Enloe, *Does Khaki Become You? The Militarization of Women's Lives* (1983; London: Pandora, 1988).

46. See Walter L. Williams, "Amazons of America: Female Gender Variance," chapter 11 of *The Spirit and the Flesh: Sexual Diversity in American Indian Culture* (Boston: Beacon, 1986), 233–51.

47. For example, by the close of the insurrectionary period, some 30 percent of Sandinista combatants in Nicaragua were women. Maxine Molyneaux, "Mobilization without Emancipation? Women's Interests. The State, and Revolution in Nicaragua," *Feminist Studies* 11, no. 2 (1985): 227. See also Cynthia Enloe, "Women in Liberation Armies," in *Does Khaki Become You?* 160–72.

48. On the social construction and historical specificity of warfare, see R. Brian Ferguson, ed., *Warfare, Culture, and Environment* (Orlando, Fla.: Academic Press, 1984), esp. Ferguson's discussion of "human agressiveness" as the "cause" of war (8–14) in the introduction, and R. Brian Ferguson and Neil L. Whitehead, eds., *War in the Tribal Zone: Expanding States and Indigenous Warfare* (Sante Fe, N.M.: School of American Research Press, 1992). Of special interest, in the light of Napoleon Chagnon's claims about the sociobiologically instructive "fierceness" of the Yanomami, is Ferguson's *Yanomami Warfare: A Political History* (Santa Fe, N.M.: School of American Research Press, 1995).

49. Bernardino de Sahagún, *General History of the Things of New Spain: Florentine Codex. Book Six: Rhetoric and Moral Philosophy,* trans. Arthur J. O. Anderson and Charles E. Dibble (Santa Fe, N.M.: School of American Research Press; Salt Lake City: University of Utah, 1969), 164, 167.

50. Katha Pollitt, "Marooned on Gilligan's Island: Are Women Morally Superior to Men?" in *Reasonable Creatures: Essays on Women and Feminism* (New York: Alfred A. Knopf, 1994), 42–62. See Carol Gilligan, *In a Different Voice: Psychological Theory and Women's Development* (Cambridge, Mass.: Harvard University Press, 1982), and Sara Ruddick, *Maternal Thinking: Toward a Politics of Peace* (New York: Ballantine, 1990).

51. Fatima Mernissi, *Beyond the Veil: Male-Female Dynamics in a Modern Muslim Society* (Cambridge, Mass.: Schenkman, 1975). Janice Boddy, *Wombs and Alien Spirits: Women, Men and the Zar Cult in Northern Sudan* (Madison: University of Wisconsin Press, 1989).

52. Kathleen Gough, "Nayar: Central Kerala," in *Matrilineal Kinship,* ed. David M. Schneider and Kathleen Gough (Berkeley: University of California Press, 1961), 328–29 and 357–63.

53. Dennis Werner, "A Cross-Cultural Perspective on Theory and Research on Male Homosexuality," *Journal of Homosexuality* 4, no. 4 (summer 1979): 345–62, esp. 356–60. Marvin Harris, "Why the Gays Came Out of the Closet," chapter 6 of *America Now: The Anthropology of a Changing Culture* (New York: Simon and Schuster, 1981), 98–115.

54. Arve Sørum, "Growth and Decay: Bedamini Notions of Sexuality," in *Ritualized Homosexuality in Melanesia,* ed. Gilbert Herdt (Berkeley: University of California Press, 1984), 325. See also Bruce Bagemihl, *Biological Exu-*

berance: *Animal Homosexuality and Natural Diversity* (New York: St. Martin's Press, 1999), 259.

55. Raymond C. Kelly, "Witchcraft and Sexual Relations: An Exploration of the Social and Semantic Implications of the Structure of Belief," in *Man and Woman in the New Guinea Highlands*, ed. Paula Brown and Georgeda Buchbinder (Washington, D.C.: American Anthropological Association, 1976), 43–44, 46.

56. David F. Greenberg, *The Construction of Homosexuality* (Chicago: University of Chicago Press, 1988), 10–11, 77–88.

57. Clellan S. Ford and Frank Beach, *Patterns of Sexual Behavior* (New York: Harper and Brothers, 1951), 125–43. Gilbert Herdt, *Same Sex, Different Cultures: Gays and Lesbians across Cultures* (Boulder, Colo.: Westview, 1997). See also Stephen O. Murray, *Homosexualities* (Chicago: University of Chicago Press, 2000).

58. Bronislaw Malinowski, *The Sexual Life of Savages in North-Western Melanesia: An Ethnographic Account of Courtship, Marriage, and Family Life among the Natives of the Trobriand Islands, British New Guinea* (New York: Harcourt, Brace and World, 1929), 179–86.

59. Karen Ericksen Paige and Jeffrey M. Paige provide a wide-ranging survey of the anthropological literature on these practices in *The Politics of Reproductive Ritual* (Berkeley: University of California Press, 1981), 34–41 and 189–90.

60. Bruno Bettelheim, *Symbolic Wounds: Puberty Rites and the Envious Male* (London: Thames and Hudson, 1955), 211.

61. See, for example, Ian Hogbin, *The Island of Menstruating Men: Religion in Wogeo, New Guinea* (Scranton, Pa.: Chandler, 1970).

62. F. E. Williams, *Papuans of the Trans-Fly* (Oxford: Clarendon Press, 1936), 200–202.

63. Thomas Gregor, *Anxious Pleasures: The Sexual Lives of an Amazonian People* (Chicago: University of Chicago Press, 1985), 60.

64. See, for instance, Gilbert Herdt, *Guardians of the Flutes: Idioms of Masculinity* (1981; New York: Columbia University Press Morningside Edition, 1987), 255–94.

65. Alan Dundes, "Earth-Diver: Creation of the Mythopoeic Male," *American Anthropologist* 64 (1964): 1032–51. See also Dundes's "Couvade and Genesis," in *Parsing through Customs: Essays by a Freudian Folklorist* (Madison: University of Wisconsin Press, 1987).

66. Alan Dundes, "A Psychoanalytic Study of the Bullroarer," in *Interpreting Folklore* (Bloomington: Indiana University Press, 1980), 167–98.

67. See various essays in Thomas Buckley and Alma Gottlieb, eds., *Blood Magic: Explorations in the Anthropology of Menstruation* (Berkeley: University of California Press, 1988). The editors' introduction (3–50) is entitled "A Critical Appraisal of Theories of Menstrual Symbolism." See esp. 25–40.

68. Bruce M. Knauft, "Bodily Images in Melanesia: Cultural Substances and Natural Metaphors," in *Fragments for a History of the Human Body, Part Three,*

ed. Michel Feher, with Ramona Naddaff and Nadia Tazi (New York: Zone, 1989), 204–5, with references to various sources.

69. Cynthia Eller, *The Myth of Matriarchal Prehistory: Why an Invented Past Won't Give Women a Future* (Boston: Beacon, 2000), 94–96.

70. Natalie Angier reviews research by anthropologists Kim Hill and Hillard Kaplan with the Ache and by Stephen Beckerman among the Barí. Natalie Angier, *Woman: An Intimate Geography* (New York: Anchor, 1999), 382–83. See Stephen Beckerman et al., "The Barí Partible Paternity Project: Preliminary Results," *Current Anthropology* 39 (1998): 164–67. Kim Hill and Hillard Kaplan, "Tradeoffs in Male and Female Reproductive Strategies among the Ache, Pt. 2," in *Human Reproductive Behavior: A Darwinian Perspective*, ed. Laura Betzig et al. (Cambridge: Cambridge University Press, 1988), 298–99. See also Kim Hill and Hillard Kaplan's general report, "On Why Male Foragers Hunt and Share Food," *Current Anthropology* 34, no. 5 (1993): 701–7.

71. William D. Hamilton, "The Genetical Evolution of Social Behaviour, I, II," *Journal of Theoretical Biology* 7, no. 1 (1964): 1–52. See also Hamilton, "Selfish and Spiteful Behaviour in an Evolutionary Model," *Nature* (London) 228 (1970): 1218–20, and "Selection of Selfish and Altruistic Behavior in Some Extreme Models," in *Man and Beast: Comparative Social Behavior*, ed. J. F. Eisenberg and W. S. Dillon (Washington, D.C.: Smithsonian Institution Press, 1971), 57–91.

72. Max Weber's classic formulation makes *meaning* the pivotal component of action: "We shall speak of 'action' insofar as the acting individual attaches a subjective meaning to his behavior. . . . Action is 'social' insofar as its subjective meaning takes account of the behavior of others and is thereby oriented in its course." Max Weber, *Economy and Society: An Outline of Interpretive Sociology*, ed. Guenther Roth and Claus Wittich (New York: Bedminster Press, 1968), 4.

73. Sahlins calculates r to be as little as $\frac{1}{32,768}$ in nine generations. (Siblings share $\frac{1}{2}$ of their genes, grandsiblings share $\frac{1}{4}$, great-grandsiblings $\frac{1}{8}$, and so on.) By my count, r is actually as little as $\frac{1}{1,024}$ in nine generations— still a very unimpressive degree of relatedness among cooperative kin. (It takes fourteen generations to derive an r as low as $\frac{1}{32,768}$.)

74. Marshall Sahlins, *The Use and Abuse of Biology: An Anthropological Critique of Sociobiology* (Ann Arbor: University of Michigan Press, 1976), 17–67, 27–28, 52–54, 47.

75. I gloss Marshall Sahlins's arguments from his *Culture and Practical Reason* (Chicago: University of Chicago Press, 1976).

CHAPTER 15. THE SOCIAL BODY

1. Natalie Angier, *Woman: An Intimate Geography* (New York: Anchor Books, 1999), xvii–xviii.

2. Clifford Geertz, "The Impact of the Concept of Culture on the Concept of Man," in *The Interpretation of Cultures* (New York: Basic Books, 1973), 49.

3. Ibid., 48.

4. Ibid., 49.

5. H. Ronald Pulliam, *Programmed to Learn: An Essay on the Evolution of Culture* (New York: Columbia University Press, 1980).

6. Maurice Merleau-Ponty, *Phenomenology of Perception* (London: Routledge and Kegan Paul, 1962), 170.

7. Geertz, "The Impact of the Concept of Culture," 49.

8. I take my cue, and this chapter's title, from Mary Douglas's summary of Marcel Mauss's classic essay on techniques of the body, from which she derives the concept, "the social body": "Every kind of [human] action carries the imprint of learning, from feeding to washing, from repose to movement and, above all, sex." *Natural Symbols* (New York: Vintage, 1970), 93.

9. See Paul Ekman, "Universal Facial Expressions of Emotion," in *Culture and Personality: Contemporary Readings,* ed. Robert A. LeVine (Chicago: Aldine, 1974), 8–15.

10. Norbert Elias, "On Human Beings and Their Emotions: A Process-Sociological Essay," in *The Body: Social Process and Cultural Theory,* ed. Mike Featherstone, Mike Hapworth, and Bryan S. Turner (London: Sage, 1991), esp. 120–25.

11. See Lawrence Hirschfeld's brief discussion of the "reading" of facial expressions in his review essay, "The Inside Story," *American Anthropologist* 102, no. 3 (September 2000): 627.

12. Even this statement may be too narrow for certain cases. In practice, some uses of the smile do in fact convey sadness. Kenneth Burke (whose clever perspectives and quirky turns on questions of form and universality were much invoked in the early years of the interpretive turn in anthropology, especially by Clifford Geertz) notes that "in Japan it is customary to smile on mentioning the death of a close friend." He goes on to add that "we might say, not 'he smiled,' but 'his face fell,' as the Western equivalent." Kenneth Burke, *Counter-Statement* (1931; Chicago: University of Chicago Press, 1953), 149. It's not such a stretch. I well remember the odd faces and strained smiles at funerals in my native South. Especially in the years before the loosening of emotional conventions, it was considered bad form to show anyone your unmodified grief. Thus the funeral smile: a way of maintaining face.

13. Charles Larson discusses his Nigerian students' responses to Thomas Hardy's *Far from the Madding Crowd*—"Excuse me, sir, what does it mean 'to kiss'?"—in "Heroic Ethnocentrism: The Idea of Universality in Literature," in *The Post-Colonial Studies Reader,* ed. Bill Ashcroft, Gareth Griffiths, and Helen Tiffin (New York: Routledge, 1995), 62–65.

14. Nancy Scheper-Hughes, *Death without Weeping: The Violence of Everyday Life in Brazil* (Berkeley: University of California Press, 1992), 326.

15. Merleau-Ponty, *Phenomenology of Perception,* 88.

16. Silvan Tomkins, *The Positive Affects,* vol. 1 of *Affect, Imagery, Consciousness* (New York: Springer, 1962), and *The Negative Affects,* vol. 2 of *Affect, Imagery, Consciousness* (New York: Springer, 1963).

17. Ruth Benedict, *Patterns of Culture* (1934; Boston: Houghton Mifflin, 1989), 23–24.

18. I paraphrase a somewhat different argument by Richard Wollheim, *On the Emotions* (New Haven, Conn.: Yale University Press, 2000).

19. Marshall Sahlins, *The Use and Abuse of Biology: An Anthropological Critique of Sociobiology* (Ann Arbor: University of Michigan Press, 1976), 7, 10.

20. Brent Berlin and Paul Kay, *Basic Color Terms: Their Universality and Evolution* (1969; Berkeley: University of California, 1991), 22–23.

21. John B. Carroll and Benjamin Lee Whorf, eds., *Language, Thought, and Reality: Selected Writings of Benjamin Lee Whorf* (Cambridge, Mass.: MIT Press, 1956). Laura Martin, " 'Eskimo Words for Snow': A Case Study in the Genesis and Decay of an Anthropological Example," *American Anthropologist* 88, no. 2 (June 1986): 418–23. Geoffrey K. Pullum and James D. McCawley, *The Great Eskimo Vocabulary Hoax and Other Irreverent Essays on the Study of Language* (Chicago: University of Chicago Press, 1991). Stephen Pinker, *The Language Instinct: How the Mind Creates Language* (New York: William Morrow, 1994), 59–67.

22. Merleau-Ponty, *Phenomenology of Perception*, 13–51.

23. Berlin and Kay, *Basic Color Terms*, 10.

24. Marshall Sahlins, "Colors and Cultures," in *Symbolic Anthropology: A Reader in the Study of Symbols and Meanings*, ed. Janet L. Dolgin, David S. Kemnitzer, and David M. Schneider (New York: Columbia University Press, 1977), 167. All emphases in the original.

25. Sahlins, "Colors and Cultures," 172.

26. Umberto Eco puts the relativity of color this way in "How Culture Conditions the Colors We See," in *On Signs*, ed. Marshall Blonsky (1977; Baltimore: Johns Hopkins University Press, 1985), 171, 173:

> There can be no units without a system [e.g., no colors without a color classification scheme]. The different ways in which cultures make the continuum of colors pertinent, thereby categorizing and identifying hues or chromatic units, correspond to different content systems. The semiotic phenomenon is not independent of perception and discrimination ability; it interacts with these phenomena and frequently overwhelms them. . . . We are animals who can discriminate colors, surely, but we are, above all, cultural animals.

27. Ferdinand de Saussure, *Course in General Linguistics*, ed. Charles Bally and Albert Sechehaye in collaboration with Albert Riedlinger, trans. Wade Baskin (1915; New York: McGraw-Hill), 120–22. The import of this concept is explicated in detail by Jonathan Culler in *Ferdinand de Saussure*, rev. ed. (1976; Ithaca, N.Y.: Cornell University Press, 1986), 36–39.

28. Merleau-Ponty, *Phenomenology of Perception*, 153.

29. I distill, perhaps too abruptly, Merleau-Ponty's approach to vision in *Phenomenology of Perception*.

30. See Simon Garfield, *Mauve: How One Man Invented a Color That Changed the World* (New York: W. W. Norton, 2001). On the broader question of how technologies affect the way we see, see Jonathan Crary, *Suspensions of*

Perception: Attention, Spectacle, and Modern Culture (Cambridge, Mass.: MIT Press, 2000).

31. See Steven Pinker's synopsis of Chomsky's arguments from various texts in *The Language Instinct: How the Mind Creates Language* (New York: Morrow, 1994), 21–24, and 38–45, and perhaps especially chapter 4, "How Language Works," 83–125.

32. Pinker, *The Language Instinct*, 59–67 and 404–30, and *How the Mind Works* (New York: W. W. Norton, 1997), 425–520. More recently, *The Blank Slate: The Modern Denial of Human Nature* (New York: Viking, 2002).

33. Terence Hawkes aptly quotes Giambattista Vico's *New Science* (1725) as a precursor of the structuralist idea: "There must in the nature of human institutions be a mental language common to all nations which uniformly grasps the substance of things feasible in human social life and expresses it with as many diverse modifications as these same things may have diverse aspects." *Structuralism and Semiotics* (Berkeley: University of California Press, 1977), 15.

34. Giorgio Agamben, *Infancy and History: Essays on the Destruction of Experience* (1978; London: Verso, 1993), 57.

35. Arthur C. Danto, *The Body/Body Problem: Selected Essays* (Berkeley: University of California Press, 1999), 214.

36. See, for instance, Lévi-Strauss's arguments at the end of *Totemism* (Boston: Beacon, 1962), esp. 99–104. See also Edmund Leach, *Claude Lévi-Strauss*, rev. ed. (1970; New York: Penguin, 1976), 36–37 and 43–44.

37. Claude Lévi-Strauss, *Elementary Structures of Kinship*, trans. James Harle Bell and John Richard von Sturmer and trans. and ed. Rodney Needham, rev. ed. (Boston: Beacon Press, 1969), and "Structural Analysis in Linguistics and in Anthropology," in *Structural Anthropology* (New York: Basic Books, 1963), 47–49. See also Lévi-Strauss's introduction to *A History of the Family, Volume One: Distant Worlds, Ancient Worlds,* ed. André Burgière, Christiane Klapisch-Zuber, Martine Segalen, and François Zonabend, trans. Sarah Hanbury Tenison, Rosemary Morris, and Andrew Wilson (Cambridge, Mass.: The Belknap Press of Harvard University Press, 1996), 1–7.

38. To make recourse once again to Marshall Sahlins, whose key works make elegant recourse to Claude Lévi-Strauss: "A limit is only a negative determination; it does not positively specify how the constraint was realized." *Use and Abuse of Biology,* 64.

39. I paraphrase one of Lévi-Strauss's interlocutors and friends: Human beings "somehow secrete culture without even wanting to." Maurice Merleau-Ponty, *Sense and Non-Sense,* trans. Hubert L. Dreyfus and Patricia Allen Dreyfus (1948; Evanston, Ill.: Northwestern University Press, 1964), 118.

40. I am grateful to Howard Hastings for this analogy.

CHAPTER 16. THE PRACTICES OF SEX

1. Researchers today express some disagreement over how to understand gender systems that allow options other than *either/or* (either male or female). Are there "third" (or fourth, and fifth) genders? Or are these alternative gen-

ders to be best understood as "crossed-over," "mixed," "intermediary," or "fluid" genders? (Note that this latter query gives not one, but four different options.) My own wording, I hope, is flexible enough to take in systems that are themselves likely to be variable and in many cases ambiguous while still keeping to the point: Not every gender system gives either/or as the only option.

2. Will Roscoe, *The Zuni Man-Woman* (Albuquerque: University of New Mexico Press, 1991), 123–46. Harriet Whitehead, "The Bow and the Burden Strap: A New Look at Institutionalized Homosexuality in Native North America," in *Sexual Meanings: The Cultural Construction of Gender and Sexuality,* ed. Sherry Ortner and Harriet Whitehead (New York: Cambridge University Press, 1981), 80–115. Charles Callender and Lee M. Kochems, "The North American Berdache," *Current Anthropology* 24, no. 4 (October 1983): 1–76. Walter L. Williams, *The Spirit and the Flesh: Sexual Diversity in American Indian Culture* (Boston: Beacon, 1986).

3. Serena Nanda, *Neither Man nor Woman: The Hijras of India* (Belmont, Calif.: Wadsworth, 1990).

4. Judith Butler, *Bodies That Matter: On the Discursive Limits of "Sex"* (New York: Routledge, 1993).

5. M. Kay Martin and Barbara Voorhies, *Female of the Species* (New York: Columbia University Press, 1975), 86.

6. Gilbert Herdt, "Introduction: Third Sexes and Third Genders," in *Third Sex, Third Gender: Beyond Sexual Dimorphism in Culture and History,* ed. Gilbert Herdt (New York: Zone, 1994), 25–33 and 78.

7. Nancy Scheper-Hughes and Margaret Locke, "The Mindful Body: A Prolegomenon to Future Work in Medical Anthropology," *Medical Anthropology Quarterly* 1, no. 1 (March 1987): 6–41.

8. For early versions of this argument, see Sylvia Yanagisako and Jane F. Collier, "The Mode of Reproduction in Anthropology," in *Theoretical Perspectives on Sexual Difference,* ed. Deborah L. Rhode (New Haven, Conn.: Yale University Press, 1990), 131–41. See also Moira Gatens, "A Critique of the Sex/Gender Distinction," *Intervention,* Special Issue (1983): 143–61.

9. Captain Laudonnière reported that there were "many hermaphrodites" in Florida in the 1560s, and Le Moyne concurred that they were "quite common." Later European reports on the practices of two-spirit people transmute "hermaphroditism" into "sodomy." In 1702, Liette maintained that "the sin of sodomy prevails more among [the Miamis of Illinois] than in any other nation." And Bossu claimed that "most of [the Choctaws] are addicted to sodomy." See Jonathan Ned Katz's excerpts in *Gay American History: Lesbians and Gay Men in the USA. A Documentary History,* chapter 4, "Native Americans/Gay Americans" (New York: Meridian, 1976), 285, 286, 288, and 291. See also Callender and Kochems, "The North American Berdache," 9–10.

10. Of course, it is not altogether clear what might count as intersexed or abnormal genitalia. Baby boys' penises are said to average between 1 and 1.5 inches long. Doctors regard infantile penises less than .6 inches long as "abnormal," even if they are fully formed and functional, and surgically remove

them (along with the testes) to build vaginas in their places. In cases like these, the number of "abnormal" genitalia (and medically intersexed persons) could be revised upward or downward, depending on what doctors agree to as the "cut-off" point, if they see need for surgical intervention at all. See Anne Fausto-Sterling, "How to Build a Man," in *The Gender/Sexuality Reader: Culture, History, Political Economy,* ed. Roger N. Lancaster and Micaela di Leonardo (New York: Routledge, 1997), 244–48. See also her subsequent discussion, "Should There Be Only Two Sexes?" in *Sexing the Body: Gender Politics and the Construction of Sexuality* (New York: Basic Books, 2000), 78–114.

11. Alfred Kroeber cannot quite bring himself to credit reports that Juaneno parents "selected" infant boys to rear as two-spirit people. *Handbook of the Indians of California* (Washington, D.C.: Bureau of American Ethnology Bulletin 78, 1925), 647. But the practice is consistent with other reports from other Native American cultures. William A. Hammond, "The Disease of the Skythians (Morbus Feminarum) and Certain Other Analogous Conditions," *American Journal of Neurology and Psychiatry* 1 (1882): 339–55. John H. Honigmann, *The Kaska Indians: An Ethnographic Reconstruction* (New Haven, Conn.: Yale University Publications in Anthropology 51, 1954), 130. Hubert Howe Bancroft, *The Native Races of the Pacific States of North America,* vol. 1 (New York: Appleton, 1874), 82.

12. Marjorie Garber, *Vested Interests: Cross-Dressing and Cultural Anxiety* (New York: Routledge, 1992).

13. Clifford Geertz, *Local Knowledge* (New York: Basic Books, 1984), 85.

14. Anne Fausto-Sterling, "The Five Sexes: Why Male and Female Are Not Enough," *The Sciences* (March–April 1993): 20–24. Fausto-Sterling has since ceased using the phrase "five sexes" not so as to re-dualize sex, but so as to avoid reifying any fixed number at all. See *Sexing the Body,* 78, 95–114.

15. Maurice Merleau-Ponty, *Phenomenology of Perception* (London: Routledge and Kegan Paul, 1962), 146.

16. Clifford Geertz, "Deep Play: Notes on the Balinese Cockfight," in *The Interpretation of Cultures* (New York: Basic Books, 1973), 421.

17. Gilbert Herdt, *The Sambia: Ritual and Gender in New Guinea* (Fort Worth, Tex.: Holt, Rinehart and Winston, 1987), 139–45.

18. F. E. Williams, *Papuans of the Trans-Fly* (Oxford: Clarendon Press, 1936), 307–8.

19. Lana Thompson, *The Wandering Womb: A Cultural History of Outrageous Beliefs about Women* (New York: Prometheus Books, 1999), 33–34.

20. Stith Thompson, *Motif-Index of Folk-Literature: A Classification of Narrative Elements in Folktales, Ballads, Myths, Fables, Mediaeval Romances, Exempla, Fabliaux, Jest-Books, and Local Legends,* rev. ed., vol. 3 (Bloomington: Indiana University Press, 1966), 165.

21. Wendy Doniger O'Flaherty, *Tales of Sex and Violence: Folklore, Sacrifice, and Danger in the Jaiminiya Brahmana* (Chicago: University of Chicago Press, 1985), 101–3. Indian legends often place vaginas in unusual places: in the forehead, in the armpit. Thompson, *Motif-Index of Folk-Literature,* vol. 3, 165.

22. Claude Lévi-Strauss, *The Raw and the Cooked: Introduction to a Science of Mythology: I,* trans. John Weightman and Doreen Weightman (1964; New York: Harper Colophone, 1975), 126 n. 9.

23. Luce Irigaray, *This Sex Which Is Not One,* trans. Catherine Porter (1977; Ithaca, N.Y.: Cornell University Press, 1985), 28.

24. Lévi-Strauss, *Raw and the Cooked,* 113. Thompson, *Motif-Index of Folk-Literature,* vol. 3, 164.

25. Terry Eagleton, "Self-Realization, Ethics, and Socialism," *New Left Review,* no. 237 (September–October 1999): 152, 153.

26. On the "taking in hand," see Merleau-Ponty, *Phenomenology of Perception,* 171–73.

CHAPTER 17. THIS QUEER BODY

1. Or, as Sartre puts it: *"Desire is consent to desire." The Philosophy of Jean-Paul Sartre,* ed. Robert Denoon Cumming (New York: Vintage, 1965), 214.

2. Romans 1:26–27.

3. Jennine Davis-Kimball and C. Scott Littleton, "Warrior Women of the Eurasian Steppes," *Archaeology* 50, no. 1 (January–February 1997): 44–48.

4. This is one of the recurring threads of Micaela di Leonardo's survey of the history of twentieth-century anthropology, *Exotics at Home: Anthropologies, Others, American Modernity* (Chicago: University of Chicago Press, 1998).

5. Colin Stewart, "An Udder Way of Making Lambs" (769–71), and I. Wilmut et al., "Viable Offspring Derived from Fetal and Adult Mammalian Cells" (810–13), *Nature* 385 (February 27, 1997).

6. Lisa Keen, "DNA Discoverer Backs Aborting 'Gay' Fetuses," *Washington Blade,* February 28, 1997, 1, 21.

7. See John J. O'Connor, *TV Weekend:* "Amniocentesis for the Gay Factor," *New York Times,* March 21, 1997, B33.

8. It hardly seems necessary to insert the usual academic "*[sic]*s" here, where gays have become a species and a fetus has become a child. Ted Gideonse, "Are We an Endangered Species?" *The Advocate,* May 27, 1997, 28–30.

9. The classic distillation of this historical argument is provided by David M. Halperin in *One Hundred Years of Homosexuality and Other Essays on Greek Love* (New York: Routledge, 1990), 15–40 and 41–53.

10. See Jonathan Ned Katz, " 'Homosexual' and 'Heterosexual': Questioning the Terms," and "Coming to Terms: Conceptualizing Men's Erotic and Affectional Relations with Men in the United States, 1820–1892," in *A Queer World: The Center for Lesbian and Gay Studies Reader,* ed. Martin Duberman (New York: New York University Press, 1997), 177–80, 216–35.

11. Jonathan Goldberg, *Sodometries: Renaissance Texts, Modern Sexualities* (Stanford, Calif.: Stanford University Press, 1992), 1–26. Michel Foucault, *The History of Sexuality, Volume 1: An Introduction,* trans. Robert Hurley (1976; New York: Vintage, 1980), 37, 101.

12. Foucault, *The History of Sexuality,* 43.

13. Ibid., 40, 43.

14. Siobhan Somerville, "Scientific Racism and the Construction of the Homosexual Body," in *The Gender/Sexuality Reader: Culture, History, Political Economy,* ed. Roger N. Lancaster and Micaela di Leonardo (New York: Routledge, 1997), 41–56. Jennifer Terry, "Anxious Slippages between 'Us' and 'Them': A Brief History of the Scientific Search for Homosexual Bodies," in *Deviant Bodies,* ed. Jennifer Terry and Jacqueline Urla (Bloomington: Indiana University Press, 1995), 129–69. See also Ann Laura Stoler, "Carnal Knowledge and Imperial Power: Gender, Race, and Morality in Colonial Asia," in *The Gender/ Sexuality Reader,* 13–36, esp. 22–27.

15. Eve Kosofsky Sedgwick, *Epistemology of the Closet* (Berkeley: University of California Press, 1990), 2. Jonathan Ned Katz, *Gay/Lesbian Almanac: A New Documentary* (New York: Harper and Row, 1983), 147–50. See also Katz's majestic study *The Invention of Heterosexuality* (New York: Dutton, 1975).

16. See Jamake Highwater's brilliant discussion in *The Mythology of Transgression: Homosexuality as Metaphor* (New York: Oxford University Press, 1997), esp. 93–97.

17. Gail Vines, *Raging Hormones: Do They Rule Our Lives?* (Berkeley: University of California Press, 1994), 116.

CHAPTER 18. THE BIOLOGY OF THE HOMOSEXUAL

1. Thomas H. Maugh II and Nora Zamichow, *Los Angeles Times,* August 30, 1991. Malcolm Gladwell, *Washington Post,* December 17, 1991. Jamie Talan, *Newsday,* December 9, 1991. Kim Painter, *USA Today,* December 17, 1991. Natalie Angier, *New York Times,* July 16, 1993. Curt Suplee, *Washington Post,* October 31, 1995.

2. Anne Fausto-Sterling, *Myths of Gender: Biological Theories about Women and Men,* rev. ed. (1985; New York: Basic Books, 1992), 257.

3. Simon LeVay, "A Difference in Hypothalamic Structure between Heterosexual and Homosexual Men," *Science* 253 (1991): 1034–37.

4. See Edward Stein's calculations in *The Mismeasure of Desire: The Science, Theory, and Ethics of Sexual Orientation* (Oxford: Oxford University Press, 1999), 200–201.

5. Gail Vines, *Raging Hormones: Do They Rule Our Lives?* (Berkeley: University of California Press, 1994), 112. John Maddox, "Is Homosexuality Hardwired?" *Nature* 353 (1991): 13.

6. Gilbert Zicklin, "Media, Science, and Sexual Ideology: The Promotion of Sexual Stability," in *A Queer World: The Center for Lesbian and Gay Studies Reader,* ed. Martin Duberman (New York: New York University Press, 1997), 383.

7. See Simon LeVay, *The Sexual Brain* (Cambridge, Mass.: MIT Press, 1993), 121–22. William Byne, "LeVay's Thesis Reconsidered," in *A Queer World,* 325, and Stein, *The Mismeasure of Desire,* 201.

8. Stein, *The Mismeasure of Desire,* 210. Simon LeVay and Dean Hamer,

"Evidence for a Biological Influence in Male Homosexuality," *Scientific American* 270 (May 1994): 44–49.

9. LeVay, *The Sexual Brain*, 122.

10. David Gelman with Donna Foote, Todd Barrett, and Mary Talbot, "Born or Bred?" *Newsweek*, February 24, 1992, 49. See also Ruth Hubbard and Elijah Wald, *Exploding the Gene Myth: How Genetic Information Is Produced and Manipulated by Scientists, Physicians, Employers, Insurance Companies, Educators, and Law Enforcers* (Boston: Beacon, 1993), 97–98.

11. Byne, "LeVay's Thesis Reconsidered," 318–27. See also William Byne, "Science and Belief: Psychobiological Research on Sexual Orientation," in *Sex, Cells, and Same-Sex Desire: The Biology of Sexual Preference*, ed. John P. De Cecco and David Allen Parker (New York: Harrington Park Press, 1995), 303–44.

12. In two subsequent papers, William Byne suggests that the INAH3 is sexually dimorphic in human beings and in rhesus monkeys. Byne et al., "The Interstitial Nuclei of the Human Anterior Hypothalamus: An Investigation of Sexual Variation in Volume and Cell Size, Number and Density," *Brain Research* 856, nos. 1–2 (February 21, 2000): 254–58, and Byne, "The Medial Preoptic and Anterior Hypothalamic Regions of the Rhesus Monkey: Cytoarchitectonic Comparison with the Human and Evidence for Sexual Dimorphism," *Brain Research* 793, nos. 1–2 (May 18, 1998): 346–50. Byne notes, however, the difficulties of securing comparable brain samples: "Because my samples from women had been in fixitive longer, on average, than the samples from men, and because fixitives can cause tissue to shrink, the apparent sex differences in my preliminary studies may merely be a fixation shrinkage artifact." Byne, "LeVay's Thesis Reconsidered," 325.

13. Byne, "LeVay's Thesis Reconsidered," 322–23.

14. See Michael Ruse's extensive review of this literature, *Homosexuality: A Philosophical Inquiry* (Oxford: Basil Blackwell, 1988), 103–12.

15. I first noted the preposterous nature of these labeling procedures in the immediate aftermath of the LeVay study. See *Life Is Hard: Machismo, Danger, and the Intimacy of Power in Nicaragua* (Berkeley: University of California Press, 1992), 314–15 n. 10 and 315–16 n. 11. See also William Byne and Bruce Parsons, "Sexual Orientation: The Biologic Theories Reappraised," *Archives of General Psychiatry* 50 (1993): 228–39, and William Byne, "Science and Belief: Psychobiological Research on Sexual Orientation," in *Sex, Cells, and Same-Sex Desire*, 311–12.

16. Michael J. Bailey and Richard Pillard, "A Genetic Study of Male Sexual Orientation," *Archives of General Psychiatry* 48 (1991): 1089–96. See also Michael J. Bailey and Richard Pillard, "Are Some People Born Gay?" *New York Times*, December 17, 1991.

17. Hubbard and Wald, *Exploding the Gene Myth*, 97.

18. Actually, adoption procedures tend to select for relatively homogeneous, middle-class environments, even for twins separated at birth. And it turns out that many twins called "separated at birth" were not really so separated after

all. Many such twins are actually reared by different sets of relatives in the same town.

19. Hubbard and Wald, *Exploding the Gene Myth*, 97.

20. Zicklin, "Media, Science, and Ideology," 385.

21. David Gelman et al., "Born or Bred: The Origins of Homosexuality," *Newsweek*, February 24, 1992, 46.

22. I leave aside here a discussion of all those terms that give away more than they need divulge of the author's presuppositions, for example, "nurturing" parents, a "lively interest" in sports, and "appropriate relations with women."

23. See the section entitled "Homosexual Outlet" in Alfred C. Kinsey et al., *Sexual Behavior in the Human Male* (Philadelphia: W. B. Saunders, 1948), 610–66.

24. Zicklin, "Media, Science, and Sexual Ideology," 384.

25. See Dean Hamer and Peter Copeland, *The Science of Desire: The Search for the Gay Gene and the Biology of Behavior* (New York: Simon and Schuster, 1994), 21.

26. Dean Hamer, Stella Hu, Victoria Magnuson, Nan Hu, and Angela Pattatucci, "A Linkage between DNA Markers on the X Chromosome and Male Sexual Orientation," *Science* 261 (1993): 321–27.

27. Hamer and Copeland, *The Science of Desire*, 203–4, 17–38.

28. Stein, *The Mismeasure of Desire*, 217. Neil Risch, E. Squires-Wheeler, and B. J. B. Keats, "Male Sexual Orientation and Genetic Evidence," *Science* 262 (December 24, 1993): 2063–65.

29. On the matrilateral skewing of American and English kinship systems, especially but not exclusively patterns of kinship in the lower classes, see David M. Schneider and Raymond T. Smith, *Class Differences in American Kinship* (1973; Ann Arbor: University of Michigan Press, 1978), 9, 40–43, 53–55. On the significance of maternal kin work, see Micaela di Leonardo, "The Female World of Cards and Holidays: Women, Families, and the Work of Kinship," *Signs* 12, no. 3 (1987): 440–53.

30. See Zicklin, "Media, Science, and Sexual Ideology," 385.

31. In *The Science of Desire* and in response to Hamer's critics, Hamer and Copeland report that the Hamer team *did* attempt other checks: the first was to ponder the distribution of *lesbian* relatives of the gay male subjects. Theoretically, if the maternal links simply reflected better knowledge of one's maternal kin, then there ought to also be elevated reportage of lesbianism along maternal lines. Hamer and Copeland report that the research team found no such pattern. The second check was to review lesbian informants' reportage of *gay male* relatives from a separate study. The authors report that there was no significant difference between maternal and paternal links for lesbian subjects (103–4). Of course, these "checks" assume that communication about relatives' sex lives occurs in a transparent environment unaffected by either sexual intolerance or gender inequalities—that talk about sex is uninflected by different maternal as opposed to paternal (and male as opposed to female, or mother-son, as opposed to mother-daughter, etc.) strategies of revelation and conceal-

ment. . . . It is by no means unthinkable that such factors could differentially distribute family knoweldge about gays and lesbians. As Edward Stein demonstrates in *The Mismeasure of Desire* (218), it remains altogether plausible that the elevated maternal pattern of homosexuality reported by gay subjects is a *strictly* sociological effect, derived from partial knowledges, selectively revealed and asymetrically conveyed.

32. Hubbard and Wald, *Exploding the Gene Myth*, 75.

33. Stella Hu, Angela Pattatucci, C. Patterson, L. Li, D. W. Fulker, S. S. Cherny, L. Kruglak, and Dean Hamer, "Linkage between Sexual Orientation and Chromosome Xq28 in Males but Not in Females," *Nature Genetics* 11, no. 3 (1995): 248–56.

34. Stein, *The Mismeasure of Desire*, 220.

35. Jonathan Marks, "Behavioral Genetics," chapter 5 in *What It Means to Be 98 Percent Chimpanzee* (Berkeley: University of California Press, 2002).

36. Natalie Angier, "Variant Gene Tied to a Love of New Thrills," *New York Times*, January 2, 1996. See Angier's follow-up story later the same year, which reports a failure to replicate the original studies: "Maybe It's Not a Gene behind a Person's Thrill-Seeking Ways," *New York Times*, November 1, 1996.

37. Although the shorthand that refers to genetic "causes" is appealing when simple Mendelian traits such as eye color are under discussion, the idea of a genetic "cause" founders when polygenic traits are in question. Simple Mendelian traits account for only a small percentage of human traits. See Hubbard and Wald's discussion in *Exploding the Gene Myth*, 40–42.

38. R. C. Lewontin, Steven Rose, and Leon J. Kamin, *Not in Our Genes: Biology, Ideology, and Human Nature* (New York: Pantheon, 1984), 97.

39. To be more precise, it means that 90 percent of the variance in height for a population is accounted for by genetic variance. See Lewontin, Rose, and Kamin, *Not in Our Genes*, 97.

40. Ibid.

41. This important point is meticulously illustrated by Richard Lewontin, from whom I draw the following example, in *The Triple Helix: Gene, Organism, and Environment* (Cambridge, Mass.: Harvard University Press, 2000), 20–24.

42. Jens Clausen, David Keck, and William Heisey, *Experimental Studies on the Nature of Species, Vol. 3: Environmental Responses of Climatic Races of Achillea*, Carnegie Institution of Washington Publication 581 (1958), 1–129.

43. I leave aside certain well-known paradoxes of the scientific approach to heritability. Since heritability is a measure of *variance*, certain traits that are absolutely genetic show no variation—hence, zero heritability. (Imagine a population in which everyone has brown eyes.) Correlatively, if certain other traits "run in families" (because of where the families live) or are *socially* attached to a genetic trait (like skin color), they display high heritability, despite having plainly environmental origins. See Edward Stein's discussion in *The Mismeasure of Desire*, 142–44.

44. See Richard Lewontin's short masterpiece of science criticism, *Biology as Ideology: The Doctrine of DNA* (New York: HarperPerennial, 1992).

45. See John Horgan, *The End of Science: Facing the Limits of Knowledge in the Twilight of the Scientific Age* (Reading, Mass.: Addison-Wesley, 1996), and *The Undiscovered Mind—How the Human Mind Defies Replication, Medication, and Explanation* (New York: The Free Press, 1999).

CHAPTER 19. DESIRE IS NOT A "THING"

1. Dean H. Hamer and Peter Copeland, *The Science of Desire: The Search for the Gay Gene and the Biology of Behavior* (New York: Simon and Schuster, 1994), 180–86, esp. 186. Hamer and Copeland, *Living with Our Genes: Why They Matter More Than You Think* (New York: Doubleday, 1998).

2. Simon LeVay, *The Sexual Brain* (Cambridge, Mass.: MIT Press, 1993).

3. Plutarch describes the sacred Band of Thebes in *The Life of Pelopidas*. On the lifeways of classical Hellenic aristocrats, see David M. Halperin, "One Hundred Years of Homosexuality," in *One Hundred Years of Homosexuality and Other Essays on Greek Love* (New York: Routledge, 1990), 15–40.

4. Gilbert Herdt, *Guardians of the Flutes: Idioms of Masculinity* (New York: Columbia University Press Morningside Edition, 1987), and *The Sambia: Ritual and Gender in New Guinea* (Fort Worth, Tex.: Holt, Rinehart and Winston, 1987). Raymond C. Kelly, "Witchcraft and Sexual Relations: An Exploration of the Social and Semantic Implications of the Structure of Belief," in *Man and Woman in the New Guinea Highlands*, ed. Paula Brown and Georgeda Buchbinder (Washington, D.C.: American Anthropological Association, 1976), 36–53.

5. Roger N. Lancaster, "Subject Honor, Object Shame," in *Life Is Hard: Machismo, Danger, and the Intimacy of Power in Nicaragua* (Berkeley: University of California Press, 1992), 235–78. Richard Parker, "Masculinity, Femininity, and Homosexuality: On the Anthropological Interpretation of Sexual Meanings in Brazil," *Journal of Homosexuality* 11, nos. 3–4 (1985): 155–63, and *Bodies, Pleasures, and Passions: Sexual Culture in Contemporary Brazil* (Boston: Beacon, 1991), 35–54.

6. In *The Triple Helix: Gene, Organism, and Environment* (Cambridge, Mass.: Harvard University Press, 2000), Richard Lewontin discusses just what is wrong with the notion of a "genetic tendency" this way:

> In everyday language we say that Bill "tends to be fat" while Ronald "tends to be thin," but it is not clear how this notion is to be used for genotypes and environments. In some environments Bill will be thin and in others, fat. . . . Often the notion of "tendency" carries with it an implicit idea of "normal" conditions. . . . But in general we do not know how to specify the ideal "normal" environment in which the tendencies of genotypes are to be compared, nor does such an idealized "normal" environment exist. (29–30)

7. These lines are from Randy Cohen's review of Stephen Fry's book, *Moab Is My Washpot: An Autobiography* (New York: Random House, 1999), in *The New York Times Book Review*, June 13, 1999, 6.

8. Michael Warner, *The Trouble with Normal: Sex, Politics, and the Ethics of Queer Life* (New York: The Free Press, 1999), 9. F. Whitam, M. Diamond, and J. Martin, "Homosexual Orientation in Twins: A Report on 61 Pairs and Three Triplet Sets," *Archives of Sexual Behavior* 22 (1993): 187–206.

9. Not since the days of Krafft-Ebing, at any rate. The early classification schemes and their purported links to biology were, in the nineteenth century, complicated and ultimately untenable. What has come down to us today is the simplified version.

10. Warner, *The Trouble with Normal*, 10.

11. See Freud's old distinction between "aim" and "object" in *Three Essays on the Theory of Sexuality*, trans. and ed. James Strachey (1962; New York: Basic Books, 1975), esp. chapter 1, "The Sexual Aberrations" (1–38).

12. Henry James, "The Beast in the Jungle," in *The Complete Tales of Henry James*, ed. Leon Edel (London: Rupert Hart-Davis, 1964), vol. 11, 401.

13. Eve Kosofsky Sedgwick, *Epistemology of the Closet* (Berkeley: University of California Press, 1990), 182–212.

14. Henry James, quoted by Sedgwick, in *Epistemology of the Closet*, 211.

15. Sedgwick, *Epistemology of the Closet*, 210.

16. James, quoted in Sedgwick, *Epistemology of the Closet*, 212.

17. Maurice Merleau-Ponty, "Indirect Language and the Voices of Silence," in *The Merleau-Ponty Aesthetics Reader: Philosophy and Painting*, ed. Galen Johnson (Evanston, Ill.: Northwestern University Press, 1993), 113.

18. See especially chapter 6, "Men Court Men: Initiations and Secret Societies," in Lionel Tiger, *Men in Groups* (New York: Random House, 1969), 126–55.

19. Freud was certainly on to the logic of substitutions, whereby one person or thing might represent or substitute for another. As Freud construed everyday queerness in a footnote to the 1915 edition of the *Three Essays:* "All human beings are capable of making a homosexual object-choice and have in fact made one in their unconscious. Indeed, libidinal attachments to persons of the same sex play no less a part as factors in normal mental life, and a greater part as a motive force for illness, than do similar attachments to the opposite sex" (11 n. 1). See also Eve Kosofsky Sedgwick, *Between Men: English Literature and Male Homosexual Desire* (New York: Columbia University Press, 1985), and *Epistemology of the Closet*.

20. Maurice Merleau-Ponty, *Phenomenology of Perception* (London: Routledge and Kegan Paul, 1962), 168–69.

21. Ibid. Here, in contrast with Merleau-Ponty's phenomenological conception of desire, I refer to Georg Lukács's theory of reification (from the Latin *res*, or "thing"), derived from Marx's ideas about alienation and fetishism. Genetic fetishism is reification in the strictest sense of the term: It ascribes human powers, actions, and attributes to chains of nucleotides. But more profoundly, the idea that desire can be dislodged from its cultural context, demarcated from other perceptual practices, and abstracted or frozen from the ongoing flow of social life also belongs to the order of reification and fetishism. Georg Lukács,

"Reification and the Consciousness of the Proletariat," in *History and Class Consciousness: Studies in Marxist Dialectics* (Cambridge, Mass.: MIT Press, 1971), 46–222.

22. Maurice Merleau-Ponty, "Cézanne's Doubt," in *The Merleau-Ponty Aesthetics Reader*, 73, 75.

23. Alexandre Kojève, *Introduction to the Reading of Hegel: Lectures on the Phenomenology of Spirit*, assembled by Raymond Queneau, ed. Allan Bloom, trans. James H. Nichols Jr. (1939; New York: Basic Books, 1969), 3–4.

CHAPTER 20. FAMILIAR PATTERNS, DANGEROUS LIAISONS

1. Marc Breedlove, "Another Important Organ" (15–16), and Jiang-Ning Zou, Michel A. Hofman, Louis J. G. Gooren, and Dick F. Swaab, "A Sex Difference in the Human Brain and Its Relation to Transsexuality" (68–70), *Nature* 378 (November 2, 1995). Christine Gorman, "Women Trapped in Men's Bodies," *Time* (Science Section), November 13, 1995.

2. Robert Finn, "Biological Determination of Sexuality Heating Up as a Research Field," *The Scientist* 10, no. 1 (January 8, 1996): 13–16. William Byne, personal communications, June 6–7, 2001. William Byne et al., "The Interstitial Nuclei of the Human Anterior Hypothalamus: An Investigation of Variation with Sex, Sexual Orientation and HIV Status," *Hormones and Behavior* 40 (2001), 86–92.

3. J. Michael Bailey, Michael P. Dunne, and Nicholas G. Martin, "Genetic and Environmental Influences in Sexual Orientation and Its Correlates in an Australian Twin Sample," *Journal of Personality and Social Psychology* 78, no. 3 (March 2000): 524–36, esp. 533–34. The authors note (533) that flawed subject recruitment methods used by Bailey and Pillard in the earlier study tend to inflate MZ concordance rates. Edward Stein cites a paper by the same authors under the working title "The Distribution, Correlates, and Determinants of Sexual Orientation in an Australian Twin Sample" (n.d.), but no paper under this title has as yet appeared. See *The Mismeasure of Desire: The Science, Theory, and Ethics of Sexual Orientation* (Oxford: Oxford University Press, 1999), 193–94.

4. Eliot Marshall, "NIH's 'Gay Gene' Study Questioned," *Science* 268 (June 30, 1995): 1841. John Crewdson, front page of the *Chicago Tribune*, June 25, 1995. Finn, "Biological Determination of Sexuality Heating Up as a Research Field."

5. Sciencescope: "No Misconduct in 'Gay Gene' Study," *Science* 275 (February 28, 1997): 1251.

6. Marshall, "NIH's 'Gay Gene' Study Questioned." Finn, "Biological Determination of Sexuality Heating Up as a Research Field." Sciencescope: "No Misconduct in 'Gay Gene' Study."

7. *New York Times*, April 23, 1999. George Rice, Carol Anderson, Neil Risch, and George Ebers, "Male Homosexuality: Absence of Linkage to Microsatellite Markers at Xq28," *Science* 284 (April 23, 1999): 665–67.

8. John C. Crabbe, Douglas Wahlsten, Bruce C. Dudek, "Genetics of Mouse

Behavior: Interactions with Laboratory Environment" (1670–72), and Martin Enserink, "Behavioral Genetics: Fickle Mice Highlight Test Problems" (1599–1600), *Science* 284 (June 4, 1999). (Many thanks to Lynne Constantine for bringing these references to my attention.)

9. Stein, *The Mismeasure of Desire*.

10. See Steven Rose's review of Dean Hamer and Peter Copeland, *Living with Our Genes: Why They Matter More Than You Think* (New York: Macmillan, 1999), in *The Guardian*, May 8, 1999.

11. It bears noting that the author exploits the rhetorical ploy that safe sex and prophylaxis are mere "technological fixes" at the very moment when physis and techne, sex and artifice, culture and technology all seem inextricably interlinked, even—especially?—in that inner sanctum of nature, heterosexual reproduction (see chapter 21 of this book). Gabriel Rotello, *Sexual Ecology: AIDS and the Destiny of Gay Men* (New York: Dutton, 1997).

12. Larry Kramer, "Our Bodies, Ourselves," *The Advocate*, issue 734, May 27, 1997, 59–70.

13. See, for instance, coverage in the *Washington Blade:* " 'I did it for my own truth': Comic Ellen DeGeneres surprises no one by coming out" (April 11, 1997), followed by "Unearthing the Roots of Sexual Orientation" (May 9, 1997), which covered the *Oprah* show's post-*Ellen* roundtable discussion among the usual lineup of bioreductionists (LeVay, Pillard, and Hamer) and critic William Byne.

14. Radicalesbians, "The Woman-Identified Woman" (1970), in *Radical Feminism*, ed. Anne Koedt, Ellen Levine, and Anita Rapone (New York: Quadrangle Books, 1973), 240–45. Dennis Altman, *Homosexual Oppression and Liberation* (1971; New York: New York University Press, 1993), esp. chapter 3, "Liberation: Towards the Polymorphous Whole," 80–116.

15. Elisabeth Bumiller, "Elite Manhattan Schools Confront the Gay Issue: Parents' Concerns at Same-Sex Academies," *New York Times*, June 13, 1997, A29.

16. Lisa Keen, "DNA Discoverer Backs Aborting 'Gay' Fetuses," *Washington Blade*, February 28, 1997, 21.

17. Michel Foucault, *History of Sexuality, Volume 1: An Introduction*, trans. Robert Hurley (1976; New York: Vintage, 1980), 101.

18. Altman, *Homosexual Oppression and Liberation*. Guy Hocquenghem, *Homosexual Desire*, trans. Diniella Dangoor (1978; Durham, N.C.: Duke University Press, 1993).

19. See Lisa Keen and Suzanne B. Goldberg's fine survey, *Strangers to the Law: Gay People on Trial* (Ann Arbor: University of Michigan Press, 1998). Kenji Yoshino argues that the gay-rights movement's emphasis on sexual orientation as an "immutable characteristic" promotes a numbing conformity in ways both subtle and not so subtle: Basically, the emphasis on essential "identity" leaves unchallengeable all those forms of discrimination based on overt (especially excessive) "behavior" (e.g., gender nonconformity, promiscuity, the display of "inappropriate" behavior). Kenji Yoshino, "Covering," *The Yale Law Journal* 111, no. 4 (January 2002): 769–939. See Kristin Eliasberg, "Making a

Case for the Right to Be Openly Different," *New York Times* (Arts and Ideas), June 16, 2001.

20. Chandler Burr, "Homosexuality and Biology," *Atlantic Monthly* 272, no. 3 (March 1993): 47–65.

21. Chandler Burr, *A Separate Creation: The Search for the Biological Origins of Sexual Orientation* (New York: Hyperion, 1996), 306.

22. I borrow Michel Foucault's apt phrase "technologies of the self." *Technologies of the Self: A Seminar with Michel Foucault*, ed. Luther H. Martin, Huck Gutman, and Patrick H. Hutton (Amherst: University of Massachusetts Press, 1988).

23. And why not? Everything in modern advertising culture, which is to say modern culture, makes increasingly open use of such a rhetorical flip: claiming the naturalness of something so as to establish the authenticity of what is in fact contrived, manufactured, and social. . . . A wink and a nod. We're all in on the joke.

24. For an up-close, sometimes funny, sometimes harrowing account of one man's experience with therapists intent on "curing" him of homosexuality, see Martin Duberman's *Cures: A Gay Man's Odyssey* (New York: E. P. Dutton, 1991).

25. Again, it is notable that every one of the major innatist texts on the "biology of homosexuality" spends at least as much time revisiting sociobiological fables and evolutionary-psychological just-so stories purporting to explain how it is that men and women got to be the way they allegedly are as they do visiting the question of homosexual exceptionalism.

26. Erica Goode, "Study Questions Gene Influence on Male Homosexuality," *New York Times*, April 23, 1999.

27. On the stakes involved in the present naturalizations of sex, the tensions between "rights" and "liberation," and the consequent "identity crisis" of the gay movement, I can think of no book more capacious and discerning than Richard Goldstein's slim analytical polemic, *The Attack Queers: Liberal Society and the Gay Right* (New York: Verso, 2002).

28. See Hamer and Copeland, *The Science of Desire*, 156–59. J. A. Y. Hall and D. Kimura, "Dermatoglyphic Asymmetry and Sexual Orientation in Men," *Behavioral Neuroscience* 108, no. 6 (1994): 1203–6.

29. M. Jane Taylor, " 'Dr. Laura' Lashes Out at Gay Activists," *Washington Blade* (National News), January 8, 1999.

30. William J. Turner, "Homosexuality, Type 1: An Xq28 Phenomenon," *Archives of Sexual Behavior* 24, no. 2 (1995): 109–34.

CHAPTER 21. "NATURE" IN QUOTATION MARKS

1. Foucault coined the term "bio-power" to describe the array of disciplines, observatory techniques, and sciences organized around "the administration of bodies and the calculated management of life." The productive effects Foucault attributes to bio-power are simultaneously material (e.g., the application of sci-

entific knowledge as medical treatments), institutional (the application of medical and sociological models to administrative techniques), and ideological (the production of new subjects and subjectivities out of material and institutional matrices). For Foucault, the formation of these new disciplines, techniques, and sciences represents "nothing less than the entry of life into history." Michel Foucault, *The History of Sexuality. Volume I: An Introduction*, trans. Robert Hurley (1976; New York: Vintage, 1980), 140, 141. On the politics of "life," see Giorgio Agamben, *Homo Sacer: Sovereign Power and Bare Life* (1995; Stanford, Calif.: Stanford University Press, 1998).

2. Here, as in what follows, I mean to evoke that old-fashioned Marxist distinction between social "relations" and technological "means."

3. Gina Kolata, "Reproductive Revolution Is Jolting Old Views," *New York Times*, January 11, 1994.

4. Faye Ginsburg and Rayna Rapp, *Conceiving the New World Order: The Global Politics of Reproduction* (Berkeley: University of California Press, 1995), xi.

5. Gina Kolata, "Babies Born in Experiments Have Genes from Three People," *New York Times*, May 5, 2001.

6. Gina Kolata, *Clone: The Road to Dolly and the Path Ahead* (New York: William Morrow, 1998). See also Gina Maranto, *Quest for Perfection: The Drive to Breed Better Human Beings* (New York: Scribner, 1996), and Elsimar M. Coutinho, *Is Menstruation Obsolete?* trans. Sheldon J. Segal (Oxford: Oxford University Press, 1999).

7. Gina Kolata, "U.S. Approves Sale of Impotence Pill; Huge Market Seen. 70% of Patients Helped. Eagerly Awaited Drug Found to Relieve Condition That Afflicts Millions of Men," *New York Times*, March 28, 1998.

8. This section is titled with apologies to Neil Young. On the signature traits of postmodernity, see Fredric Jameson, *Postmodernism, or, the Cultural Logic of Late Capitalism* (Durham, N.C.: Duke University Press, 1991), 1–54.

9. See Paulina Borsook, *Cyberselfish: A Critical Romp through the Terribly Libertarian Culture of High Tech* (New York: PublicAffairs, 2000).

10. Anne Fausto-Sterling, "Nature Is the Human Heart Made Tangible," in *Reinventing Biology: Respect for Life and the Creation of Knowledge*, ed. Lynda Birke and Ruth Hubbard (Bloomington: Indiana University Press, 1995), 121–36. Marilou Awiakta, "Trail Warning," in *Selu: Seeking the Corn-Mother's Wisdom* (Golden, Colo.: Fulcrum, 1993), 39.

11. Marilyn Strathern, *After Nature: English Kinship in the Late Twentieth Century* (Cambridge: Cambridge University Press, 1992); see also her *Reproducing the Future: Anthropology, Kinship, and the New Reproductive Technologies* (New York: Routledge, 1992).

12. Raymond Williams, *Keywords: A Vocabulary of Culture and Society*, rev. ed. (1976; New York: Oxford University Press, 1983), 219–24. Fredric Jameson, *Postmodernism*. Donna Haraway, "Manifesto for Cyborgs: Science, Technology, and Socialist Feminism in the 1980s," *Socialist Review* 80 (1985): 65–108.

13. Ian Hacking, *Representing and Intervening: Introductory Topics in Phi-*

losophy and Natural Science (Cambridge: Cambridge University Press, 1983), esp. 130–46.

14. Evelyn Fox Keller, *Secrets of Life, Secrets of Death: Essays on Language, Gender, and Science* (New York: Routledge, 1992), 3–6, 33.

15. David Rothenberg, *Hand's End: Technology and the Limits of Nature* (Berkeley: University of California Press, 1993), xii, xiv.

16. Karl Marx, *Capital: A Critique of Political Economy, Volume 1: The Process of Capitalist Production*, ed. Friedrich Engels (1887; New York: International, 1967), 177.

17. The term "production" might be too straightforward here. Instead of referring to the "production" of nature, it might be better to use Donna Haraway's decentered term "artifactuality" (which is "askew of productionism") to mark the kinkiness of the convolutions in play, the crisscrossing of multiple agencies, actions, and reactions—and the concomitant uncertainty of any clear "outcome." "The Promise of Monsters: A Regenerative Politics for Inappropriate/d Others," in *Cultural Studies*, ed. Lawrence Grossberg, Cary Nelson, and Paula Treichler (New York: Routledge, 1992), 296–300.

CHAPTER 22. MONEY'S SUBJECT

1. Natalie Angier, "Sexual Identity Not Pliable after All, Report Says," *New York Times*, March 14, 1997. Milton Diamond and Keith Sigmund, "Sex Reassignment at Birth," *Archives of Pediatric and Adolescent Medicine* (March 1997).

2. Dr. Milton Diamond, quoted in the *New York Times*, March 14, 1997.

3. John Colapinto, *As Nature Made Him: The Boy Who Was Raised as a Girl* (New York: HarperPerennial, 2000).

4. Louis J. G. Gooren, "Biomedical Concepts of Homosexuality: Folk Belief in a White Coat," in *Sex, Cells, and Same-Sex Desire: The Biology of Sexual Preference*, ed. John P. De Cecco and David Allen Parker (New York: Harrington Park Press, 1995), 237–46.

5. Suzanna Danuta Walters, *All the Rage: The Story of Gay Visibility in America* (Chicago: University of Chicago Press, 2001).

6. Michael Musto, "La Dolce Musto," *Village Voice*, July 2, 2002, 12.

7. *Ellen's* well-realized coming-out extravaganza topped the Nielsens, but the show's ratings progressively sank over the season that followed as the writers slogged through the familiar pageant of gay identity. Perhaps if the writers had followed an alternative course, less attuned to coming-out narratives established in the seventies and more in step with the queer nineties, the sitcom's history might have followed a different trajectory.

8. *Two Guys and a Girl*, formerly *Two Guys, a Girl, and a Pizza Place*, is itself indicative of how domesticity is imagined and complicated in modern entertainment media representations. The original "two guys" were roommates whose best friend, the "girl," lived in the same apartment building. Ad promos offered viewers a series of sexy come-ons that showed the actors entwined in

various playful scenes evocative of a ménage à trois. Eventually, the show's writers dropped the "pizza place" and introduced a boyfriend for the girl and a girlfriend for one of the guys, thus effectively expanding the cast to the five-person ensemble treated in this example.

9. Mikhail Bakhtin, *Rabelais and His World,* trans. Hélène Iswolsky (1965; Bloomington: University of Indiana Press, 1984).

10. See Herbert Muschamp's brilliant and subversive little essay, "Beefcake for the Masses," *New York Times Magazine,* November 14, 1999.

11. Andy Cole, "New Television Ads Appeal to Gays," *Washington Blade,* May 25, 2001, 16–17.

12. "More than 40 percent of the early news reports on the investigation of President Clinton's relationship with Monica S. Lewinsky consisted of analysis, opinion and speculation, not factual reporting," according to the Committee of Concerned Journalists, the media watchdog organization that published "The Clinton Crisis and the Press: A New Standard of American Journalism." The study concluded that "roughly half the time there was no evidence offered" to substantiate empirical claims in reportage. Felicity Barringer, *New York Times* (The Press), "Study Finds More Views Than Facts," February 19, 1998.

13. In this quote, Baudrillard means "transsexuals of the political sphere," but his treatment of the manipulation and switching of sexual signs makes it clear that he also treats transsexuality proper as the degree zero of postmodern culture. *The Transparency of Evil: Essays on Extreme Phenomena,* trans. James Benedict (1990; London: Verso, 1993), 7, 21.

14. Martha Bayles, *Hole in Our Soul: The Loss of Beauty and Meaning in American Popular Music* (Chicago: University of Chicago Press, 1994).

CHAPTER 23. HISTORY AND HISTORICITY

1. Antonio Gramsci, *Selections from the Prison Notebooks,* ed. and trans. Quintin Hoare and Geoffrey Nowell Smith (New York: International, 1971), 279, 296–97.

2. These cadences were later echoed in the key works of Frankfurt School Marxism. See especially Herbert Marcuse, *One-Dimensional Man: Studies in the Ideology of Advanced Industrial Society* (Boston: Beacon, 1964), and *Eros and Civilization: A Philosophical Inquiry into Freud* (Boston: Beacon, 1955).

3. Gramsci, *Selections,* 297–98, 299, 299–300.

4. Gramsci, *Selections,* 304–5.

5. Everything Gramsci writes about, from theories of ideological hegemony to strategies of class alliance, turns on connectivity. Purists might well argue that I have not quite employed Gramsci (or Marx) in the usual sense—that is, I have not developed an analysis of how ruling groups "hegemonize" their power through a state apparatus (nor have I consistently kept my eye on the play of class conflict and compromises in the material at hand). But it seems to me that

Gramsci's text could be used in more than one way, or, at any rate that theories of hegemony ought to attend more rigorously to how an institutional nexus articulates with and reproduces (or fails to reproduce) a given regime of accumulation.

6. David Harvey, *The Condition of Postmodernity* (Cambridge, Mass.: Basil Blackwell, 1990).

7. Here and in the following paragraphs, I telegraph a broad current of feminist scholarship on gender relations and family life in changing political-economic contexts. See Elaine Tyler May's study *Homeward Bound: American Families in the Cold War Era* (New York: Basic Books, 1988) and Judith Stacey's ethnography *Brave New Families: Stories of Domestic Upheaval in Late Twentieth Century America* (New York: Basic Books, 1990), esp. 3–19. The tradition that "never really was" is perhaps best debunked in Stephanie Coontz's *The Way We Never Were: American Families and the Nostalgia Trap* (New York: Basic Books, 1992), which undermines conservative nostalgia for lost origins on so many fronts. Also recommended are Stacey's *In the Name of the Family* (Boston: Beacon, 1996) and Coontz's *The Way We Really Are: Coming to Terms with America's Changing Families* (New York: Basic Books, 1997).

8. Perhaps my thumbnail sketch will seem like bad form: cheerleading the decline of a venerable institution in veritable parody of that conservative bogeyman, "the homosexual agenda." If so, so be it. Gays have nothing to fear from the growing realization that there are many ways to live, love, and bring children into the world. We have everything to gain from the progressive disentanglement of those conceptions of norm, nature, and ideal that are woven together in the gloomy history of heteronormativity.

9. Milton Friedman, with the assistance of Rose D. Friedman, *Capitalism and Freedom* (1962; Chicago: University of Chicago Press, 1974).

10. Taken from a *New York Times* letter to the editor, August 13, 1999, Larry Penner, "Is the G.O.P. Moving Closer to Tolerance?"

11. Marcuse, *One-Dimensional Man* and *Eros and Civilization*.

12. John D'Emilio, "Capitalism and Gay Identity" (1983), reprinted in *The Gender/Sexuality Reader: Culture, History, Political Economy*, ed. Roger N. Lancaster and Micaela di Leonardo (New York: Routledge, 1997), 169–78. I update D'Emilio's arguments here.

13. I invoke Max Weber's old distinctions here. See "Class, Status, Party," in *From Max Weber: Essays in Sociology*, ed. H. H. Gerth and C. Wright Mills (New York: Oxford University Press, 1946), 180–95.

14. Youth movements played major roles in transforming twentieth-century popular culture. Their successes in mass-producing new identities drew in no small part on a pattern of consumer economics linking lifestyle, subculture, and identification: the recurring template for the youth subculture of the 1950s, the counterculture of the 1960s, and gay, lesbian, and other sexual communities from the 1970s onward. See Stuart Hall and Tony Jefferson, eds., *Resistance through Rituals: Youth Subcultures in Post-War Britain* (1975; Lon-

don: Routledge, 1998), Dick Hebdidge, *Subculture: The Meaning of Style* (1979; New York: Routledge, 1987), and, of course, Dennis Altman's gay-studies classic, *The Homosexualization of America, the Americanization of the Homosexual* (New York: St. Martin's Press, 1982). On gay capitalism and the development of the gay subculture, see three pioneering books: David T. Evans, *Sexual Citizenship: The Material Construction of Sexualities* (London: Routledge, 1993), Daniel L. Wardlow, ed., *Gays, Lesbians, and Consumer Behavior: Theory, Practice, and Research Issues in Marketing* (New York: Harrington Park Press, 1996), and Amy Gluckman and Betsy Reed, eds., *Capitalism, Community, and Lesbian and Gay Life* (New York: Routledge, 1997).

15. Katha Pollitt, "Did Someone Say 'Hypocrites'?" *The Nation*, April 13, 1998, 9.

16. Alexandra Chasin, *Selling Out: The Gay and Lesbian Movement Goes to Market* (New York: St. Martin's Press, 2000).

17. Rosemary Hennessey, *Profit and Pleasure: Sexual Identities in Late Capitalism* (New York: Routledge, 2000), 108–9.

18. Karl Marx and Friedrich Engels, *The Communist Manifesto*, in *Karl Marx: Selected Writings*, ed. David McLellan (1977; Oxford: Oxford University Press, 2000), 247–55.

19. Dennis Altman, *Global Sex* (Chicago: University of Chicago Press, 2001).

20. Richard Goldstein, "Attack of the Homocons," *The Nation*, July 1, 2002, 11.

21. This claim, following from the arguments developed here and in the preceding chapter, represents an explicit critique of the positions taken by both Judith Butler and Nancy Fraser in their *Social Text* debate on the relationship of the new social movements—particularly feminism and lesbigay liberation—to questions of a political-economic order. Butler argues that gender, sexuality, and identity are not "merely cultural" phenomena, but are part of the "economic base" of society. I am in considerable sympathy with the gist of what Butler sets out to demonstrate, but, regrettably, Butler's argument derails on both logical and factual grounds. Butler announces her intention to analyze the culture of contemporary capitalism, but her key arguments draw on a (mis)reading of classical anthropological texts about *archaic* societies. She thus inadvertently ends up demonstrating that the heteronormative regulation of sexuality is necessary for the functioning of *precapitalist societies* (not late capitalist political economies)—an empirically unsustainable point not taken seriously in anthropology since the heyday of socialist feminism in the mid-1970s.

Fraser replies that although all injustices are "morally indefensible," it remains useful to distinguish between cultural and economic dimensions of inequality. She appeals to Weberian distinctions between "economic class" and "social status," arguing that what she calls "injustices of distribution" (e.g., class inequalities) are distinct from "injustices of recognition" (e.g., hatred of gays). Fraser concludes with a proper call for theoretical "historicization" (as against Butler's invocations of Lévi-Strauss and Mauss to "destabilize" or "deconstruct" the culture/economy distinction). But one might have wished for a more admirable historicity in Fraser's own staging of arguments. Her starting point in moral judgment scarcely lends itself to an adequate historicization of political theory—and hardly seems appropriate in connection with the sexual question. (Think here of Marx's unwavering rejection of Christian socialism, and his aversion in principle to moral arguments for political positions.) Her emphasis on the "politics of recognition" seems to imply that social identities are pregiven (in "nature"?), but somehow "misrecognized"—an implicitly essentialist framing that sidesteps an enormous current of lesbigay and feminist scholarship on the historical contingency of social identities. As Rosemary Hennessey puts it, Fraser "presents a historical effect—the emergence of sexual identities—as an ontological given." *Profit and Pleasure: Sexual Identities in Late Capitalism* (New York: Routledge, 2000), 223. The invocation of Max Weber to sustain a distinction between culture and economy, too, is problematic, precisely because Weber counts both class *and* status as forms of economic stratification (class having to do with production, and status with consumption), and since Weber also shows how the one can be translated, over either long or short periods of time, into the other. Fraser is no doubt correct when she concludes, contrary to Butler's lazy conflations, that "contemporary capitalism seems not to require heterosexism." But her muted cavil, at a moment when advertising culture shamelessly circulates homoerotic images on a global scale, is a case of too little insight, too late. It also gives considerably less traction than it might in understanding current struggles of a cultural *and* economic order.

See Judith Butler, "Merely Cultural," and Nancy Fraser, "Heterosexism, Misrecognition, and Capitalism: A Response to Judith Butler," *Social Text* 52–53 (fall–winter 1997): 265–77 and 279–89. See my arguments on sexuality and political economy in *Life Is Hard: Machismo, Danger, and the Intimacy of Power in Nicaragua* (Berkeley: University of California Press, 1992), esp. xvii–xix, 19–21, 221–30, 235–37, 279–82, 292–93; "That We Should All Turn Queer? Homosexual Stigma in the Making of Manhood and the Breaking of a Revolution in Nicaragua," in *Conceiving Sexuality: Approaches to Sex Research in a Postmodern World*, ed. Richard Parker and John Gagnon (New York: Routledge, 1995), 135–56; and [with Micaela di Leonardo] "Gender, Sexuality, Political Economy," *New Politics*, n.s., 21, no. 6 (summer 1996): 29–43, reprinted in *The Socialist Feminist Project: A Reader*, ed. Nancy Holmstrom (New York: Monthly, forthcoming).

22. David Harvey, *The Condition of Postmodernity: An Enquiry into the Origins of Cultural Change* (Cambridge, Mass.: Blackwell, 1990). See especially Harvey's use of tables on 174–79 and 340–41.

23. See Paul Smith's incisive review of some recent scholarship on masculinity, "Millennial Man," *American Literary History* 10, no. 4 (winter 1998): 733–52.

24. I thus leave largely aside the question of whether heterosexuality is "always" in a crisis mode, a notion that claims to plumb psychoanalytic depths and

that has become something of a cliché in queer theory. This sort of claim strikes me as ahistorical.

CHAPTER 24. THE POLITICS OF DREAD AND DESIRE

1. I've cribbed the title of this chapter from Valerie Hartouni, who surveys the "changing geography of dread and desire" in her *Cultural Conceptions: On Reproductive Technologies and the Remaking of Life* (Minneapolis: University of Minnesota Press, 1997), 9.

2. Antonio Gramsci, *Selections from the Prison Notebooks*, ed. and trans. Quintin Hoare and Geoffrey Nowell Smith (New York: International, 1971), 300.

3. "Everybody's Free (to Wear Sunscreen)," from the CD *Baz Luhrmann Presents Something for Everybody* (EMI Group's Capitol Records), delivers neatly balanced and commonsensical advice on matters both profound and trivial in a deadpan style. The spoken words are taken from Mary Schmich's *Chicago Tribune* column of June 1, 1997. Before Luhrmann, the director of *William Shakespeare's Romeo and Juliet*, acquired recording rights to the words, an Internet hoax attributed the words to Kurt Vonnegut as the text of his graduation speech at MIT. "Why So Many People Know to Wear Sunscreen," *New York Times,* May 10, 1999, C12.

4. Debbie Nathan, "Sodomy for the Masses," *The Nation,* April 19, 1999, 16–22, 22.

5. I quote here from Timothy Egan's exceptional journalism, "Hard Time: Less Crime, More Criminals," *New York Times* (Week in Review), March 7, 1999. See also Joel Dyer, *The Perpetual Prisoner Machine: How America Profits from Crime* (Boulder, Colo.: Westview, 2000).

6. David Wagner, *The New Temperance: The American Obsession with Sin and Vice* (New York: Westview, 1997).

7. Jane E. Brody, "Coping with Cold, Hard Facts on Teen-Age Drinking," *New York Times* (Science Times, Health and Fitness), April 6, 1999.

8. Homage, of course, to Herbert Marcuse, whose pessimistic suspicion was that the sixties counterculture would not so much eliminate repression as bring about an era of "repressive desublimation" perfectly articulated with a new type of consumer society.

9. Barbara Ehrenreich, *Nickel and Dimed: On (Not) Getting By in America* (New York: Metropolitan, 2001), 213, 3.

10. Christopher Lasch, *Haven in a Heartless World: The Family Besieged* (1977; New York: Norton and Company, 1995). See Wini Breines, Margaret Cerullo, and Judith Stacey's early, well-crafted critique in "Social Biology, Family Studies, and Antifeminist Backlash," *Feminist Studies* 4, no. 1 (February 1978): 43–67, esp. 57–62.

11. Judith Stacey, *Brave New Families: Stories of Domestic Upheaval in Late Twentieth-Century America* (New York: Basic Books, 1990).

12. Arlene Stein, *The Stranger Next Door: The Story of a Small Community's Battle over Sex, Faith, and Civil Rights* (Boston: Beacon, 2001).

13. George Yúdice, "What's a Straight White Man to Do" (267–83), and Barbara Ehrenreich, "The Decline of Patriarchy" (284–90), in *Constructing Masculinity*, ed. Maurice Berger, Brian Wallis, and Simon Watson (New York: Routledge, 1995).

14. Sylvia Ann Hewelett and Cornel West, *The War against Parents: What We Can Do for America's Beleaguered Moms and Dads* (Boston: Houghton Mifflin, 1998).

15. Patrick Buchanan, quoted at length by Doug Ireland, "Pat the Red Menace," *The Nation*, May 4, 1998, 7.

CHAPTER 25. SEX AND CITIZENSHIP

1. See Diana Jean Schemo's somewhat deceptively titled report, "Virginity Pledges by Teenagers Can Be Highly Effective, Federal Study Finds," *New York Times*, January 4, 2001.

2. Kathleen Kelleher, "Birds and Bees: More Teens Are Having Sex, but Don't Always Think They Are," *Los Angeles Times*, January 22, 2001. Karen S. Peterson, "Younger Kids Trying It Now, Often Ignorant of Disease Risks," *USA Today*, November 16, 2000. Damian Whitworth, "Oral Sex Becomes the Norm among U.S. Teens," *London Times*, July 9, 1999. Tamar Levin, "Inside America: Teenagers of AIDS Era Find New Pleasures," *London Guardian*, April 8, 1997.

3. Dudley Clendinen, "AIDS Spreads Pain and Fear among Ill and Healthy Alike," *New York Times*, June 17, 1983. Richard D. Lyons, "Sex in America: Conservative Attitudes Prevail," *New York Times*, October 4, 1983. David Gelman et al., "The Social Fallout from an Epidemic," *Newsweek*, August 12, 1985, 28. Lynn Simross, "Singles Mull Sexual Behavior as Fears of AIDS Intensify," *Los Angeles Times*, December 13, 1985.

4. Stacey D'Erasmo suggests that such TV shows as *Sex in the City* and *Ally McBeal*—which "seem like the dominant narrative of life on earth right now"—embody "in very slender form the argument that not only is feminism over. It failed." But her conclusion—that these fictional characters "really do just want to get married"—seems very wrong. "Single File," *New York Times Magazine* (The Way We Live Now), August 29, 1999, 13–14. Quite the contrary, it seems to me that the heroines of the 1990s struggle to balance love and career, commitment and freedom, in a world undeniably and irreversibly changed. If these TV characters are so often discontent, they show little sign of longing for a return to prefeminist ways of life.

5. Sheryl Gay Stolberg, "President Decides against Financing Needle Programs; Bitter Internal Debate; Ban Stays, Even as Scientists Cite Chance to Stem Spread of AIDS among Addicts," *New York Times*, April 21, 1998.

6. Debbie Nathan and Michael Snedeker, *Satan's Silence: Ritual Abuse and the Making of a Modern American Witch Hunt* (New York: Basic Books, 1995).

7. "Mozart for Baby? Some Say, Maybe Not," *New York Times* (Science Times), August 3, 1999. "Pediatricians Urge Limiting TV Watching," *New York Times*, August 4, 1999.

8. Lauren Berlant, *The Queen of America Goes to Washington City: Essays on Sex and Citizenship* (Durham, N.C.: Duke University Press, 1997).

9. With the release of its report documenting the torture and abuse of children, *Hidden Scandal, Secret Shame* (2000), Amnesty International called upon the United States to ratify the U.N. Convention on the Rights of the Child. AI's web site lists the United States and Somalia as "the only two countries who have not ratified the convention."

10. The National Abortion and Reproductive Rights Action League's fact sheet "Clinic Violence, Intimidation and Terrorism" documents the murder of three doctors, two clinic employees, a clinic escort, and a security guard, as well as seventeen attempted murders, and "2,500 reported acts of violence against abortion providers," including fire bombings, arson, and death threats.

11. Carey Goldberg and Janet Elder, "Public Still Backs Abortion, but Wants Limits, Poll Says," *New York Times,* January 16, 1998.

12. David Firestone, "At 29, an Elder Statesman of Internet Erotica," *New York Times,* November 6, 1998.

13. I borrow the phrase "fissioning public sphere" from Micaela di Leonardo's *Exotics at Home: Anthropologies, Others, American Modernities* (Chicago: University of Chicago Press, 1998), 263.

14. A set of open-ended questions, no doubt. For an excellent history of how these questions have been treated, from Theodor Adorno to Jürgen Habermas, from Raymond Williams to Stuart Hall, and from Marshall McLuhan to Jean Baudrillard, see Paul Marris and Sue Thornham, eds., *Media Studies: A Reader* (Edinburgh: Edinburgh University Press, 1996).

15. Berlant, *The Queen of America Goes to Washington City,* 3.

16. Mikhail Bakhtin, *Rabelais and His World,* trans. Hélène Iswolsky (Bloomington: Indiana University Press, 1984).

17. My understanding of this problem is considerably indebted to discussions with Marcial Godoy-Anativia, whose research in Chile surveys the close connection between, on the one hand, state terror, neoliberal privatization, and structural adjustment programs initiated by the Pinochet dictatorship and, on the other, the promulgation of policies favoring faith and family as "stabilizing" elements in a post-coup society. In more ways than one, the Chile model continues to inspire reactionaries the world over.

AN OPEN-ENDED CONCLUSION

1. Actually, the story first broke in the *London Observer,* whose editors had obtained advance copy of forthcoming articles in *Science* and *Nature.* In the scramble for news that ensued, the two scientific journals lifted their usual embargo on prereleasing the contents of their publications to the press. See "The Human Genome," *Nature* 409 (February 15, 2001): 813–958, and *Science* 291 (February 16, 2001).

2. Nicholas Wade, "Genome Analysis Shows Humans Survive on Low Number of Genes," *New York Times,* February 11, 2001. See also Nicholas Wade,

"Long-Held Beliefs Are Challenged by New Human Genome Analysis," *New York Times*, February 12, 2001, and "Genome's Riddles: Few Genes, Much Complexity," *New York Times* (Science Times), February 13, 2001, and Natalie Angier, "Genome Shows Evolution Has an Eye for Hyperbole," *New York Times* (Science Times), February 13, 2001.

3. See Stephen Jay Gould, "Humbled by the Genome's Mysteries," *New York Times* (Op Ed), February 19, 2001.

4. Nicholas Wade, "The Other Secrets of the Genome: The Story of Us," *New York Times* (Week in Review), February 18, 2001.

5. Natalie Angier, "The Sandbox: Bully for You: Why Push Comes to Shove," *New York Times* (Week in Review), May 20, 2001. Ironically, socio-biologist Richard Dawkins gives one of the least reductive perspectives in the article.

6. John Tierney, "The Big City: Fantasies of Vengeance, Fed by Fury," *New York Times*, September 18, 2001.

7. Natalie Angier, "Of Altruism, Heroism and Evolution's Gifts," *New York Times* (Science Times), September 18, 2001.

8. I invoke here, and over the last couple of chapters, Jeffrey Weeks's *Invented Moralities: Sex Values in an Age of Uncertainty* (New York: Columbia University Press, 1995), Kath Weston's *Families We Choose: Lesbians, Gays, Kinship* (New York: Columbia University Press, 1991), and Alan Wolfe's *Moral Freedom: The Impossible Idea That Defines the Way We Live Now* (New York: W. W. Norton, 2001).

Index

Lombroso, Cesare, 26
London Observer, 419n1
Loreal advertising, 303
Los Angeles Times, 323
"Lucy" and mate, 44 fig.
Luhrmann, Baz, 417n3
Lukács, Georg, 351n13, 375n3, 407n21
Lutz, Catherine, xi

MacCormack, Carole, 36
Madness and Civilization: A History of Insanity in the Age of Reason (Foucault), 366n15
Madonna, 296
Mae Enga society, 194
male animals. *See* animal behaviors
The Male Body (Bordo), 133
male bonding, 264, 407n19
males. *See* men
Malinowski, Bronislaw, 124–26, 127–28, 193, 357–58n8
Man Made Monster (film), 277 fig.
Mansfield, Harvey C., 153–54
Man the Hunter (Lee and DeVore), 363n18
Man with Dog (Witkin), 224 fig.
Mapplethorpe, Robert, 343–44
Maranto, Gina, xii, 365n7
March on Washington (1963), 6–7
Marcuse, Herbert, 413n2, 414n11, 417n8
Marks, Jonathan, xii, 60, 252
marriage: cultural variations in, 181–82, 183–84; DOMA's regulation of, 315; economic function of, 325; Fordist regulation of, 308, 309–10; as homosocial bonding, 264, 407n19; sociobiology's advocacy of, 160–63; whimsical bioreductive stories about courtship rituals, "pair-bonding," and, 41–45, 108–10, 160–61, 161–63. *See also* family, the; kinship systems; nuclear family
Martin, Emily, 109
Martin, M. Kay, 222
Martin, Nicholas G., 269, 408n3
Marx, Karl, 26, 106, 217–18; on com-

modity fetishism, 102–3, 375n3; on culture/nature relationship, 78–79, 175–76, 291, 387n15; on forms of family, 350–51n10
masculinism, 142–44, 150, 307, 330. *See also* masculinity; men; men's movement; Promise Keepers
masculinity: comparative studies of, 184–85, 206; crisis of, 151–54, 157–58, 297, 321. *See also* masculinism; men's movement
masks, 129
mass media: on aborting gay fetus, 235, 236 fig., 237; biology of beauty in, 118–20, 119 fig., 134, 135; constructionism exposé in, 71–72; on depression, 171, 387n4; gender essentialism of, 12, 163–64, 169–70, 293–94, 296, 304, 305; on geneticized homosexuality, 240, 246–47, 269–70; on heritability, 254, 405n37; heteronormalizing role of, 335–36; heteronormativity slippage in, 138; on human genome, 99–100, 342–43, 374n23, 419n1; on marriage benefits, 161–63; on masculinity crisis, 54–57, 151–52, 153–54, 360n3, 384n3; national political topics in, 334–35; nature focus of, 46–48, 288–89; on prison populations, 323–24; on racism, 57–58, 360n8; replication of themes in, 268, 269–71; on reproductive technology, 286; on September 11, 344–45; sociobiological discourse of, 11, 117, 160–61, 304; transformed values in, 4–6, 323, 349n5, 350n8; two sectors of, 336–38. *See also* advertising; *New York Times* articles; television shows
maternal kin: family knowledge about, 250, 404–5n31; as gay transmission route, 248–49, 251, 252, 281
Maternal Thinking (Ruddick), 139–40
mating position: with complementary genitalia, 36–37, 38–39; evolutionary tale of, 41–42; missionary model of, 43, 357–58n8

Turner, Frederick, 41–42, 43–44, 45, 47, 58
Turner, William, 281
TV shows. *See* television shows
The Twilight of the Golds (TV movie), 235, 237
twins: adoption of, 403–4n18; gay concordance rates in, 245–46, 269, 408n3; questionable research on, 246–48; with same sexual preference, 262, 407n9
Two Guys and a Girl (TV show), 301–3, 412–13n8
Two Men (Holland), 265 fig.
Tyson, Mike, 55

Ulrichs, Karl, 9–10, 16
"Unconventional Wisdom" (Morin), 57–58
universal woman motif, 25–26, 145–47, 190
Untitled (Haring), 228 fig.
Updike, John, 42–43
Upper Paleolithic ivory carving, 132 fig.
upright stance, 41–42, 50
Use of Abuse of Biology (Sahlins), xii, 62
Uterus of Vesalius, 37 fig.

vaginas, 226, 227–28, 400n21
Vaid, Urvashi, 21, 22
Van Den Berghe, Pierre, 110
Variety, 141, 142
Vedder, Eddie, 305
Ventura, Jesse, 152, 384n4
"venus observa" (term), 357–58n8
Vermont's gay civil union bill, 25
Veronica's Closet (TV show), 298
Viagra, 287–88
Vico, Giambattista, 398n33
Victorian culture, 56, 86–90, 146, 150, 371n26
Village Voice, 355n55
Vines, Gail, 239
virginity movement, 331
Vogel, Steven, 351n13
Voorhies, Barbara, 222

Waddington, C. H., 57
Wade, Nicholas, 99, 100–101, 342, 343, 374n23
wages, 325
Waite, Linda, 161–63, 165
Wald, Elijah, 97, 245, 246
Walters, Suzanna, xii, 299
The War against Parents (West), 328
warfare, female participation in, 189–90, 235, 393n47
Warner, Michael, xii, 38, 261–62
Washington Blade, 234–35, 273, 409n13
Washington Post, 57–58, 374n23
Watson, James, 235, 273
Weber, Bruce, 303
Weber, Max, 395n72, 415–16n21
Weed, Susan, 132
Weeks, Jeffrey, 26, 107, 354n43
West, Cornel, 328–29
Westphal, Karl, 16
Whitman, Walt, 351n11
Whorf, Lee, 210–11
Why Men Are the Way They Are (Farrell), 142
Wilde, Oscar, 55, 127, 165
wildness motif, 56–57, 88–90, 92, 158, 371n26
Will and Grace (TV show), 300–301, 301 fig.
Williams, F. E., 193
Williams, Raymond, 121, 290
Wilson, Edward O., 30–31, 51–52, 62, 343, 362n4; on courtship ritual, 108; on genetic code, 92; on nuclear family, 109; on sexual difference, 382n10; transformed sociobiology of, 145, 381–82n9
Witkin, Joel-Peter, *Man with Dog*, 224 fig.
Wittig, Monique, 104
Wolfe, Tom, 294
The Woman That Never Evolved (Hrdy), 383n15
women: attractiveness research on, 118, 136, 378n5; as cognitively specialized, 163–64, 386n9; cultural

Compositor:	Integrated Composition Systems, Inc.
Text:	10/13 Aldus
Display:	Aldus
Printer and Binder:	Friesens Corporation
Indexer:	Patricia Deminna